T0418910

VOLUME SEVENTY FOUR

ADVANCES IN QUANTUM CHEMISTRY

Löwdin Volume

VOLUME SEVENTY FOUR

ADVANCES IN QUANTUM CHEMISTRY

Löwdin Volume

Edited by

JOHN R. SABIN
University of Florida, Gainesville, FL, United States

ERKKI J. BRÄNDAS
Uppsala University, Uppsala, Sweden

Academic Press is an imprint of Elsevier
50 Hampshire Street, 5th Floor, Cambridge, MA 02139, United States
525 B Street, Suite 1800, San Diego, CA 92101-4495, United States
The Boulevard, Langford Lane, Kidlington, Oxford OX5 1GB, United Kingdom
125 London Wall, London, EC2Y 5AS, United Kingdom

First edition 2017

Notices

Knowledge and best practice in this field are constantly changing. As new research and experience broaden our understanding, changes in research methods, professional practices, or medical treatment may become necessary.

Practitioners and researchers must always rely on their own experience and knowledge in evaluating and using any information, methods, compounds, or experiments described herein. In using such information or methods they should be mindful of their own safety and the safety of others, including parties for whom they have a professional responsibility.

To the fullest extent of the law, neither the Publisher nor the authors, contributors, or editors, assume any liability for any injury and/or damage to persons or property as a matter of products liability, negligence or otherwise, or from any use or operation of any methods, products, instructions, or ideas contained in the material herein.

ISBN: 978-0-12-809988-9
ISSN: 0065-3276

Publisher: Zoe Kruze
Acquisition Editor: Poppy Garraway
Editorial Project Manager: Shellie Bryant
Production Project Manager: Vignesh Tamil
Cover Designer: Greg Harris

Typeset by SPi Global, India

CONTENTS

CONTRIBUTORS

Dževad Belkić
Karolinska Institute, Stockholm, Sweden

Karen Belkić
Karolinska Institute, Stockholm, Sweden; School of Community and Global Health, Claremont Graduate University, Claremont; Institute for Health Promotion and Disease Prevention Research, University of Southern California School of Medicine, Los Angeles, CA, United States

Anael Ben-Asher
Schulich Faculty of Chemistry, Technion—Israel Institute of Technology, Haifa, Israel

Manuel Berrondo
Brigham Young University, Provo, UT, United States

Debarati Bhattacharya
Schulich Faculty of Chemistry, Technion—Israel Institute of Technology, Haifa, Israel

Erkki J. Brändas
Institute of Theoretical Chemistry, Angstrom Laboratory, Uppsala University, Uppsala, Sweden

Carlos F. Bunge
Instituto de Física, Universidad Nacional Autónoma de México, Mexico City, Mexico

Sylvio Canuto
Instituto de Física, Universidade de São Paulo, Cidade Universitária, São Paulo, SP, Brazil

Marcelo Hidalgo Cardenuto
Instituto de Física, Universidade de São Paulo, Cidade Universitária, São Paulo, SP, Brazil

Héctor H. Corzo
Auburn University, Auburn, AL, United States

Kaline Coutinho
Instituto de Física, Universidade de São Paulo, Cidade Universitária, São Paulo, SP, Brazil

Hazel Cox
University of Sussex, Brighton, United Kingdom

Lawrence J. Dunne
School of Engineering, London South Bank University; Imperial College London, London; University of Sussex, Brighton, United Kingdom

Idan Haritan
Schulich Faculty of Chemistry, Technion—Israel Institute of Technology, Haifa, Israel

Arik Landau
Schulich Faculty of Chemistry, Technion—Israel Institute of Technology, Haifa, Israel

Sven Larsson
Chalmers University of Technology, Göteborg, Sweden

Jan Linderberg
Aarhus University, Aarhus, Denmark

Ingvar Lindgren
University of Gothenburg, Gothenburg, Sweden

David A. Micha
University of Florida, Gainesville, FL, United States

Nimrod Moiseyev
Schulich Faculty of Chemistry, Technion—Israel Institute of Technology, Haifa, Israel

Cleanthes A. Nicolaides
Theoretical and Physical Chemistry Institute, National Hellenic Research Foundation, Athens, Greece

Yngve Öhrn
QTP, University of Florida, Gainesville, FL, United States

J. Vince Ortiz
Auburn University, Auburn, AL, United States

Jose Récamier
Instituto de Ciencias Físicas, Universidad Nacional Autónoma de México, Cuernavaca, Morelos, Mexico

John R. Sabin
University of Florida, Gainesville, FL, United States; University of Southern Denmark, Odense, Denmark

Suchitra Sampath
Centre for Neural and Cognitive Sciences, University of Hyderabad, Hyderabad, Telangana, India

Vipin Srivastava
Centre for Neural and Cognitive Sciences; School of Physics, University of Hyderabad, Hyderabad, Telangana, India

Orlando Tapia
Chemistry Ångström, Uppsala University, Uppsala, Sweden

PREFACE

In this particular volume of *Advances in Quantum Chemistry* we celebrate the 100th anniversary of the birth of Per-Olov Löwdin. He was appointed as the first chair in Quantum Chemistry at Uppsala University in 1960, and is one of the Founders of the Science of Quantum Chemistry—a subdiscipline that grew out of the synthesis and the integration of knowledge and work in physics, chemistry, biology, and applied mathematics. He established the Quantum Chemistry Group in Uppsala, Sweden, and the Quantum Theory Project in Gainesville, Florida, and founded the present channel for scientific dissemination, *Advances in Quantum Chemistry* and a few years later also the Wiley academic journal the *International Journal of Quantum Chemistry*.

In this volume we attempt to reflect the many varied scientific accomplishments of Per-Olov Löwdin, emphasizing his introduction of fundamental concepts and the addition of rigor and consistency to the field, which today are inherent parts of the technological development of quantum chemistry and theoretical chemical physics.

The invited writers involve Per-Olov Löwdin's first recruits to the new field and continue with PhD students supervised in Uppsala and Florida along with international summer school acolytes and collaborators. As already stated, the field of quantum chemistry has emerged as a subject of its own, and the chapters in this volume will display the multifaceted overlaps with various other fields with contributions from analytic work, accurate QED perturbations, via strong correlations in superconductors, time-dependent approaches, partitioning techniques, photonic schemes, thermodynamic conditions, and learning–memory to fundamental applications in medicine.

As presented, the contents of this volume are multivarious, involving both fundamental theory and innovative applications, but also reflecting the human nature of the master. The contributing authors have made great strides to share their views of a unique and fascinating era of Quantum Chemistry. As series editors, we hope that the Löwdin memorial volume will convey the same pleasure and enjoyment as we, and also the contributors, felt during the planning of this volume.

John R. Sabin
Erkki J. Brändas

CHAPTER ONE

Per-Olov Löwdin

Jan Linderberg*, Yngve Öhrn†, Erkki J. Brändas‡, John R. Sabin†,§,1

*Aarhus University, Aarhus, Denmark
†University of Florida, Gainesville, FL, United States
‡Uppsala University, Uppsala, Sweden
§University of Southern Denmark, Odense, Denmark
[1]Corresponding author: e-mail address: sabin@qtp.ufl.edu

This volume celebrates the centennial of the birth of one of the major figures in the world of quantum chemistry, Per-Olov Löwdin. As one of the founders of the science of quantum chemistry, he contributed more than 60 years to the field. His many varied scientific accomplishments include the introduction of novel basic concepts, which are today inherent parts of the development of theoretical chemical physics and quantum chemistry. He advocated and developed linear algebra, partitioning technique, orthogonalization procedures, and established basic notions such as correlation energy, natural spin orbitals, general configuration interaction (CI) methods, and reduced density matrices, and he pioneered interesting applications to neighboring fields such as quantum biology.

He contributed to the sociology and the international spread of quantum chemistry through Summer Schools in the Scandinavian Mountains, Winter Institutes at the Sanibel Island in the Gulf of Mexico; all known for their mix of intense lectures schedules, adventurous mountain hikes, hard-fought soccer, and water polo games. He established the Quantum Chemistry Institute at Uppsala University, and the Quantum Theory Project (QTP) at the University of Florida at Gainesville. Both have for many years been influential research centers for the study of novel molecular and material systems.

Uppsala and Gainesville were the nodal points in Per-Olov Löwdin's World Wide Web of quantum chemistry. He was a scientist, who acknowledged the importance of a deductive approach to science and introduced new and lasting education and wisdom through schooling and lecturing, while maintaining an unusual zeal in spreading the good word due to extensive travel and indefatigable concern about teaching quantum chemistry

Advances in Quantum Chemistry, Volume 74
ISSN 0065-3276
http://dx.doi.org/10.1016/bs.aiq.2016.04.001

through all—embracing lecture tours. He took new initiatives for making scientists from Latin America, Africa, and Asia members of his community.

Löwdin was born in Uppsala on October 28, 1916 as the son of the musician Erik Wilhelm Löwdin and his wife Eva Kristina, nee Östgren. He showed early mathematical proficiency in his schoolwork, and when he entered Uppsala University in 1935, his plans were to major in mathematical physics. His first scientific paper in 1939, on the Lorentz-transformation and the kinematical principle of relativity, was published in Swedish in the journal *Elementa*. The years of war that followed meant interruptions in communications and research opportunities and Löwdin, like most young Swedish men, spent time in the military, defending his country.

The new quantum electrodynamics presented serious challenges to a theoretician and Löwdin's research notes from the 1940s on this topic show that he spent much time and effort working on these problems. An extended stay with Wolfgang Pauli at Zurich in 1946 became a turning point in Löwdin's scientific interests. At about this time, he started work in solid-state physics and began his dissertation research with Professor Ivar Waller at Uppsala. The topic of his thesis work was a quantum mechanical study of ionic crystals and the examination of the Cauchy relations. These relations arise from the assumption of pairwise interactions between spherical ions, but were not satisfied in real crystals such as sodium chloride. This pioneering application of quantum mechanics, before the advent of electronic computers, was accomplished with the aid of a cadre of doctoral students using desk calculators and ingenious numerical algorithms designed by Löwdin.

A Theoretical Investigation into Some Properties of Ionic Crystals. A Quantum Mechanical Treatment of the Cohesive energy, the Interionic Distance, the Elastic Constants, and the Compression at High Pressures with Numerical Applications to some Alkali Halides (Almqvist & Wiksell, Uppsala, 1948) is the full title of Löwdin's dissertation for the degree of Doctor of Philosophy. The preface is marked May 1, 1948, the successful defense took place some 3 weeks later, and the ring, laurel, and diploma, the Swedish insignia of the doctorate, were awarded at the commencement ceremony on May 31.

It is interesting to read Löwdin's discussion of his method of investigation and the numerical results from Chapter XI. He specifies the fixed nuclei Hamiltonian in the Schrödinger form where no relativistic effects are considered. He specifies the one-electron approximation with the free ion orbitals and shows that polarization and van der Waals effects are small compared to the terms from exchange and the *S-energy*. This latter concept was to remain a key feature in Löwdin's future work. He considered all overlap integrals, denoted $S_{\mu\nu}$, and chose to evaluate the formal energy expression to second order. Only nearest neighbor overlap was included since he found the others to be smaller than the general tolerance level he adopted. The detailed numerical calculations were based on the so-called α-function expansions. Orbitals on one atomic center were expanded in terms of spherical harmonics on a neighboring center. The resulting radial factors are the Löwdin α-functions.

The failure of the Cauchy relations derives from the three- and four-body interactions, which stem from the overlap terms. The description of the properties of ionic crystals was brought to a new and improved level by Löwdin's thesis and he developed an arsenal of tools, which were sharpened and extended, throughout his career.

A 5-month stay with Neville Mott at Bristol in 1948, and extended periods with Robert Mulliken at Chicago, and with the group of Hertha Sponer at Duke University in the early 1950s set the tone for life filled with international travel. Particularly significant was a stay with the Solid State and Molecular Theory Group of John C. Slater at the Massachusetts Institute of Technology. There developed a close association between Slater and Löwdin, which was to last until Slater's death in 1976.

Ram's Head Inn on Shelter Island, in Gardiners Bay at the Eastern end of Long Island, was the venue for a conference that was considered epoch-making for the emerging field of quantum chemistry by the participants. The National Academy of Sciences sponsored the Conference on Quantum-Mechanical Methods in Valence Theory with the financial support of the Office of Naval Research.

It was held September 8–10, 1951, and was attended by the leading figures in the field.[1] Emphasis was given to numerical work. Slater and Ufford reported on the use of electronic computers for self-consistent field calculations, Coulson and Barnett were implementing the ζ-function technique in London, Roothaan, Ruedenberg, and Shull at Chicago and Ames were pursuing diatomic integrals, Kotani's table project was well under way, and Löwdin could present experiences with direct numerical integration methods, which he was developing together with Stig Lundqvist. Per-Olov Löwdin was now an established scientist in the new field of quantum chemistry and he came to play a decisive role in forming the program for quantum chemistry in the postwar world.

Three publications are particularly noteworthy in defining the scientist Per-Olov Löwdin in the first half of the 1950s. His concern with overlap led him to formulate *symmetric orthogonalization* as a means of forming an orthonormal basis from an overlapping one.[2] This paper may still be among his most cited works[3] and has been an essential element in the justification of *neglect of differential* overlap.[4] The trilogy *Quantum Theory of Many-Particle Systems*[5] was worked out during Löwdin's stay with Slater's group and contains a detailed analysis of the configuration space method for dealing with the correlation problem. Density matrices were used as the principle vehicle for the advancement of interpretation of many-electron wave functions. Detailed prescriptions were offered for the evaluation of density matrices from general wave functions in the form of superposition of configurations. General orbital basis sets with overlaps were used. The *natural spin orbitals* were defined and the first efforts to formulate *extended Hartree–Fock* methods appeared here. Löwdin also initiated the use of projection operators in order to handle degenerate states. Spin multiplets were of primary concern and the foundation was laid for the later development of the *alternant molecular orbital* approach.[6] *Quantum Theory of Cohesive Properties of Solids*[7] exhibits, in its 172 pages, the structure of the theoretical approach toward the description of electronic properties of matter as Löwdin saw it at the time. The numerical procedures, density matrices, natural orbitals, and projection operators were developed in these papers and a thorough review of the literature was included.

Löwdin received substantial grants that enabled the creation of the Quantum Chemistry Group at Uppsala University in 1955. The *King Gustaf VI Adolf's 70-years Fund for Swedish Culture* and the *Knut and Alice Wallenberg's Foundation* provided funding for the project, while Löwdin's position as a research professor was supported by the *Swedish Natural Sciences Research Council.* Jointly with Inga Fischer-Hjalmars, Löwdin arranged the first symposium on quantum chemistry in Sweden at Uppsala and Stockholm in 1955 with participation from the leading experts.[8]

Extension of the group was made possible by a research grant starting in 1957 from *Aerospace Research Laboratories, OAR*, through the *European Office of Aerospace Research (OAR), United States Air Force.* An agreement with the Swedish agencies opened the possibility for purchasing a state-of-the-art electronic computer, the *ALWAC IIIE.* This was the first device of its kind at Uppsala and was formally inaugurated in conjunction with the group's new offices at Rundelsgränd on April 23, 1958, the centennial of the birth of Max Planck.

The electronic correlation problem was a primary research theme in the Quantum Chemistry Group when Jan Linderberg and Yngve Öhrn joined the group in 1957 and 1958, respectively. The review[9] and Yoshizumi's bibliography[10] had been submitted, natural orbitals had been determined,[11] and the Quantum Chemistry Group was emerging as a center for quantum chemical investigation. Further enhancement came through the Summer Schools. Vålådalen in the mountains of Sweden was the location for the first Summer School, held in August 1958. Löwdin managed to gather several prominent lecturers such as Robert Mulliken, Linus Pauling, Kenneth Hedberg, F.A. Matsen, and others for the final symposium week. Ruben Pauncz, a participant then, returned as a valued lecturer for more than 20 years in these annual Summer Schools.

Quantum Chemistry became recognized as an academic discipline in its own right at Uppsala University with the establishment of a chair in the subject. Löwdin was appointed to the chair in 1960, the same year that he founded an interdisciplinary research institute, the Quantum Theory Project for Research in Atomic, Molecular, and Solid-State Theory (informally known as QTP), at the University of Florida. Dean Linton E. Grinter and the heads of the departments of chemistry and physics at UF, Professor Harry S. Sisler and Professor Stanley S. Ballard, respectively, provided the local support for Löwdin and a small contingent of young Swedish scientists[11] as the original staff of the new QTP. With academic homes on both sides of the Atlantic the number of quantum chemists connected to Löwdin grew and was sustained by the Scandinavian Summer Schools and Florida Winter Schools in Quantum Chemistry. It is estimated that about 4000 scientists have attended a summer school, winter institute, or conference organized by Löwdin and his colleagues.

There were certain areas of theory that were not embraced by Löwdin in the early years. Group theory was not given a satisfactory form to his liking until John Coleman had lectured on group algebra and Löwdin could cast this in the form of linear algebra.[12] Second quantization had been familiar to him since the 1940s when he worked with quantum electrodynamics and he wanted to connect the use of the Fock space in many-electron theories with the configuration space formulation. In spite of asking his junior colleagues to study the early papers of V. Fock, and others on the subject, no concrete project arose from these efforts and Löwdin seemed to share Slater's feelings[13] that field operators were of little use in the study of electronic systems. It was the influence of Stig Lundqvist that inspired the use of field theoretical methods in the Uppsala and Florida groups.

In 1964, POL started the series *Advances in Quantum Chemistry* and later, in 1967, he founded the Wiley academic journal the *International Journal of Quantum Chemistry*. In its first issue he presented the program and outlined in detail the nature of the new field.[14]

Through 1960, Löwdin had authored and coauthored some 50 papers. He added more than 200 in the following 40 years. These covered topics in quantum genetics, proton tunneling, and science in society in addition to further pursuits in quantum theory. The series *Studies in Perturbation Theory* I–XIV demonstrated a search for economy and elegance in presentation, which was important to Löwdin and became one of his trademarks as a scientist. Löwdin frequently referred to "the economy of thinking" as shorthand for elegance and compactness of a mathematical derivation. This characteristic manifests itself in his book *Linear Algebra for Quantum Theory* (J. Wiley & Sons, New York 1998), which summarizes over five decades of Löwdin's teaching efforts. Even after his official retirement from the University of Florida faculty in 1992, he and his wife Karin continued to travel between Uppsala and Gainesville. Their Florida stays coincided with the annual Sanibel Symposia, where Löwdin always attended all sessions, and from his first row seat continued his habit of making insightful and often complimentary remarks after many of the plenary presentations.

Uppsala University bestows a *Jubilee doctorate* on those alumni who have survived 50 years after their doctoral commencement. Thus it was that Löwdin received a new laurel and an additional diploma on May 29, 1998, as well as a two-shot salute by the artillery cannon outside the University Aula. The president of the student union saluted the five recipients with a speech that was answered by Löwdin in his characteristically youthful style. Five months later he had a serious heart operation. He never recovered fully. His health deteriorated, and he passed away quietly on October 6, 2000. A memorial symposium was arranged at Uppsala on the day before the funeral, on October 26. Many of Löwdin's colleagues and friends also gathered on the University of Florida campus to honor his memory with eulogies by Löwdin's longtime collaborator and friend Harrison Shull, and by Robert A. Bryan, who held many top administrative positions at UF, including that of President, during most of Löwdin's tenure at Florida. The 42nd Sanibel Symposium held in 2002 was dedicated to his memory.

We remember Löwdin as a teacher, an indefatigable lecturer, a helpful colleague, and a friendly competitor. Perhaps his most important characteristic is that of a true internationalist. The Scandinavian Summer Schools, the Florida Winter Schools, and the Sanibel Symposia have, through the vision of Per-Olov Löwdin, furthered international cooperation and friendship between scientists from all continents. He was quoting Harrison Shull,[15] "a veritable giant in the scientific Universe, a dynamo of energy, full of *joie de vivre* and a kind mentor to us all". Per-Olov, or Pelle as he wanted to be called, was the most colorful, exciting and fascinating person, always maintaining that we all worked in quantum chemistry "JUST FOR FUN"!

ACKNOWLEDGMENTS

Most of the text in this chapter is *Reprinted from* Jan Linderberg, J.; Öhrn, N. Y.; Sabin, J. R. Advancing Quantum Chemistry: Per-Olov Löwdin 1916–2000. *Adv. Quant. Chem.* **2002,** *41*, x–xv. Copyright 2002, with permission from Elsevier.

REFERENCES

1. Parr, R. G.; Crawford, B. L., Jr. On Quantum-Mechanical Methods in Valence Theory. *Proc. Natl. Acad. Sci. U.S.A.* **1952**, *38*, 547.
2. Löwdin, P. O. On the Non-orthogonality Problem Connected with the Use of Atomic Wave Functions in the Theory of Molecules and Crystals. *J. Chem. Phys.* **1950**, *18*, 365.
3. Manne, R. In *Quantum Science. Methods and Structure*; Calais, J.-L., Goscinski, O., Linderberg, J., Öhrn, N. Y., Eds.; Plenum Press: New York, 1976. p. 25.
4. Parr, R. G. *The Quantum Theory of Molecular Electronic Structure*; W.A. Benjamin, Inc.: New York, 1963. p. 51; Parr, R. G. *J. Chem. Phys.* **1960**, *33*, 1184.
5. Löwdin, P. O. Quantum Theory of Many-Particle Systems. 1. Physical Interpretations by Means of Density Matrices, Natural Spin-Orbitals, and Convergence Problems in the Method of Configurational Interaction. *Phys. Rev.* **1955**, *97*, 1474; Löwdin, P. O. Quantum Theory of Many-Particle Systems. 2. Study of the Ordinary Hartree-Fock Approximation. *Phys. Rev.* **1955**, *97*, 1490; Löwdin, P. O. Quantum Theory of Many-Particle Systems. 3. Extension of the Hartree-Fock Scheme to Include Degenerate Systems and Correlation Effects. *Phys. Rev.* **1955**, *97*, 1509.
6. Calais, J.-L. *Abstracts of Uppsala Dissertations in Science*, 52; 1965; Pauncz, R. *Alternant Molecular Orbital Method*; W.B. Saunders: Philadelphia, 1967.
7. Löwdin, P. O. Quantum Theory of Cohesive Properties of Solids. *Adv. Phys.* **1956**, *5*, 1.
8. Fischer-Hjalmars, I.; Löwdin, P. O. Report from the Symposium on Quantum Theory of Molecules, Stockholm and Uppsala, 1955. *Sv. Kem. Tidskrift* **1955**, *67*, 365.
9. Löwdin, P. O. Correlation Problem in Many-Electron Quantum Mechanics. 1. Review of Different Approaches and Discussion of Some Current Ideas. *Adv. Chem. Phys.* **1959**, *2*, 207.
10. Yoshizumi, H. Correlation Problem in Many-Electron Quantum Mechanics. 2. Bibliographical Survey of the Historical Development with Comments. *Adv. Chem. Phys.* **1959**, *2*, 323.
11. Löwdin, P. O. In *Partners in Progress*; Kastrup, A., Olsson, N. W., Eds.; Swedish Council of America: Minneapolis, MN, 1977. p. 255.
12. Löwdin, P. O. Group Algebra Convolution Algebra and Applications to Quantum Mechanics. *Rev. Mod. Phys.* **1967**, *39*, 259.
13. Slater, J. C. (Review of) Second Quantization and Atomic Spectroscopy. *Am. J. Phys.* **1968**, *36*, 69.1.
14. Löwdin, P. O. Program. *Int. J. Quant. Chem.* **1967**, *1*, 1.
15. Shull, H. In *Fundamental World of Quantum Chemistry*; Brändas, E. J., Kryachko, E. S., Eds.; Kluwer Academic Publishers: Dordrecht, 2003. p. 1.

CHAPTER TWO

From Numerical Orbitals to Analytical Ones and Back

Jan Linderberg[1,2]
Aarhus University, Aarhus, Denmark
[1]Corresponding author: e-mail address: boforsarn@mail.dk

Contents

Abstract

Attempting to survey and review some developments in the design and use of atomic basis sets for molecular electronic structure calculations from the perspective of Per--Olov Löwdin's contributions this chapter is offered as a contribution to the celebration of the centennial of his birth.

1. INTRODUCTION

Numerical tables of the radial factors in the atomic and ionic orbitals were the basis at disposal for Per-Olov Löwdin when he was assigned his thesis problem in the mid-1940s. He had been, like all able-bodied young men in Sweden during World War II, serving his country through extended periods in remote areas along the borders of Sweden. His mind was set upon the problems of quantum electrodynamics and he finished his licentiate

[2] Born (1934) and raised at Karlskoga, Sweden. Graduate of Karlskoga Kommunala Gymnasium 1954. Doctor's degree earned at Uppsala University 1964, Appointed chair of theoretical chemistry at Aarhus 1968, Emeritus status since 2003. Awarded the dignity of Jubeldoktor at Uppsala University in 2014.

Advances in Quantum Chemistry, Volume 74
ISSN 0065-3276
http://dx.doi.org/10.1016/bs.aiq.2016.04.003

degree with a thesis in this domain. There were also periods of teaching interspersed among his efforts until he could resume his graduate work. A visiting period in Zürich with Wolfgang Pauli did not result in tangible records. The professor of mechanics and mathematical physics, Ivar Waller, suggested upon Löwdin's return to Uppsala that he applied himself to a quantum mechanical investigation of some properties of ionic crystals. Evidence from measurements of the elastic features of these crystals was contradictory to the simple model of pair-wise interactions between the ions. Thus it rested with Löwdin to provide a more general description where many-body couplings appeared.

Waller[1] had his interest in condensed matter physics from his early work on the influence of the thermal motion in crystals on the scattering of X-rays. His work is remembered through the Debye–Waller factor and applies to the scattering of neutrons as well.

Löwdin[2] developed, in his thesis project, an energy expression for the ionic crystals from a single Slater determinant state wave function built from doubly occupied Bloch crystal orbitals formed from isolated ion orbitals. It was imperative to treat overlap accurately and to form the various parts of the energy expression in combinations that were convergent and minimized the accumulation of rounding errors in the numerical work. During the course of the numerical effort, there were several functions that only were available as tabulated values at a discrete mesh of points. Such tables were put into analytical forms as a recourse for simplifications. Large-scale numerical calculations required assistants, and Löwdin could employ a number of *parallel processors* as was known both at Cambridge, Massachusetts and Cambridge, England. The young students who were engaged by Löwdin and paid by grants from the Research Council later assumed positions as *parallel professors* in the Swedish University system.

Halogen orbitals were available for the anions of Fluorine[3] and Chlorine[4] from calculations with exchange accounted for. Alkali metal orbitals could similarly be retrieved from the literature for the cations of Lithium,[5] Sodium,[6] and Potassium.[7] There were in these papers several additional functions tabulated, and these could be included in Löwdin's formalism directly or by analytical fittings.

Slater[8] introduced a set of screening constants to be used in exponential-type radial function to represent the more accurate numerical ones. He built his values from a range of atomic and ionic properties and did not use the energy minimization principle like Zener[9] and Guillemin.[10] Early comprehensive studies by Pauling[11] led him to suggest screening constants for hydrogenic orbitals. Slater's orbitals were excellent vehicles for numerous numerical estimates of atomic and molecular parameters, such as overlap. Tables of essential integrals were produced by Mulliken's group[12] and in book form by Kotani's group.[13] The latter provided, in addition, tables for spin couplings through application of the symmetric group and its irreducible representations. Such material was valuable at a time when electronic computational devices were few and far between.

Analytical representations of numerical functions were still a matter of importance to Löwdin when an able computor, Klaus Appel, joined his group. Löwdin's earlier efforts[14] were then supplemented with analytical orbitals also for transition metals.[15] The paper with Appel contains the basic formulae for finding optimal exponents and amplitudes for fitting an exponential type function.

Knut and Alice Wallenberg's Foundation provided Löwdin with a generous grant that allowed him to purchase the first electronic computer at Uppsala for the Quantum Chemistry Group.[16] A Swedish agency, Matematikmaskinnämnden, had initiated a development toward electronic computation[17] and BESK, for *binär elektronisk sekvenskalkylator*, was made available for scientists in the mid 1950s. Its location in Stockholm and the limited capacity made it obvious to Löwdin that the Uppsala group needed a machine dedicated to the needs of the group.

It was an exciting time when I was inducted into the Uppsala Quantum Chemistry Group in Aug. 1957. I had some experience with electrical desk calculators from summer jobs at AB Bofors in my hometown Karlskoga and got into Boolean algebra and programming in hexadecimal machine code with considerable enthusiasm.

2. EXPONENTIALS

Four young members of the Uppsala Quantum Chemistry Group were summoned to Löwdin's office in January 1959 and were assigned topics to be presented at the first Nordic conference on electronic computing, NordSAM, held at Karlskrona, Sweden in May that year. Mine was *Interpolation with Exponentials*, and I was directed to the previously cited papers and to the numerical analysis literature. I found Lanczos's monograph[18] particularly readable.

Consider a map

$$X:\{x_j = x_0 + jh | j = 0, 1, \ldots n\} \rightarrow Y:\{y_j | j = 0, 1, \ldots n\}$$

where values of the function y are given at equidistant points of the independent variable x. The tenet that y can be represented as a sum of exponentials implies that

$$y_j = \sum_{i=1}^{m} a_i k_i^j$$

and that the Hankel determinants

$$H_p = \begin{vmatrix} y_0 & y_1 & \cdots & y_p \\ y_1 & y_2 & & y_{p+1} \\ \vdots & & \ddots & \vdots \\ y_p & y_{p+1} & \cdots & y_{2p} \end{vmatrix}$$

differ from zero when $p < m$ but do equal zero for $p \geq m$. The rank of the expansion is then possible to determine from the series of values for the Hankel determinants. It follows that the zeroes of the function

$$H_m(z) = \begin{vmatrix} 1 & z & \cdots & z^m \\ y_0 & y_1 & & y_m \\ \vdots & & \ddots & \vdots \\ y_{m-1} & y_m & \cdots & y_{2m-1} \end{vmatrix}$$

give the set of exponentials $\{k_i | i = 1, 2, \ldots m\}$.

These formal relations are useful but there is a general experience that rounding errors and limited accuracy of the set Y limit the rank of the forms to a noticeable extent. Input data with a six-figure accuracy can often be represented by six exponentials. Near linear dependence among a set of exponentials is readily demonstrated by considering two sets of exponentials of rank m and $m+1$ where the first set is included in the second one. The ratio between the two Hankel determinants is[19]

$$\frac{H_{m+1}}{H_m} = a_m \prod_{i=1}^{m} (k_i - k_{m+1})^2$$

and shows that an additional exponential within the domain spanned by others gives little improvement.

Very similar results obtain for the optimization problem of finding the best fit to an exponential when expanded in a set of exponentials. Consider the basis set B for decay-type functions on the interval $[0, \infty)$

$$B: \{k_i^x | 0 < k_i < 1, i = 1, 2, \ldots m\}; \; 0 \leq x < \infty$$

and the functional

$$J(\{a_i\}) = \int_0^\infty dx \left| k^x - \sum_{i=1}^{m} a_i k_i^x \right|^2$$

with the explicit form

$$J(\{a_i\}) = -\frac{1}{\ln(k^2)} + \sum_{i=1}^{m} \frac{2a_i}{\ln(kk_i)} - \sum_{i,i'=1}^{m} \frac{a_i a_{i'}}{\ln(k_i k_{i'})}$$

Simplifying notations are introduced so that

$$A_{00} = -\frac{1}{\ln(k^2)};\ A_{0i} = -\frac{1}{\ln(kk_i)};\ A_{ii'} = -\frac{1}{\ln(k_i k_{i'})};$$

and

$$J(\{a_i\}) = A_{00} - \sum_{i=1}^{m} 2A_{0i}a_i + \sum_{i,i'=1}^{m} a_i A_{ii'} a_{i'}$$

A lower bound on the functional is the quotient of two determinants,

$$J \geq \frac{\begin{vmatrix} A_{00} & -A_{01} & -A_{02} & & -A_{0m} \\ -A_{01} & A_{11} & A_{12} & \cdots & A_{1m} \\ -A_{02} & A_{21} & A_{22} & & A_{2m} \\ & \vdots & & \ddots & \vdots \\ -A_{0m} & A_{m1} & A_{m2} & \cdots & A_{mm} \end{vmatrix}}{\begin{vmatrix} A_{11} & A_{12} & \cdots & A_{1m} \\ A_{21} & A_{22} & & A_{2m} \\ \vdots & & \ddots & \vdots \\ A_{m1} & A_{m2} & \cdots & A_{mm} \end{vmatrix}}$$

with the explicit form

$$J \geq A_{00} \prod_{i=1}^{m} \left(1 - \frac{A_{0i}}{A_{00}}\right)^2 = -\frac{1}{\ln(k^2)} \prod_{i=1}^{m} \left(\frac{\ln(k/k_i)}{\ln(kk_i)}\right)^2$$

The error in the fit will be small when k is the range of the basis exponentials. An illustration of the factor

$$\sigma = \prod_{i=1}^{m} \left(\frac{\ln(k/k_i)}{\ln(kk_i)}\right)$$

is given below for $m=4$ and 6 with exponentials given by the recipe $k_{i+1}=\sqrt{k_i}$.

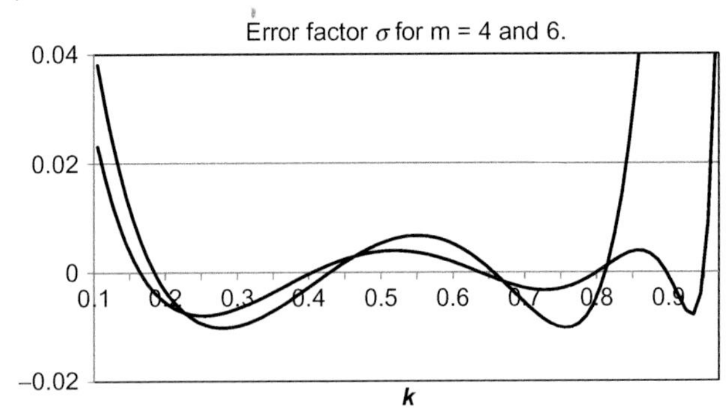

The relative error in the norm of the difference function measured by $|\sigma|$ is seen to be less than 1% when $\{0.15<k<0.825,\ m=4;\ 0.125<k<0.95,\ m=6\}$.

Experience with exponential interpolation was useful during my military service at the Research Institute for National Defence where I was assigned a study of the inhour equation for the reactivity and the critical neutron flux in a nuclear reactor. The delayed neutrons play a significant role but their origin and the rate of appearance was determined from measurements of the half-lives from decay data. Several isotopes are created in the fission, and their properties were still being chartered in the fifties. Analysis of decay data was one route to knowledge but was plagued by the numerical difficulties of fitting exponentials.

3. GENERAL BASES ON ONE-DIMENSIONAL DOMAINS

Electronic computers impacted the theory of electronic structure profoundly at the end of the fifties and search for numerical methods had priority in most groups. Boys[20] pioneered the use of Gaussian basis sets. He realized that the advent of the new computing devices made it feasible to map an electronic structure problem onto a basis of functions that offered simple integral evaluations albeit the representation of atomic and molecular orbitals required more extensive sets of basis functions. Thus was initiated a development that is still going on.

Preuss, who produced a set of integral tables for exponential type orbitals,[21] took an interest in Gaussian functions during his year at Uppsala where he explored the notion of distributed basis functions centered at points other than atomic nuclei. The limited capacity of the available

ALWAC IIIE computer prevented any conclusive results. Preuss created a group at Stuttgart where Gaussian basis functions formed the foundation for significant contributions to molecular theory.

A similar effort was undertaken by Frost[22] who used floating Gaussians. He optimized the locations for the basis functions as well as the molecular geometry. Approximate linear dependencies turned out to mar applications, and these developments were eclipsed by John Pople's switch from parameterized models, eg, Pariser–Parr–Pople (*PPP*) and Complete Neglect of Differential Overlap (*CNDO*), to first principles methods. Hall's[23] and Roothaan's[24] formulations of the nonlinear, self-consistent approach were feasible at an ever increasing scale.

Replacement of the exponential-type Slater orbitals with combinations of Gaussian functions became a common practice and so called contracted Gaussians, eg, *STO3G* with three Gaussian basis functions replacing one Slater-type orbital, came into use. The figure below illustrates that a single 1s Slater orbital with the effective charge parameter ζ in the

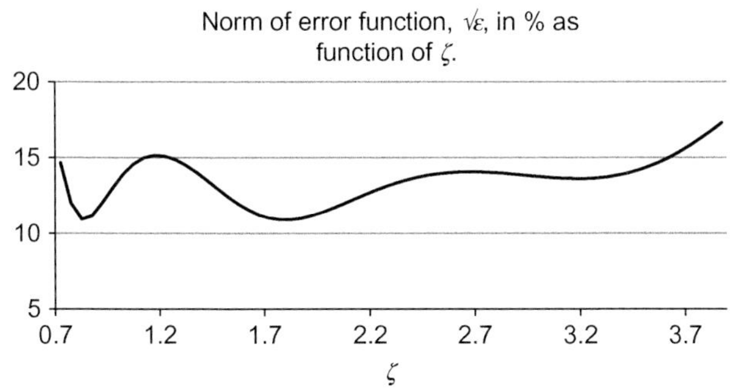

exponential can be approximated by a three term expansion in Gaussians with parameters 0.2, 1, and 5 and the norm of the difference function less than 15% for $0.7 < \zeta < 3.7$:

$$\varepsilon(\zeta) = \min_{\{a_j\}} \int_0^\infty r^2 dr \left| 2\zeta\sqrt{\zeta} e^{-\zeta r} - a_1 e^{-r^2/5} - a_2 e^{-r^2} - a_3 e^{-5r^2} \right|^2$$

Systematic efforts have been undertaken, notably by Dunning,[25] to develop bases of higher accuracy and with adaptation to various usages.

Large basis sets generate very large sets of integrals, and the search for alternatives seems to have failed in producing viable options. Friesner[26] brought in a method using point sets in configuration space, and Light[27] and collaborators realized that, when integrals were evaluated by numerical integration rules, one might use the function amplitudes at these points as the primary variational parameters.

Shull and Löwdin[28] demonstrated the inadequacy of the hydrogenic eigenstates as a basis unless the continuum was included while the Laguerre set with constant exponential parameter could cover the space. Their results were instrumental for me in a perturbation study of atomic Hartree–Fock equations,[29] and the importance of the continuum was of great concern to Löwdin over the years.

Consider the orthonormal set of functions on the domain $\{0 \leq r < \infty\}$

$$\Phi : \left\{ \phi_{n\ell}(r) = \sqrt{\frac{2\zeta n!}{(n+2\ell+2)!}} (2\zeta r)^{\ell+1} e^{-\zeta r} L_n^{(2\ell+2)}(2\zeta r) | n = 0, 1, \ldots, M \right\}$$

defined in terms of Generalized Laguerre polynomials.[30] They span a range of the function space characterized by the projector kernel

$$\begin{aligned} P(r;r') &= \sum_{n=0}^{M} \phi_{n\ell}(r)\phi_{n\ell}(r') \\ &= \frac{\sqrt{(M+1)(M+2\ell+3)}}{r'-r} \begin{vmatrix} \phi_{M+1\ell}(r) & \phi_{M+1\ell}(r') \\ \phi_{M\ell}(r) & \phi_{M\ell}(r') \end{vmatrix} \end{aligned}$$

Its rank is $M+1$, and it can be represented equally well by alternative sets of functions. The discrete variable[27] form employs the set

$$U : \left\{ u_j(r) = \frac{P(r;\, r_j)}{\sqrt{P(r_j;\, r_j)}} | \phi_{M+1\ell}(r_j) = 0, j = 0, 1, \ldots M \right\}$$

It has the property that $u_j(r_{j'}) = \delta_{jj'} \sqrt{P(r_j;\, r_j)} = \delta_{jj'} / \sqrt{w_j}$. The weights and abscissas for Laguerre quadrature over $M+1$ points constitute the set $\{w_j,\, r_j | j = 0, 1, \ldots, M\}$. Matrix elements of a local potential $V(r)$ in the basis U:

$$\left\{ V_{jj'} = \int_0^{\infty} dr u_j(r) V(r) u_{j'}(r) = \delta_{jj'} V(r_j) + \mathcal{R}_{M+1} \right\}$$

are quite simple when the error term is negligible. This is so when the potential is a linear function of r. Kinetic energy matrix elements couple the different basis functions of the set but there is a precise expansion for the form

$$r^{\ell+1} \frac{d}{dr} \left[\frac{u_j(r)}{r^{\ell+1}} \right] = \sum_{j'}^{M} u_{j'}(r) D_{j'j}$$

and no numerical errors need occur in the evaluation.

A general variational procedure based on the use of the set U needs a viable control of the error in the potential integrals and possibly additional constraints on the expansion coefficients of the trial functions. These would be of the form that a given trial function should be contained in a basis of lower rank than M.

Only modest fits of exponentials can be expected in terms of polynomials such as the Laguerre ones. The figure below illustrates a Laguerre type basis with $M=5$

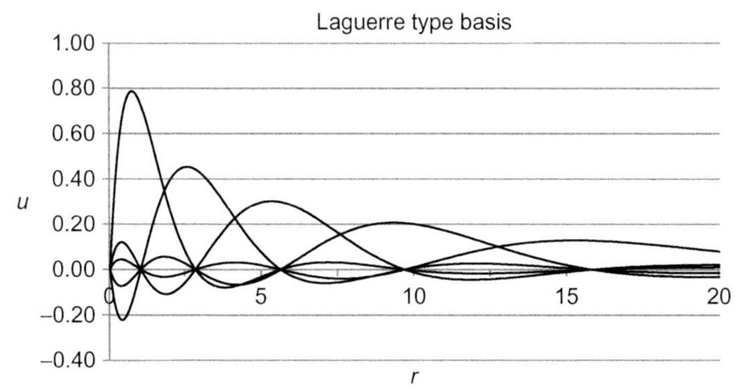

and the following picture gives the norm $\sqrt{\eta}$ of the difference function for a 1s Slater-type orbital, with orbital exponent ζ, expanded in the basis $\Phi(\zeta=0.5)$ for $M=4$ and $M=5$:

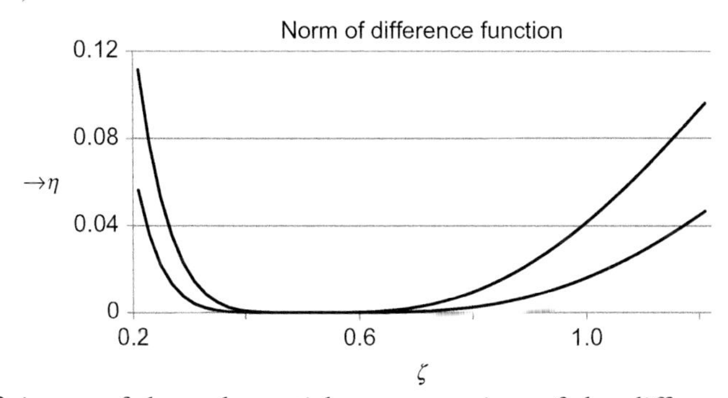

The deficiency of the polynomial representation of the difference in the exponential factor is evident also for a modest deviation.

4. MOMENTUM SPACE REPRESENTATIONS

Fourier transform methods were revived by Prosser and Blanchard[31] by showing options for integral evaluations in momentum space. An orbital of the form

$$\psi\left(\vec{r}\right) = \frac{P_\ell(r)}{r} Y_{\ell m}(\theta, \phi)$$

in standard polar coordinates has the Fourier transform

$$\tilde{\psi}\left(\vec{k}\right) = \frac{1}{2\pi\sqrt{2\pi}} \int d\vec{r} e^{-i\vec{k}\cdot\vec{r}} \psi\left(\vec{r}\right) = \frac{(-)^\ell H_\ell(k)}{k} Y_{\ell m}(\theta_k, \phi_k)$$

and the integral representation

$$\psi\left(\vec{r}\right) = \frac{1}{2\pi\sqrt{2\pi}} \int d\vec{k} e^{i\vec{k}\cdot\vec{r}} \tilde{\psi}\left(\vec{k}\right)$$

The function $H_\ell(k)$ is the Hankel transform of $P_\ell(r)$

$$H_\ell(k) = \int_0^\infty dr P_\ell(r) J_{\ell+\frac{1}{2}}(kr)\sqrt{kr}$$

Transforms of Slater orbitals are expressed in terms of Gegenbauer polynomials while Gaussian orbitals transform into Gaussian forms.

Schrödinger's functional in configuration space

$$J(\psi) = \int d\vec{r} \left\{ \left[\varepsilon - U\left(\vec{r}\right)\right] \left|\psi\left(\vec{r}\right)\right|^2 - \frac{\hbar^2}{2m} \left|\nabla\psi\left(\vec{r}\right)\right|^2 \right\}$$

takes, in momentum space, the form

$$J(\tilde{\psi}) = \int d\vec{k} \left\{ \left[\varepsilon - \frac{\hbar^2 k^2}{2m}\right] \left|\tilde{\psi}\left(\vec{k}\right)\right|^2 - \tilde{\psi}^*\left(\vec{k}\right) \int d\vec{k'}\ \tilde{U}\left(\vec{k} - \vec{k'}\right) \tilde{\psi}\left(\vec{k'}\right) \right\}$$

with the potential integral kernel

$$\tilde{U}\left(\vec{k} - \vec{k'}\right) = \left(\frac{1}{2\pi}\right)^3 \int d\vec{r}\ U\left(\vec{r}\right) e^{ir\cdot\left(\vec{k'} - \vec{k}\right)}$$

A simple screened central potential, $U\left(\vec{r}\right) = Ze^2 \exp(-r/\rho)/r$, gives the transform

$$\tilde{U}\left(\vec{k} - \vec{k'}\right) = \frac{Ze^2}{2\pi^2} \frac{\rho^2}{1 + \rho^2\left|\vec{k} - \vec{k'}\right|^2}$$

where the unscreened Coulomb potential is represented by an infinite screening length ρ.

Optimization of the functional in momentum space leads to an integral equation with a kernel that is not readily separable to admit standard Fredholm methods toward a solution. Silverstone,[32] in particular, has studied the evaluation of molecular integrals through the Fourier method.

Fock[33] observed that the integral equation in momentum space was equivalent to the equations for the irreducible representations of the three-dimensional rotation group, the Wigner functions. The change of variable

$$k = k\tan\left(\frac{\alpha}{2}\right);\; 0 \leq \alpha < \pi$$

and its consequence in the volume element

$$k^2\frac{dk}{d\alpha} = k^3\frac{\sin^2\alpha}{(1+\cos\alpha)^3}$$

admits the form

$$\int d\vec{k} f\left(\vec{k}\right) = \int d\mathbf{x}\frac{k^3 f\left(\vec{k}\right)}{(1+\cos\alpha)^3};|\mathbf{x}| = 1;$$

$$\mathbf{x} = (\chi \;\; \xi \;\; \eta \;\; \zeta); \vec{k} = \frac{k}{1+\chi}(\xi \;\; \eta \;\; \zeta)$$

or, in polar form

$$\mathbf{x} = (\cos\alpha \quad \sin\alpha\sin\theta\cos\phi \quad \sin\alpha\sin\theta\sin\phi \quad \sin\alpha\cos\theta)$$

The four-vector $\mathbf{x}$ is restricted to the unit sphere in four dimensions, and a mapping from an infinite to a finite domain is accomplished. A scaling parameter k allows focus to be placed on a relevant part of momentum space. Schrödinger's functional comes out as, for the special case of the Hydrogen atom,

$$J(\hat{\psi}) = \int d\mathbf{x}\left\{\left[(1+\chi)\frac{\varepsilon}{k} - \frac{\hbar^2 k}{2m}(1-\chi)\right]|\hat{\psi}(\mathbf{x})|^2 - \frac{e^2}{2\pi^2}\hat{\psi}^*(\mathbf{x})\int d\mathbf{x}'\frac{\hat{\psi}(\mathbf{x}')}{|\mathbf{x}'-\mathbf{x}|^2}\right\}$$

where the wave function is

$$\hat{\psi}(\mathbf{x}) = \left(\frac{k}{1+\chi}\right)^2 \tilde{\psi}\left(\vec{k}\right)$$

The first term of the integrand in the functional simplifies, for $\varepsilon = -\frac{\hbar^2 k^2}{2m}$, to $-\frac{\hbar^2 k}{m}\int d\mathbf{x}|\hat{\psi}(\mathbf{x})|^2$ and gives an indication that the Wigner functions might offer an appropriate basis.[34]

A set of Wigner functions, $\mathcal{W}: \left\{ D^j_{m'm}(\mathbf{x}) | j = 0, \frac{1}{2}, \ldots J, \ -j \leq m', m \leq j \right\}$, in the Euler–Rodrigues parameterization,[35] is characterized by the projector kernel

$$\mathcal{P}(\mathbf{x}'; \mathbf{x}) = \sum_{j=0}^{J} \frac{2j+1}{2\pi^2} \sum_{m'm=-j}^{j} D^{j*}_{m'm}(\mathbf{x}') D^j_{m'm}(\mathbf{x}) = \sum_{j=0}^{J} \frac{2j+1}{2\pi^2} U_{2j}(\mathbf{x}' \cdot \mathbf{x}).$$

The final summation over the Chebyshev polynomials of the second kind is given by the Christoffel–Darboux formula as

$$\mathcal{P}(\mathbf{x}'; \mathbf{x}) = \frac{1}{4\pi^2(1 - \mathbf{x}' \cdot \mathbf{x})} \begin{vmatrix} U_{2J+1}(1) & U_{2J+1}(\mathbf{x}' \cdot \mathbf{x}) \\ U_{2J}(1) & U_{2J}(\mathbf{x}' \cdot \mathbf{x}) \end{vmatrix}$$

These functions span a space with dimension

$$\int d\mathbf{x} \mathcal{P}(\mathbf{x}; \mathbf{x}) = \lim_{\epsilon \to 0} \frac{1}{2\epsilon} \begin{vmatrix} U_{2J+1}(1) & U_{2J+1}(1-\epsilon) \\ U_{2J}(1) & U_{2J}(1-\epsilon) \end{vmatrix} = \frac{J(2J+1)(4J+1)}{3}$$

Any function within this space can be spanned by its amplitudes at a point set:

$$\hat{\psi}(\mathbf{x}) = \sum_{n=1}^{N} \hat{\psi}(\mathbf{x}_n) u_n(\mathbf{x})$$

where the set must have the property that $u_n(\mathbf{x}_{n'}) = \delta_{n'n}$. This cannot generally be accomplished in several dimensions while it is feasible in one, as shown above. The properties of a projection kernel, such as $\mathcal{P}(\mathbf{x}'; \mathbf{x})$, admits the option of an expansion in the form

$$\hat{\psi}(\mathbf{x}) = \sum_{n=1}^{N} \widehat{\psi}_n \frac{\mathcal{P}(\mathbf{x}_n; \mathbf{x})}{\mathcal{P}(\mathbf{x}_n; \mathbf{x}_n)}$$

When the rank of the kernel is larger than N, there exist a unique solution of the equation system

$$\widehat{\psi}_n = \hat{\psi}(\mathbf{x}_n) - \sum_{n' \neq n}^{1,N} \widehat{\psi}_{n'} \frac{\mathcal{P}(\mathbf{x}_{n'}; \mathbf{x}_n)}{\mathcal{P}(\mathbf{x}_n; \mathbf{x}_n)}$$

that may be found iteratively when the projector kernel elements fall off rapidly with distance between the points.

An illustration of the possibility of using a pointwise representation of atomic orbitals on the unit sphere in four dimensions is offered by a point set associated with a particular polytope called the 600 cell.[36] The polytope {5,3,3} has 120 vertices, 720 edges, 1200 faces, and 600 equivalent tetrahedra. Each vertex defines an element of the binary icosahedral group, and the full group has the order 14,400. Its dual, the polytope {3,3,5}, has 600 vertices that are associated with the centers of the tetrahedra of {5,3,3}. Their coordinates are listed by Coxeter,[36] and each point has four nearest neighbors, the centers of the four neighboring tetrahedra. The figure below displays the elements

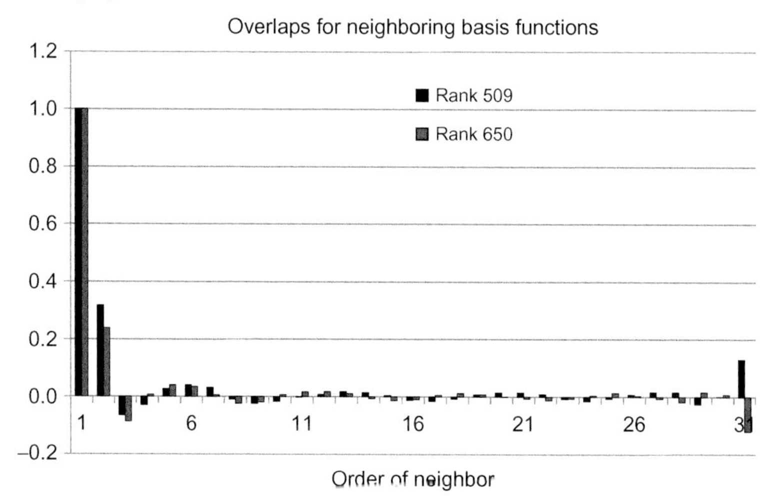

$\mathcal{P}(\mathbf{x}_{n'}; \mathbf{x}_n)/\mathcal{P}(\mathbf{x}_n; \mathbf{x}_n)$ as functions of order of neighbor for $J=5.5$ and $J=6$. The ranks of the kernels are 506 and 650, respectively. An increase of the rank to 1240 ($J=7.5$) results in a very small positive value at first neighbors while an increase to 1496 ($J=8$) gives a small negative value there.

A more detailed picture of the functions is provided in the next picture where two functions on neighboring points are plotted as functions of the arc length between the two. The arc itself is also included and is nearly linear over the interval [0 degree, 15 degree]. The finite element method uses, in its simplest form, linear interpolants as basis functions that appear quite similarly but cannot, in contrast to the Wigner functions, give a continuous gradient.

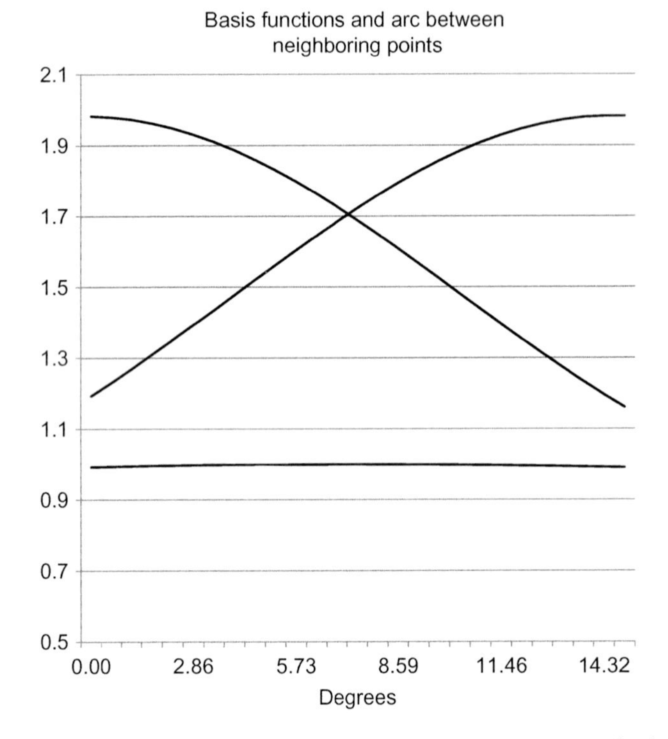

No implementation of a molecular or atomic structure calculation has appeared but exploratory efforts on the determination of Wigner functions from an isoparametric finite element are available.[37]

Talman[38] exhibited how to combine the momentum representation and the configuration space form through the use of fast Fourier transform techniques and how this can give some advantages for the evaluation of four-center Coulomb interaction integrals from numerically defined orbitals. His procedure uses the commonly employed logarithmic grids in the radial variable and its counterpart in momentum space. Thus, for the orbital Hankel transform,

$$H_\ell(k) = \int_0^\infty dr P_\ell(r) J_{\ell+\frac{1}{2}}(kr)\sqrt{kr}$$

$$\Downarrow$$

$$H_\ell(k_0 q^m) = r_0 \ln q \sum_{n=0}^{N-1} q^n P_\ell(r_0 q^n) J_{\ell+\frac{1}{2}}(k_0 r_0 q^{m+n})\sqrt{k_0 r_0 q^{m+n}}$$

Talman obtains a convolution where the Fourier summation over the spherical Bessel function is available in closed form. Some additional manipulations can be used to emphasize accuracy in the inner or outer region of an atomic domain.

Wavelets are associated with the Fourier analysis and Calais[39] initiated a study of these as basis functions. This piece of work ended with the untimely death of its author.

5. MIXED REPRESENTATIONS

Schrödinger's functional and the associated Euler equations have implicit boundary conditions, particularly for bound states. Scattering solutions are conveniently included through the Wigner–Eisenbud R-matrix formulation. Thus one considers, in a spherical atomic context, the functional

$$
\begin{aligned}
J_R(P_\ell) = & \int_a^b \mathrm{d}r \left\{ [\varepsilon - U(r)]|P_\ell(r)|^2 - \frac{\hbar^2}{2m}\left| r^{\ell+1} \frac{d}{dr}\left[\frac{P_\ell(r)}{r^{\ell+1}}\right]\right|^2 \right\} \\
& + P_\ell^*(a) Q_{\ell,a} + Q_{\ell,a}^* P_\ell(a) + P_\ell^*(b) Q_{\ell,b} + Q_{\ell,b}^* P_\ell(b) \\
& - \begin{pmatrix} Q_{\ell,a}^* & Q_{\ell,b}^* \end{pmatrix} \begin{bmatrix} \mathcal{R}_{aa} & \mathcal{R}_{ab} \\ \mathcal{R}_{ba} & \mathcal{R}_{bb} \end{bmatrix} \begin{pmatrix} Q_{\ell,a} \\ Q_{\ell,b} \end{pmatrix}
\end{aligned}
$$

and the Euler equations

$$
\begin{aligned}
[\varepsilon - U(r)] P_\ell(r) + \frac{\hbar^2}{2mr^{\ell+1}} \frac{d}{dr} r^{2\ell+2} \frac{d}{\mathrm{d}r}\left[\frac{P_\ell(r)}{r^{\ell+1}}\right] &= 0 \\
Q_{\ell,a} + \frac{\hbar^2}{2m}\left[\frac{dP_\ell(a)}{dr} - \frac{(\ell+1)P_\ell(a)}{a}\right] &= 0 \\
Q_{\ell,b} - \frac{\hbar^2}{2m}\left[\frac{dP_\ell(b)}{dr} - \frac{(\ell+1)P_\ell(b)}{b}\right] &= 0 \\
\begin{bmatrix} P_\ell(a) \\ P_\ell(b) \end{bmatrix} = \begin{bmatrix} \mathcal{R}_{aa} & \mathcal{R}_{ab} \\ \mathcal{R}_{ba} & \mathcal{R}_{bb} \end{bmatrix} \begin{pmatrix} Q_{\ell,a} \\ Q_{\ell,b} \end{pmatrix}
\end{aligned}
$$

These equations hold for both linearly independent solutions of the differential equation. Such a pair of solutions is

$$P_\ell^a(r)\wedge\frac{dP_\ell^a(a)}{dr}-\frac{(\ell+1)P_\ell^a(a)}{a}=0$$

$$P_\ell^b(r)\wedge\frac{dP_\ell^b(b)}{dr}-\frac{(\ell+1)P_\ell^b(b)}{b}=0$$

so that the R-matrix comes out as

$$\begin{bmatrix}\mathcal{R}_{aa} & \mathcal{R}_{ab}\\ \mathcal{R}_{ba} & \mathcal{R}_{bb}\end{bmatrix}$$

$$=\frac{2m}{\hbar^2}\begin{bmatrix}P_\ell^a(a) & P_\ell^b(a)\\ P_\ell^a(b) & P_\ell^b(b)\end{bmatrix}\begin{bmatrix}0 & -\frac{dP_\ell^b(a)}{dr}+\frac{(\ell+1)P_\ell^b(a)}{a}\\ \frac{dP_\ell^a(b)}{dr}-\frac{(\ell+1)P_\ell^a(b)}{b} & 0\end{bmatrix}^{-1}$$

Elements of the R-matrix define the propagation of a solution:

$$\begin{bmatrix}P_\ell(b)\\ Q_{\ell,b}\end{bmatrix}=\begin{bmatrix}1 & -\mathcal{R}_{bb}\\ 0 & \mathcal{R}_{ab}\end{bmatrix}^{-1}\begin{bmatrix}0 & \mathcal{R}_{ba}\\ 1 & -\mathcal{R}_{aa}\end{bmatrix}\begin{bmatrix}P_\ell(a)\\ Q_{\ell,a}\end{bmatrix}$$

The initial values at a determine the values at b and the cusp condition at the nucleus is contained in the R-matrix when the starting point a equals zero. Potential free domains offer solutions of the differential equation in terms of Riccati–Bessel functions,[40] $=\sqrt{2m\varepsilon}/\hbar$, and a pair of solutions are

$$P_\ell^a(r)=\begin{vmatrix}kaj_{\ell+1}(ka) & krj_\ell(kr)\\ kay_{\ell+1}(ka) & kry_\ell(kr)\end{vmatrix},\quad P_\ell^b(r)=\begin{vmatrix}kbj_{\ell+1}(kb) & krj_\ell(kr)\\ kby_{\ell+1}(kb) & kry_\ell(kr)\end{vmatrix}$$

The R-matrix is then, after a few manipulations,

$$\begin{bmatrix}\mathcal{R}_{aa} & \mathcal{R}_{ab}\\ \mathcal{R}_{ba} & \mathcal{R}_{bb}\end{bmatrix}=\frac{2m}{k\hbar^2}\frac{\begin{bmatrix}P_\ell^b(a) & 1\\ 1 & P_\ell^a(b)\end{bmatrix}}{\begin{vmatrix}kaj_{\ell+1}(ka) & kbj_{\ell+1}(kb)\\ kay_{\ell+1}(ka) & kby_{\ell+1}(kb)\end{vmatrix}}$$

and can be analytically continued to complex k-values. This is relevant for bound states where the wave function amplitude decreases exponentially toward large radial values. A requirement that the orbital should be normalized puts a condition on the values at the boundary between the inner atomic region and the asymptotic one through a relation between the amplitudes $\{P_\ell(a), Q_{\ell,a}\}$, or commonly, the logarithmic derivative $Q_{\ell,a}/P_\ell(a)$. Some algebra gives

$$P_\ell(b) = \begin{vmatrix} P_\ell(a)kbj_{\ell+1}(kb) - \frac{2m}{k\hbar^2}Q_{\ell,a}kbj_\ell(kb) & kaj_\ell(ka) \\ P_\ell(a)kby_{\ell+1}(kb) - \frac{2m}{k\hbar^2}Q_{\ell,a}kby_\ell(kb) & kay_\ell(ka) \end{vmatrix}$$

$$Q_{\ell,b} = \begin{vmatrix} Q_{\ell,a}kbj_\ell(kb) - \frac{k\hbar^2}{2m}P_\ell(a)kbj_{\ell+1}(kb) & kaj_{\ell+1}(ka) \\ Q_{\ell,a}kby_\ell(kb) - \frac{k\hbar^2}{2m}P_\ell(a)kby_{\ell+1}(kb) & kay_{\ell+1}(ka) \end{vmatrix}$$

Further examination gives that the first columns of the determinants above are proportional and have asymptotic forms for large values of the radius b:

$$\begin{pmatrix} P_\ell(a)kbj_{\ell+1}(kb) - \frac{2m}{k\hbar^2}Q_{\ell,a}kbj_\ell(kb) \\ P_\ell(a)kby_{\ell+1}(kb) - \frac{2m}{k\hbar^2}Q_{\ell,a}kby_\ell(kb) \end{pmatrix}$$

$$\times \underset{b\to\infty}{\Rightarrow} \begin{pmatrix} -P_\ell(a)\cos(kb-\ell\pi/2) - \frac{2m}{k\hbar^2}Q_{\ell,a}\sin(kb-\ell\pi/2) \\ -P_\ell(a)\sin(kb-\ell\pi/2) + \frac{2m}{k\hbar^2}Q_{\ell,a}\cos(kb-\ell\pi/2) \end{pmatrix}$$

$$= -\left|P_\ell(a) + \frac{i}{k}\left[\frac{dP_\ell(a)}{dr} - \frac{(\ell+1)P_\ell(a)}{a}\right]\right| \begin{pmatrix} \cos(kb-\ell\pi/2+\delta_\ell) \\ \sin(kb-\ell\pi/2+\delta_\ell) \end{pmatrix};$$

The phase shift of scattering theory is given as

$$\delta_\ell = \arg\left(P_\ell(a) + \frac{i}{k}\left[\frac{dP_\ell(a)}{dr} - \frac{(\ell+1)P_\ell(a)}{a}\right]\right).$$

Bound states, that have negative energy parameter and $k = i|k|$, vanish at $b = \infty$ and thus there is a quantization condition

$$\left|P_\ell(a) + \frac{1}{|k|}\left[\frac{dP_\ell(a)}{dr} - \frac{(\ell+1)P_\ell(a)}{a}\right]\right| = 0$$

This is often expressed in terms of a logarithmic derivative:

$$\frac{d}{dr}\ln\left(\frac{P_\ell(r)}{r^{\ell+1}}\right) + |k| = 0.$$

Muffin-tin orbitals are defined through numerical tables inside spheres and tails in terms of Riccati–Bessel functions for imaginary arguments outside the spheres.

Finite size atomic spheres became a vehicle for the study of crystalline materials and the study by Wigner and Seitz[41] illustrated the power of this approach. Slater built further on this by considering basis functions that had the spherical atomic features and a numerical representation close to the nuclei and were joined to plane wave functions in the interstitial regions. He named these Augmented Plane Waves[42] and numerous calculations have appeared related to systems with periodic symmetry. The technique was further extended to molecular systems where several atomic spheres were enclosed by an outer sphere, outside of which a simple potential, either zero or Coulombic, with analytical solutions to the differential equation, was applied to obtain proper asymptotic forms of the orbitals.

Systematic use of *muffin-tin orbitals* has been championed by Andersen.[43] Sets of basis orbitals have been determined with different features. Orbitals engaged in bonding are determined with energy parameters near the Fermi energy. Muffin-tin orbitals can be defined so that the atomic spheres are nonoverlapping so that the matrix element evaluations are mixtures of numerical integrals and analytical ones. Such conditions are generally easier to manage; configuration space is partitioned into disjoint domains. Molecular structures and lattices with large pockets without atoms invite the option of overlapping spherical domains. This comes at the cost of more elaborate integral evaluations, leading back to Löwdin's techniques where atomic orbitals and potential functions on one center were expanded in spherical harmonics on another one. Present computational resources exceed by far the ones that were in the hands of Löwdin so many of his notions can be and has been implemented with advantage today.

6. FOUR-COMPONENT SPINORBITALS

Several efforts are currently actively pursued into the realm of relativistic formulations of many-electron theory.[44] The basis set selection meets then with new challenges with the possible appearance of negative energy states. There must be a balance between the basis functions for the large and small components of the spinorbitals as has been pointed out commendably by Grant.[45] His analysis will be considered here from an appropriate variational functional for the radial factors of the spinorbitals for a central symmetric screened Coulomb potential. Notations are adapted from the ones used by Condon and Shortley.[46] The functional

$$J_D(F,G)=\int_0^\infty dr\left\{\begin{bmatrix}F\\G\end{bmatrix}^\dagger\begin{bmatrix}h+\sigma & -k/r\\-k/r & h-\sigma\end{bmatrix}\begin{bmatrix}F\\G\end{bmatrix}\right.$$

$$\left.+\begin{bmatrix}\frac{dF}{dr}\\ \frac{dG}{dr}\end{bmatrix}^\dagger\begin{bmatrix}\frac{G}{2}\\ -\frac{F}{2}\end{bmatrix}+\begin{bmatrix}\frac{G}{2}\\ -\frac{F}{2}\end{bmatrix}^\dagger\begin{bmatrix}\frac{dF}{dr}\\ \frac{dG}{dr}\end{bmatrix}\right\}$$

when the Euler equations apply,

$$(h+\sigma)F-\frac{dG}{dr}-\frac{kG}{r}=0\wedge\frac{dF}{dr}-\frac{kF}{r}+(h-\sigma)G=0$$

and boundary conditions are implemented to give the expected result. The spinorbital energy parameter W and the screened Coulomb potential are combined, in wave number units, as

$$h(r)=\frac{W}{\hbar c}+\frac{e^2Z(r)}{r\hbar c}$$

while $\sigma=\mu c/\hbar$ denotes the inverse Compton scattering length for the electron. The total angular momentum state is characterized by the quantum number $k=\pm1,\pm2,\ldots$. The Coulomb singularity at the origin requires detailed attention in order to maintain a Hermitian form but is occasionally circumvented by the use of a finite size nuclear charge distribution.

There is no compelling reason to use different basis sets for the F and G, thus, in a modest effort, $F(r)=u(r)f_u+v(r)f_v; G(r)=u(r)g_u+v(r)g_v$; brings the functional to a sesquilinear form

$$J_D=\begin{bmatrix}f_u\\f_v\\g_u\\g_v\end{bmatrix}^\dagger\begin{bmatrix}h_{uu}+\sigma & h_{uv} & -p_{uu} & -p_{uv}\\ h_{vu} & h_{vv}+\sigma & -p_{vu} & -p_{vv}\\ -p_{uu} & p_{uv} & h_{uu}-\sigma & h_{uv}\\ p_{vu} & p_{vv} & h_{vu} & h_{vv}-\sigma\end{bmatrix}\begin{bmatrix}f_u\\f_v\\g_u\\g_v\end{bmatrix}$$

The matrix elements are denoted in the conventional fashion with the caveat that

$$\int_0^\infty dr\,u(r)v(r)/r=0$$

and thus

$$p_{uv}=\frac{1}{2}\int_0^\infty dr\left[\frac{du(r)}{dr}v(r)-u(r)\frac{dv(r)}{dr}\right]=-p_{vu}$$

The four eigenvalues of the matrix can be expressed in the form[47]

$$\frac{1}{2}(h_{uu}+h_{vv})$$
$$+\begin{cases} \sqrt{\alpha+2\beta\cos\gamma}+\sqrt{\alpha-\beta\cos\gamma+\beta\sqrt{3}\sin\gamma}+\sqrt{\alpha-\beta\cos\gamma-\beta\sqrt{3}\sin\gamma} \\ \sqrt{\alpha+2\beta\cos\gamma}-\sqrt{\alpha-\beta\cos\gamma+\beta\sqrt{3}\sin\gamma}-\sqrt{\alpha-\beta\cos\gamma-\beta\sqrt{3}\sin\gamma} \\ -\sqrt{\alpha+2\beta\cos\gamma}+\sqrt{\alpha-\beta\cos\gamma+\beta\sqrt{3}\sin\gamma}-\sqrt{\alpha-\beta\cos\gamma-\beta\sqrt{3}\sin\gamma} \\ -\sqrt{\alpha+2\beta\cos\gamma}-\sqrt{\alpha-\beta\cos\gamma+\beta\sqrt{3}\sin\gamma}+\sqrt{\alpha-\beta\cos\gamma-\beta\sqrt{3}\sin\gamma} \end{cases}$$

The parameters of these forms derive from the matrix elements, and it follows that both α and β are dominated by the term $\frac{1}{3}\sigma^2$ and that $\gamma \cong \frac{\alpha-\beta}{\sigma^2}\sqrt{3}$, a small quantity. The first two eigenvalues give negative energy states while the last two give the relevant electronic states.

The purpose of the exercise with a basis of two functions is to demonstrate that the kinetic operator, that here manifests itself in the integral p_{uv}, disappears in a single function basis when the Hermitian functional form applies. Boundary conditions should properly be included in the functional in the way used in the Schrödinger case.

A conventional way of studying the four-dimensional matrix problem involves a Löwdin partitioning procedure[48] and gives a two-dimensional, implicit eigenvalue condition. Thus

$$\left\{\begin{bmatrix} h_{uu}-\sigma & h_{uv} \\ h_{vu} & h_{vv}-\sigma \end{bmatrix} - \begin{bmatrix} -p_{uu} & p_{uv} \\ p_{vu} & -p_{vv} \end{bmatrix}\begin{bmatrix} h_{uu}+\sigma & h_{uv} \\ h_{vu} & h_{vv}+\sigma \end{bmatrix}^{-1}\begin{bmatrix} -p_{uu} & -p_{uv} \\ -p_{vu} & -p_{vv} \end{bmatrix}\right\}$$
$$\times\begin{bmatrix} g_u \\ g_v \end{bmatrix} = 0$$

where the second term in the bracket is an inner projection[49] of the full inverse operator, and it is concluded that this representation of the kinetic energy operator is less than the full one and that detailed examination of the basis set is called for to secure accuracy.

7. EPITOME

Atomic orbitals, be they numerical, analytical, or mixed numerical/analytical, serve to discretize the configuration space and the quantum mechanical operators. Such a representation is readily brought to a comprehensive form through the algebra of second quantization, a *stylish technique* according to Slater.[50] Löwdin realized that there might be a useful link between configuration space formulations and Fock space methods and assigned this author to learn and to present the algebraic odds and ends as the first assignment upon being inducted into the Uppsala Quantum Chemistry Group. Field-theoretical methods had shown their potential in connection with studies of the degenerate electron gas and metallic conduction and were admitted anew at Uppsala through Lundqvist and Hedin.[51] Numerous advantages are offered through the second quantization algebra, but the general usage would take its good time making. Similarly as group theory had much resistance to overcome, remember *gruppenpest*, before earning its place as a most versatile tool in molecular physics and chemistry there was much reluctance within the molecular electronic structure circles to accept second quantization and its algebra as the comprehensive formulation of the quantum theory of the discretized configuration space.

Dimensionality of the dataset generated by a basis, formed as the union of many atomic bases, is a major concern for modern electronic structure efforts. It was particularly disturbing that the array of Coulomb interaction integrals increased with fourth power of dimension of the basis set. Thus it was that Beebe investigated the rank of density–density interaction array and found substantial linear dependencies and a Cholesky separable representation even for quite modest basis sets.[52] The Cholesky form is an inner projection of the full array, and thus it provides an approach from below of the total electron–electron interaction energy, with a controllable error. This latter feature is not readily secured in other approximate forms as density fittings and resolution-of-the-identity methods.

Separable representations of the Coulomb array can be constructed with error control from certain integral inequalities.[53] These should also be compared to formulations of the Coulomb and Breit interactions as integrals over field intensities

$$\int d\vec{r}\, d\vec{r}\,' \frac{q(\vec{r})q(\vec{r}\,')}{|\vec{r}-\vec{r}\,'|} = \int d\vec{r}\,\left|\vec{E}(\vec{r})\right|^2/4\pi$$

$$\int d\vec{r}\, d\vec{r}\,' \left[\frac{\vec{I}(\vec{r})\cdot\vec{I}(\vec{r}\,')}{2c^2|\vec{r}-\vec{r}\,'|} + \frac{\vec{I}(\vec{r})\cdot(\vec{r}-\vec{r}\,')(\vec{r}-\vec{r}\,')\cdot\vec{I}(\vec{r}\,')}{2c^2|\vec{r}-\vec{r}\,'|^3}\right]$$
$$= \int d\vec{r}\left|\vec{H}(\vec{r})\right|^2/4\pi c$$

The fields diminish as the second power of the inverse distance from the sources and similarly as overlap integrals for two-center sources. Multipole expansions are commonly used for these types of integrals.

The charge and current densities above contribute, in their second quantization form, to the total energy functional in the many-electron approximation to a proper and consistent field theoretical treatment.

Löwdin maintained the belief that theories could be improved at a rate comparable to the one for advances in computer technology. The main difference is though that computer science has a larger economy to draw on than electronic structure theory. Development of the theory is also dwarfed by applied theory where conventional methods are considered as *good enough* and practitioners have little background for the critical evaluation of their results. Löwdin was the eternal optimist and showed great trust in his younger co-workers by delegating not only scientific problems but also major administrative charges.

It is an honor and a pleasure to be allowed to add to this tribute to Löwdin's legacy.

REFERENCES

1. Waller, I. Zur Frage der Einwirkung der Wärmebevegung auf die Interferenz von Röntgenstrahlen. *Z. Phys.* **1923**, *17*, 398–408.
2. Löwdin, P.-O. *A Theoretical Investigation into Some Properties of Ionic Crystals. A Quantum Mechanical Treatment of the Cohesive Energy, the Interionic Distance, the Elastic Constants, and the Compression at High Pressures with Numerical Applications to Some Alkali Halides*; Almqvist & Wiksells Boktryckeri AB: Uppsala, 1948.
3. Hartree, D. R. Results of Calculations of Atomic Wave Functions IV—Results for F^-, Al^{+3}, and Rb^+. *Proc. Roy. Soc. Lond. A* **1935**, *151*, 96–105.
4. Hartree, D. R.; Hartree, W. Self-Consistent Field, with Exchange, for Cl^-. *Proc. Roy. Soc. Lond. A* **1936**, *156*, 45–62.
5. Fock, V.; Petrashen, M. Self-Consistent Field with Exchange for Lithium. *Phys. Z. Sow.* **1935**, *8*, 547–561.
6. Fock, V.; Petrashen, M. On the Numerical Solution of Generalized Equations of the Self-Consistent Field. *Phys. Z. Sow.* **1934**, *6*, 368–414.
7. Hartree, D. R.; Hartree, W. Self-Consistent Field, with Exchange, for Potassium and Argon. *Proc. Roy. Soc. Lond. A* **1938**, *166*, 450–464.
8. Slater, J. C. Atomic Shielding Constants. *Phys. Rev.* **1930**, *36*, 57–64.
9. Zener, C. Analytic Atomic Wave Functions. *Phys. Rev.* **1930**, *36*, 51–56.
10. Guillemin, V., Jr.; Zener, C. Über eine einfache Eigenfunktion für den Grundzustand des Li-Atoms und der Ionen mit drei Elektronen. *Z. Phys.* **1930**, *61*, 199–205.
11. Pauling, L. The Theoretical Prediction of Physical Properties of Many-Electron Atoms and Ions. Mole Refraction, Diamagnetic Susceptibility, and Extension in Space. *Proc. Roy. Soc. Lond. A* **1927**, *114*, 181–211.
12. Mulliken, R. S.; Rieke, C. A.; Orloff, D.; Orloff, H. Formulas and Numerical Tables for Overlap Integrals. *J. Chem. Phys.* **1949**, *17*, 1248–1267.

13. Kotani, M.; Amemiya, A.; Ishiguro, E.; Kimura, T. *Tables of Molecular Integrals*; Maruzen Company Ltd.: Tokyo, 1955.
14. Löwdin, P.-O. Studies of Atomic Self-Consistent Fields. I. Calculation of Slater Functions. *Phys. Rev.* **1953**, *90*, 120–125; Studies of Atomic Self-Consistent Fields. II. Interpolation Problems. *Phys. Rev.* **1954**, *94*, 1600–1609.
15. Löwdin, P.-O.; Appel, K. Studies of Atomic Self-Consistent Fields. Analytic Wave Functions for the Argon-Like Ions and for the First Row of the Transition Metals. *Phys. Rev.* **1956**, *103*, 1746–1755.
16. Fröman, A.; Linderberg, J. Inception of Quantum Chemistry at Uppsala. In *Acta Universitatis Upsaliensis C. Organization and History*, Vol. 78, 2006; pp 1–46.
17. Hallberg, T. J. *IT-gryning*. Studentlitteratur: Linköping, 2007.
18. Lanczos, C. *Applied Analysis*; Prentice Hall, Inc.: Englewood Cliffs, NJ, 1956.
19. Ref.[15], Eq. (10), p. 1747.
20. Boy, S. F. Electronic Wave Functions. I. A General Method of Calculation for the Stationary States of Any Molecular System. *Proc. Roy. Soc. A* **1950**, *200*, 542–554.
21. Preuss, H. *Integraltafeln für Quantenchemie*; Springer-Verlag: Berlin, 1956–1961.
22. Frost, A. A. Floating Spherical Gaussian Orbital Model of Molecular Structure. I. Computational Procedure. LiH as an Example. *J. Chem. Phys.* **1967**, *47*, 3707–3713; Floating Spherical Gaussian Orbital Model of Molecular Structure. II. One- and Two-Electron-Pair Systems. *J. Chem. Phys.* **1967**, *47*, 3714–3716.
23. Hall, G. G. The Molecular Orbital Theory of Chemical Valency. 8. A Method of Calculating Ionization Potentials. *Proc. Roy. Soc. A* **1951**, *205*, 541–552.
24. Roothaan, C. C. J. New Developments in Molecular Orbital Theory. *Rev. Mod. Phys.* **1951**, *23*, 69–89.
25. Dunning, T. H., Jr. Gaussian Basis Functions for Use in Molecular Calculations. I. Contraction of (9s5p) Atomic basis Sets for the First-Row Atoms. *J. Chem. Phys.* **1970**, *53*, 2823–2833; Gaussian basis sets for use in correlated molecular calculations. I. The atoms boron through neon and hydrogen. *J. Chem. Phys.* **1989**, *90*, 1007–1023.
26. Friesner, R. A. Solution of Self-Consistent Field Electronic Structure Equations by a Pseudospectral Method. *Chem. Phys. Lett.* **1985**, *116*, 39–43; Solution of the Hartree-Fock Equations by a Pseudospectral Method: Application to Diatomic Molecules. *J. Chem. Phys.* **1986**, *85*, 1462–1468; Solution of the Hartree-Fock Equations for Polyatomic Molecules by a Pseudospectral Method. *J. Chem. Phys.* **1987**, *86*, 3522–3531.
27. Light, J. C.; Hamilton, I. P.; Lill, J. V. Generalized discrete variable approximation in quantum mechanics. *J. Chem. Phys.* **1985**, *82*, 1400–1409.
28. Shull, H.; Löwdin, P.-O. Role of the Continuum in Superposition of Configurations. *J. Chem. Phys.* **1955**, *23*, 1362.
29. Linderberg, J. Perturbation Treatment of Hartree-Fock Equations. *Phys. Rev.* **1961**, *121*, 816–819.
30. Abramowitz, M.; Stegun, I. A. *Handbook of Mathematical Functions with Formulas, Graphs, and Mathematical Tables*; National Bureau of Standards: Washington, D.C., 1965; p 773 ff. 4th Printing.
31. Prosser, F. P.; Blanchard, C. H. On the Evaluation of Two-Center Integrals. *J. Chem. Phys.* **1962**, *36*, 1112.
32. Silverstone, H. J. Analytical Evaluation of Multicenter Integrals of r_{12}^{-1} with Slater-Type Atomic Orbitals. I. (1-2)-Type Three-Center Integrals. *J. Chem. Phys.* **1968**, *48*, 4098–4106; Silverstone, H. J. Analytical Evaluation of Multicenter Integrals of r_{12}^{-1} with Slater-Type Atomic Orbitals. II. Three-Center Nuclear Attraction Integrals. *J. Chem. Phys.* **1968**, *48*, 4106–4108.
33. Fock, V. Zur Theorie des Wasserstoffatoms. Z. *Phys.* **1935**, *98*, 145–154.
34. Biedenharn, L. C.; Louck, J. D. *Angular Momentum in Quantum Physics. Theory and Applications*; Addison-Wesley Publishing Company: Reading, MA, 1981; 335 ff.

35. Ref.[34], p. 66.
36. Coxeter, H. S. M. *Regular Polytopes*; Dover Publications, Inc.: New York, NY, 1973.p. 156. J. L. is grateful to the late A. J. Coleman for pointing me in the direction of Coxeter, to A. Weiss and T. Bisztricsky for admission to A NATO Advanced Workshop on polytopes, and R.V. Moody for giving the name of the binary icosahedral group.
37. Linderberg, J.; Öhrn, Y. Isoparametric Finite Elements on Hyperspheres. In: Bowman, J. M. Ed.; Advances in Molecular Vibrations and Chemical Dynamics, Vol. 2A; JAI Press, Inc.: Greenwich, CT, 1994; pp 33–49. Padkjær, S. B. personal communication.
38. Talman, J. D. Numerical Fourier and Bessel Transforms in Logarithmic Variables. *J. Comp. Phys.* **1978**, *39*, 35–48; Numerical Calculation of Four-Center Coulomb Integrals. *J. Chem. Phys.* **1984**, *80*, 2000–2008; Optimised Numerical Basis Functions. *Mol. Phys.* **2010**, *108*, 3289–3297.
39. Calais, J.-L. Wavelets—Something for Quantum Chemistry? *Int. J. Quant. Chem.* **1996**, *58*, 541–548.
40. Ref.[30]. p. 445.
41. Wigner, E.; Seitz, F. On the Constitution of Metallic Sodium. *Phys. Rev.* **1933**, *43*, 804–810; On the Constitution of Metallic Sodium. II. *Phys. Rev.* **1934**, *46*, 509–524.
42. Slate, J. C. An Augmented Plane Wave Method for the Periodic Potential Problem. *Phys. Rev.* **1953**, *92*, 603–608; Wave Functions in a Periodic Potential. *Phys. Rev.* **1937**, *51*, 846–851.
43. Andersen, O. K. NMTOs and Their Wannier Functions. In *Correlated Electrons: From Models to Materials*, Reihe Modeling and Simulation, Vol. 2; Forschungszentrum Jülich: Jülich, 2012. Chapter 3.
44. Saue, T. Relativistic Hamiltonians for Chemistry: A Primer. *ChemPhysChem* **2011**, *12*, 3077–3094.
45. Grant, I. P. Conditions for Convergence of Variational Solutions of Dirac's Equation in a Finite Basis. *Phys. Rev.* **1982**, *A25*, 1230–1232.
46. Condon, E. U.; Shortle, G. H. *The Theory of Atomic Spectra*. Cambridge University Press: Cambridge, 1963; 125 ff.
47. Linderberg, J. Algebraic Reductions of the Real Symmetric 4 × 4 Secular Problem. *Int. J. Quant. Chem.* **1981**, *19*, 237–249.
48. Löwdin, P. O. *Linear Algebra in Quantum Theory*; Wiley: New York, NY, 1998. p. 65.
49. Ref.[48]. p. 288.
50. Slater, J. C. Second Quantization and Atomic Spectroscopy. *Am. J. Phys.* **1968**, *36*, 69.
51. Hedin, L.; Lundqvist, S. O. *Introduction to the Field Theoretical Approach to the Many-Electron Problem*; Technical Note, Quantum Chemistry Group, Uppsala University: Uppsala, Sweden, Nov 15, 1960 (unpublished).
52. Beebe, N. H. F.; Linderberg, J. Simplifications in the Generation and Transformations of Two-Electron Integrals in Molecular Calculations. *Int. J. Quant. Chem.* **1977**, *12*, 683–705.
53. Linderberg, J. Separable Representations of the Coulomb Interaction. *Int. J. Quant. Chem.* **2009**, *109*, 2866–2871.

CHAPTER THREE

The Time-Dependent Variational Principal in Quantum Mechanics and Its Application

Yngve Öhrn[1]
QTP, University of Florida, Gainesville, FL, United States
[1]Corresponding author: e-mail address: ohrn@qtp.ufl.edu

Contents

Abstract

This account of the time-dependent variational principle is presented in memory of Per-Olov Löwdin on the occasion of the centenary of his birth. The material presented here has been published as part of a book chapter,[1] and is reintroduced here in recognition of Löwdin's interest in this topic.[2] Also the importance of the use of coherent state parameters as functions of time is emphasized, as well as the connection to the electron nuclear dynamics (END) theory.[3]

1. BASIC EQUATIONS

The time-dependent variational principle in quantum mechanics starts from the quantum mechanical action[1–6]

$$A = \int_{t_1}^{t_2} L(\psi^*, \psi)\, dt, \tag{1}$$

where the quantum mechanical Lagrangian is

$$L(\psi^*, \psi) = \left\langle \psi \middle| i\hbar \frac{\partial}{\partial t} - H \middle| \psi \right\rangle / \langle \psi | \psi \rangle, \tag{2}$$

Advances in Quantum Chemistry, Volume 74
ISSN 0065-3276
http://dx.doi.org/10.1016/bs.aiq.2016.04.002

and H is the quantum mechanical Hamiltonian of the system. When the wave function ψ is completely general and allowed to vary in the entire Hilbert space, then the TDVP yields the time-dependent Schrödinger equation. However, if the possible wave function variations are restricted in any way, such as is the case for a wave function represented in a finite basis and being of a particular functional form, then the corresponding Lagrangian will generate an approximation to the Schrödinger time evolution.

We consider a wave function expressed in terms of a set of (in general complex) parameters $\mathbf{z}$ (eg, molecular orbital coefficients, average nuclear positions and momenta). These parameters are time dependent and can be expressed as $z_\alpha \equiv z_\alpha(t)$ and thought of as arranged in a column or row array. We write

$$\psi = \psi(\mathbf{z}) = |\mathbf{z}\rangle \tag{3}$$

and employ the principle of least action

$$\delta A = \int_{t_1}^{t_2} \delta L(\psi^*, \psi)\, dt = 0 \tag{4}$$

with the Lagrangian

$$L = \left[\frac{i}{2}\langle \mathbf{z}|\dot{\mathbf{z}}\rangle - \frac{i}{2}\langle \dot{\mathbf{z}}|\mathbf{z}\rangle - \langle \mathbf{z}|H|\mathbf{z}\rangle\right] / \langle \mathbf{z}|\mathbf{z}\rangle, \tag{5}$$

where we have put $\hbar = 1$, and write the symmetric form of the time-derivative term. One way to see how this can come about is to consider

$$\int_{t_1}^{t_2} \frac{\frac{\partial}{\partial t}\langle \mathbf{z}|\mathbf{z}\rangle}{\langle \mathbf{z}|\mathbf{z}\rangle} dt = 0 \tag{6}$$

which holds if we require

$$\langle \mathbf{z}(t_2)|\mathbf{z}(t_2)\rangle = \langle \mathbf{z}(t_1)|\mathbf{z}(t_1)\rangle, \tag{7}$$

as our boundary condition.

The variation of the Lagrangian can be expressed in more detail as

$$\begin{aligned}
\delta L = &\frac{i}{2}\left[\langle \delta\mathbf{z}|\dot{\mathbf{z}}\rangle - \langle \delta\dot{\mathbf{z}}|\mathbf{z}\rangle - \langle \delta\mathbf{z}|H|\mathbf{z}\rangle\right] / \langle \mathbf{z}|\mathbf{z}\rangle \\
&- \left[\frac{i}{2}\langle \mathbf{z}|\dot{\mathbf{z}}\rangle - \frac{i}{2}\langle \dot{\mathbf{z}}|\mathbf{z}\rangle - \langle \mathbf{z}|H|\mathbf{z}\rangle\right] \times \langle \delta\mathbf{z}|\mathbf{z}\rangle / \langle \mathbf{z}|\mathbf{z}\rangle^2 \\
&+ \text{complex conjugate}
\end{aligned} \tag{8}$$

We would like to get rid of all the terms that contain the variation $\delta\dot{\mathbf{z}}$. To this end we add and subtract the total time derivative

$$\frac{d}{dt}\frac{\langle\delta\mathbf{z}|\mathbf{z}\rangle}{\langle\mathbf{z}|\mathbf{z}\rangle}=\frac{\langle\delta\dot{\mathbf{z}}|\mathbf{z}\rangle+\langle\delta\mathbf{z}|\dot{\mathbf{z}}\rangle}{\langle\mathbf{z}|\mathbf{z}\rangle}-\frac{\langle\delta\mathbf{z}|\mathbf{z}\rangle}{\langle\mathbf{z}|\mathbf{z}\rangle^2}\frac{d}{dt}\langle\mathbf{z}|\mathbf{z}\rangle, \tag{9}$$

and its complex conjugate to write

$$\begin{aligned}\delta L=&\frac{i}{2}\frac{\langle\delta\mathbf{z}|\dot{\mathbf{z}}\rangle}{\langle\mathbf{z}|\mathbf{z}\rangle}+\frac{i}{2}\frac{\langle\delta\mathbf{z}|\dot{\mathbf{z}}\rangle}{\langle\mathbf{z}|\mathbf{z}\rangle}-\frac{i}{2}\frac{\langle\delta\mathbf{z}\mid\mathbf{z}\rangle}{\langle\mathbf{z}|\mathbf{z}\rangle^2}\frac{d}{dt}\langle\mathbf{z}|\mathbf{z}\rangle-\frac{i}{2}\frac{d}{dt}\frac{\langle\delta\mathbf{z}|\mathbf{z}\rangle}{\langle\mathbf{z}|\mathbf{z}\rangle}\\&-\frac{\langle\delta\mathbf{z}|H|\mathbf{z}\rangle}{\langle\mathbf{z}|\mathbf{z}\rangle}\\&-\frac{\langle\delta\mathbf{z}|\mathbf{z}\rangle}{\langle\mathbf{z}|\mathbf{z}\rangle^2}\left[\frac{i}{2}\langle\mathbf{z}|\dot{\mathbf{z}}\rangle-\frac{i}{2}\langle\dot{\mathbf{z}}|\mathbf{z}\rangle-\langle\mathbf{z}|H|\mathbf{z}\rangle\right]\\&+\text{complex conjugate.}\end{aligned} \tag{10}$$

The time integration involved in $\delta A=0$ eliminates the total derivative terms since due to the boundary conditions they are zero, ie,

$$\langle\delta\mathbf{z}(t_2)|\mathbf{z}(t_2)\rangle-\langle\delta\mathbf{z}(t_1)|\mathbf{z}(t_1)\rangle=0, \tag{11}$$

which follows from Eq. (7) and the fact that $|\delta\mathbf{z}\rangle$ and $\langle\delta\mathbf{z}|$ are independent variations.

The surviving terms of δL can be expressed as

$$\begin{aligned}&i\frac{\langle\delta\mathbf{z}|\dot{\mathbf{z}}\rangle}{\langle\mathbf{z}|\mathbf{z}\rangle}-\frac{\langle\delta\mathbf{z}|H|\mathbf{z}\rangle}{\langle\mathbf{z}|\mathbf{z}\rangle}-\frac{\langle\delta\mathbf{z}|\mathbf{z}\rangle}{\langle\mathbf{z}|\mathbf{z}\rangle^2}\left[i\langle\mathbf{z}|\dot{\mathbf{z}}\rangle-\langle\mathbf{z}\mid H\mid\mathbf{z}\rangle\right]\\&+\text{complex conjugate.}\end{aligned} \tag{12}$$

Since $\delta\mathbf{z}$ and $\delta\mathbf{z}^*$ can be considered as independent variations one can conclude that

$$\left(i\frac{\partial}{\partial t}-H\right)|\mathbf{z}\rangle=\frac{\langle\mathbf{z}|i\partial/\partial t\quad H|\mathbf{z}\rangle}{\langle\mathbf{z}|\mathbf{z}\rangle}|\mathbf{z}\rangle, \tag{13}$$

which is the Schrödinger equation if the right hand side is zero. By explicitly considering the overall wave function phase we can eliminate the right hand side. We write

$$|\mathbf{z}\rangle\longrightarrow e^{-i\gamma}|\mathbf{z}\rangle \tag{14}$$

with γ only a function of time and obtain

$$\begin{aligned}\langle\mathbf{z}|i\partial/\partial t-H|\mathbf{z}\rangle &\longrightarrow\langle\mathbf{z}|e^{i\gamma}(i\partial/\partial t-H)e^{-i\gamma}|\mathbf{z}\rangle\\ &=\dot{\gamma}\langle\mathbf{z}|\mathbf{z}\rangle+\langle\mathbf{z}|i\partial/\partial t-H|\mathbf{z}\rangle=0,\end{aligned}\tag{15}$$

which means that the time derivative of the overall phase must be

$$-\dot{\gamma}=\frac{\langle\mathbf{z}|i\partial/\partial t-H|\mathbf{z}\rangle}{\langle\mathbf{z}|\mathbf{z}\rangle}.\tag{16}$$

We introduce the notations $S(\mathbf{z},\mathbf{z}^*)=\langle\mathbf{z}|\mathbf{z}\rangle$ and $E(\mathbf{z},\mathbf{z}^*)=\langle\mathbf{z}|H|\mathbf{z}\rangle/\langle\mathbf{z}\mid\mathbf{z}\rangle$, which leads to the equation

$$-\dot{\gamma}=\frac{i}{2}\sum_{\alpha}\left[\dot{z}_\alpha\frac{\partial}{\partial z_\alpha}-\dot{z}_\alpha^*\frac{\partial}{\partial z_\alpha^*}\right]\ln S(\mathbf{z},\mathbf{z}^*)-E(\mathbf{z},\mathbf{z}^*),\tag{17}$$

where we have used the chain rule of differentiation. Note that for a stationary state all $\dot{z}=0$, and $E(\mathbf{z},\mathbf{z}^*)=E$ yielding $\gamma=Et$ and the phase factor $e^{-iEt/\hbar}$. The above expression for δA can be similarly written as

$$0=\delta A=\int_{t_1}^{t_2}\delta L dt\tag{18}$$

$$=\int_{t_1}^{t_2}\left\{\sum_{\alpha}\left[\sum_{\beta}i\frac{\partial^2\ln S}{\partial z_\alpha^*\partial z_\beta}\dot{z}_\beta-\frac{\partial E}{\partial z_\alpha^*}\right]\delta z_\alpha^*\right.\tag{19}$$

$$\left.+\sum_{\alpha}\left[\sum_{\beta}-i\frac{\partial^2\ln S}{\partial z_\alpha\partial z_\beta^*}\dot{z}_\beta^*-\frac{\partial E}{\partial z_\alpha}\right]\delta z_\alpha\right\}dt,\tag{20}$$

where, say $\delta\mathbf{z}=\sum_\alpha\frac{\partial}{\partial z_\alpha}\delta z_\alpha$, and since δz_α and δz_α^* are independent variations we can write

$$i\sum_{\beta}C_{\alpha\beta}\dot{z}_\beta=\frac{\partial E}{\partial z_\alpha^*},\tag{21}$$

$$-i\sum_{\beta}C_{\alpha\beta}^*\dot{z}_\beta^*=\frac{\partial E}{\partial z_\alpha},\tag{22}$$

where $C_{\alpha\beta}=\partial^2\ln S/\partial z_\alpha^*\partial z_\beta$. We introduce the matrix $\mathbf{C}=\{C_{\alpha\beta}\}$ and assume it to be invertible to write

$$\begin{bmatrix}\dot{\mathbf{z}}\\ \dot{\mathbf{z}}^*\end{bmatrix}=\begin{bmatrix}-i\mathbf{C}^{-1} & \mathbf{0}\\ \mathbf{0} & i\mathbf{C}^{*-1}\end{bmatrix}\begin{bmatrix}\partial E/\partial\mathbf{z}^*\\ \partial E/\partial\mathbf{z}\end{bmatrix},\tag{23}$$

which is a matrix equation in block form. One may introduce a generalized Poisson bracket by considering two general differentiable functions $f(\mathbf{z}, \mathbf{z}^*)$ and $g(\mathbf{z}, \mathbf{z}^*)$ and write

$$\begin{aligned}\{f, g\} &= \left[\partial f/\partial \mathbf{z}^T \;\; \partial f/\partial \mathbf{z}^\dagger\right]\begin{bmatrix}-i\mathbf{C}^{-1} & \mathbf{0} \\ \mathbf{0} & i\mathbf{C}^{*-1}\end{bmatrix}\begin{bmatrix}\partial g/\partial \mathbf{z}^* \\ \partial g/\partial \mathbf{z}\end{bmatrix} \\ &= -i\sum_{\alpha,\beta}\left[\frac{\partial f}{\partial z_\alpha}\left(\mathbf{C}^{-1}\right)_{\alpha\beta}\frac{\partial g}{\partial z_\beta^*} - \frac{\partial g}{\partial z_\alpha}\left(\mathbf{C}^{-1}\right)_{\alpha\beta}\frac{\partial f}{\partial z_\beta^*}\right].\end{aligned} \tag{24}$$

It follows that

$$\begin{aligned}\dot{\mathbf{z}} &= \{\mathbf{z}, E\}, \\ \dot{\mathbf{z}}^* &= \{\mathbf{z}^*, E\},\end{aligned} \tag{25}$$

which shows that the time evolution of the wave function parameters, and thus, of the wave function, is governed by Hamilton-like equations. Such a set of coupled first-order differential equations in time plus the equation for the evolution of the overall phase can be integrated by a great variety of methods. Schematically we write $\dot{\mathbf{z}}(t) = F(\mathbf{z}(t))$ and proceed by finite steps such that $\Delta\mathbf{z}_i = F(\mathbf{z}_i)\Delta t$ and $\mathbf{z}_{i+1} = \mathbf{z}_i + \Delta\mathbf{z}_i$; $\mathbf{z}(0) = \mathbf{z}_0$.

2. COHERENT STATES

The discussion of the TDVP in the previous chapter exploits a family of state vectors $|\mathbf{z}\rangle$ labeled by a set of time dependent, and complex parameters $\mathbf{z} = \{z_1, z_2, \ldots, z_M\}$.[7] Such parameter spaces should be continuous and *complete* in the sense that as the state vector evolves in time and the complex parameters assume all possible values throughout their range, all states of the particular form $|\mathbf{z}\rangle$ are obtained. Such demands on parameter spaces are satisfied by (generalized) *coherent states*[8] which relate the parameters to a particular Lie group G. Typically one chooses a unitary irreducible representation of G and a corresponding lowest (or highest) weight state $|0\rangle$ of such a representation. A maximal subgroup H of G that leaves $|0\rangle$ invariant is called the stability group and the cosets of G by H provide a suitable nonredundant set of parameters to label the coherent state. In general one would also require the existence of a positive measure $d\mathbf{z}$ on this parameter space such that when the integral

$$\int |\mathbf{z}\rangle|\langle\mathbf{z}|d\mathbf{z} = I \tag{26}$$

is taken over the range of the parameter space one obtains the identity.

Already the notion of continuity of the labels rules out as coherent states some familiar sets of states used in quantum mechanics. For instance, a set of discrete orthogonal states, such as a set of orthonormal basis functions $\{|n\rangle\}$ cannot be coherent states.

2.1 Gaussian Wave Packet as a Coherent State

A Gaussian wave packet in one dimension can be expressed as $(\hbar = 1)$

$$\psi(x) \propto \exp\left[-\frac{1}{2}\left(\frac{x-q}{b}\right)^2 + ipx\right], \tag{27}$$

where we take the view that the parameters p and q are time dependent, while the width parameter b is time independent. The interpretation of these parameters is evident from the definition of quantum mechanical averages, ie,

$$\langle x\rangle = \int_{-\infty}^{\infty} xe^{-\left(\frac{x-q}{b}\right)^2} dx \Big/ \int_{-\infty}^{\infty} e^{-\left(\frac{x-q}{b}\right)^2} dx = \langle\psi|x\psi\rangle/\langle\psi|\psi\rangle. \tag{28}$$

Since the average value $\langle x-q\rangle = 0$, it follows that $q = \langle x\rangle$, which is the average position of the wave packet. The average momentum of the wave packet

$$\langle\hat{p}\rangle = \left\langle -i\frac{\partial}{\partial x}\right\rangle = p - \langle x-q\rangle/b^2 = p \tag{29}$$

defines the parameter p. The square of the width of the wave packet

$$\left\langle (x-q)^2\right\rangle = -\partial\langle\psi|\psi\rangle/\partial(1/b^2)/\langle\psi|\psi\rangle = \sqrt{\pi}\frac{b^3}{2}/\sqrt{\pi}b \tag{30}$$

making the width $\Delta x = \left\langle (x-q)^2\right\rangle^{1/2} = b/\sqrt{2}$.

The Gaussian wave packet has a number of interesting properties. For instance, it has a minimal uncertainty product $\Delta x \Delta p = 1/2$ in units of $\hbar$. This follows from

$$\langle\hat{p}^2\rangle = \langle -\hbar^2\frac{d^2}{dx^2}\rangle = p^2 + \frac{\hbar^2}{2b^2} \tag{31}$$

and $\Delta p = \langle\hat{p}^2 - p^2\rangle^{1/2} = 1/b\sqrt{2}$, $(\hbar = 1)$.

In addition the Gaussian wave packet can be written as a displaced harmonic oscillator ground state such that

$$\psi(x) = \exp[ipx]\exp[-iq\hat{p}]\exp\left[-\frac{1}{2}\left(\frac{x}{b}\right)^2\right], \tag{32}$$

ie, the oscillator ground state is displaced, $x \to x - q$, and boosted $0 \to p$. This can be seen from

$$\begin{aligned} e^{-iq\hat{p}}\exp\left[-\frac{1}{2}\left(\frac{x}{b}\right)^2\right] &= \left[1 - q\frac{\partial}{\partial x} + \frac{q^2}{2}\frac{\partial^2}{\partial x^2} - \cdots\right]\exp\left[-\frac{1}{2}\left(\frac{x}{b}\right)^2\right] \\ &= \exp\left[-\frac{1}{2}\left(\frac{x-q}{b}\right)^2\right]. \end{aligned} \tag{33}$$

The Gaussian wave packet is a coherent state and can be expressed as a superposition of oscillator states. This means that

$$\psi(x) = \sum_n c_n|n\rangle = \exp\left[za^\dagger - z^*a\right]|0\rangle, \tag{34}$$

where $|n\rangle$ is a harmonic oscillator eigenstate, a and $a^\dagger$ are harmonic oscillator field operators, and z is a suitable complex combination of wave function parameters.

This can be seen from the result

$$\psi(x) \propto e^{ipx}e^{-iq\hat{p}}e^{-\frac{1}{2}\left(\frac{x}{b}\right)^2} \propto e^{-i(q\hat{p}-px)}e^{-\frac{1}{2}\left(\frac{x}{b}\right)^2}, \tag{35}$$

where, since x and $\hat{p}$ do not commute, the last step is nontrivial. Introducing the complex parameter $z = (q/b + ibp)/\sqrt{2}$ and observing that the harmonic oscillator field operators can be expressed as

$$a^\dagger = -i(b\hat{p} + ix/b)/\sqrt{2}, \tag{36}$$

$$a = i(b\hat{p} - ix/b)/\sqrt{2}, \tag{37}$$

we can write $za^\dagger - z^*a = -i(qp - px)$ and

$$\psi(x) \propto e^{za^\dagger - z^*a}e^{\frac{1}{2}\left(\frac{x}{b}\right)^2} \propto e^{za^\dagger - z^*a}|0\rangle. \tag{38}$$

The last expression is the "classical" or canonical *coherent state* $|z\rangle$. The Baker–Campbell–Hausdorff (BCH) formula yields

$$e^{A+B} = \exp\left[-\frac{1}{2}[A,B]_-\right]e^Ae^B, \tag{39}$$

which is true for the case when the commutator $[A,B]_-$ commutes with A and B. When applied to

$$|z\rangle = e^{za^\dagger - z^* a}|0\rangle \tag{40}$$

the BCH formula yields

$$\begin{aligned}|z\rangle &= e^{-\frac{1}{2}|z|^2} e^{za^\dagger} e^{-z^* a}|0\rangle \\ &= e^{-\frac{1}{2}|z|^2} e^{za^\dagger}|0\rangle = e^{-\frac{1}{2}|z|^2}\sum_{n=0}^{\infty}(n!)^{-1}\left(za^\dagger\right)^n|0\rangle \\ &= e^{-\frac{1}{2}|z|^2}\sum_{n=0}^{\infty}(n!)^{-1/2}(z)^n|n\rangle.\end{aligned} \tag{41}$$

The Gaussian wave packet in this form is the original "coherent state". Generalizations of this concept have been made; in particular the work of Perelomov[9] has introduced so-called group-related coherent states. Such a state is formed by the action of a Lie group operator $\exp\left\{\sum_m z_m F_m\right\}$ acting on a reference state $|0\rangle$, where $\{z_m\}$ are the, in general complex, Lie group parameters and $\{F_m\}$ are the generators of the corresponding Lie algebra. The reference state is usually a lowest weight state and called the fiducial state. It is commonly invariant to some of the group elements, thus defining a so-called stability group of the fiducial state. The parameters labeling the coherent state are then associated with the left coset of the Lie group with respect to the stability group. This assures non-redundancy of parameters. The canonical coherent state has this form in terms of the so-called Weyl group, whose Lie algebra generators are $\{1, a, a^\dagger\}$. The one parameter stability group is just the phase factor $e^{i\alpha}$ and the coset representative is $e^{za^\dagger - z^* a}$.

The scalar product of two coherent states

$$\begin{aligned}\langle z_1|z_2\rangle &= \exp\left\{-\left(|z_1|^2 + |z_2|^2\right)/2\right\}\sum_n \frac{\left(z_1^* z_2\right)^n}{n!} \\ &= \exp\left\{-\frac{1}{2}|z_1|^2 + z_1^* z_2 - \frac{1}{2}|z_2|^2\right\},\end{aligned} \tag{42}$$

ie, a nowhere vanishing continuous function of the parameters. The canonical coherent state has the property that

$$a|z\rangle = z|z\rangle, \tag{43}$$

which can easily be seen from

$$e^{-za^\dagger} a e^{za^\dagger} = a + z\left[a, a^\dagger\right]_- + \frac{z^2}{2}\left[\left[a, a^\dagger\right]_-, a^\dagger\right]_- + \cdots = a + z, \tag{44}$$

which means that $e^{-za^\dagger} a e^{za^\dagger}|0\rangle = z|0\rangle$. Furthermore, the coherent state for all the values of the complex parameter z is a set of states satisfying the resolution of the identity

$$\begin{aligned}
\pi^{-1}\int |z\rangle\langle z| d^2z &= \frac{1}{\pi}\sum_{n,m}(n!m!)^{-1/2}\int e^{-|z|^2}(z^*)^n(z)^m|m\rangle\langle n| d^2z \\
&= \frac{1}{\pi}\sum_{n,m}(n!m!)^{-1/2}\int_0^\infty e^{-|z|^2}|z|^{n+m+1}d|z| \\
&\quad\times\int_0^{2\pi} e^{i(m-n)\phi}d\phi|m\rangle\langle n| \\
&= \sum_n (n!)^{-1}\int_0^\infty e^{-|z|^2}|z|^{2n}d|z|^2|n\rangle\langle n| \\
&= \sum_n |n\rangle\langle n| = 1.
\end{aligned} \tag{45}$$

This result permits us to write

$$|z'\rangle = \pi^{-1}\int |z\rangle\langle z|z'\rangle d^2z, \tag{46}$$

which illustrates the overcompleteness of the set $\{|z\rangle\}$. Thus, as a set of functions labeled by the continuous complex parameter z the coherent state satisfies the resolution of the identity and is inherently linearly dependent.

Considering the time evolution of a harmonic oscillator with $|z\rangle$ as the initial state, we obtain

$$\begin{aligned}
e^{-iHt}|z\rangle &= e^{-it\omega\left(n+\frac{1}{2}\right)}|z\rangle \\
&= e^{-\frac{1}{2}|z|^2}\sum_{n=0}^\infty (n!)^{-1/2}\left(ze^{-it\omega}\right)^n|n\rangle e^{-it\omega/2} \propto \left|e^{-it\omega}z\right\rangle.
\end{aligned} \tag{47}$$

This shows that the coherent state evolves into other coherent states by a time-dependent label change that follows the classical oscillator solution.

Application of the TDVP to the wave packet dynamics with the coherent state $|z\rangle$ is straightforward. We note that

$$S(z^*, z') = \langle z|z'\rangle = \exp\left(-\frac{1}{2}|z|^2 + z^*z' - \frac{1}{2}|z'|^2\right), \tag{48}$$

$$\begin{aligned} E(z^*, z) &= \langle z|H|z\rangle / \langle z|z\rangle = \omega\langle z|\left(a^\dagger a + \frac{1}{2}\right)|z\rangle / \langle z|z\rangle \\ &= \omega\left(z^*z + \frac{1}{2}\right), \end{aligned} \tag{49}$$

and that the dynamical equations become

$$\begin{bmatrix} i & 0 \\ 0 & -i \end{bmatrix} \begin{bmatrix} \dot{z} \\ \dot{z}^* \end{bmatrix} = \begin{bmatrix} \omega z \\ \omega z^* \end{bmatrix}, \tag{50}$$

since $C = \partial^2 \ln S/\partial z^* \partial z'|_{z'=z} = 1$. The equation $i\dot{z} = \omega z$ becomes in more detail

$$\frac{i}{\sqrt{2}}(\dot{q}/b + ib\dot{p}) = \omega\frac{1}{\sqrt{2}}(q/b + ibp) \tag{51}$$

assuming a constant width wave packet. One easily deduces that

$$\ddot{p} = -\omega^2 p, \tag{52}$$

$$\ddot{q} = -\omega^2 q, \tag{53}$$

ie, in an oscillator field with $b = 1/\sqrt{m\omega}$ the Gaussian wave packet is coherent and that its average position has a harmonic motion $q(t) = q_0 \cos\omega t + \frac{p_0}{m\omega}\sin\omega t$, while $p(t) = p_0 \cos\omega t - m\omega \sin\omega t$.

2.1.1 Gaussian Wave Packet with Evolving Width

A more general coherent state description of a Gaussian wave packet is required when we allow the width parameter to evolve in time. The corresponding Lie group is then $Sp(2, R)$, which is isomorphic to $SU(1, 1)$ or $SO(2, 1)$. The generators of the $sp(2, R)$ Lie algebra are

$$t_1 = \frac{i}{2}\begin{bmatrix} -1 & 0 \\ 0 & 1 \end{bmatrix}, \quad t_2 = \frac{i}{2}\begin{bmatrix} 0 & 1 \\ 1 & 0 \end{bmatrix}, \quad t_3 = \frac{i}{2}\begin{bmatrix} 0 & 1 \\ -1 & 0 \end{bmatrix}, \tag{54}$$

satisfying the relations

$$[t_1, t_2] = -it_3, \quad [t_2, t_3] = it_1, \quad [t_3, t_1] = it_2, \tag{55}$$

where the different signs on the right indicate that we are dealing with a noncompact group.

Realization of the generators in terms of a Cartesian coordinate x and its conjugate momentum p, such that $[x, p]=i\hbar$, are

$$\begin{aligned} t_1 &\rightarrow T_1 = -(xp+px)/4\hbar \\ t_2 &\rightarrow T_2 = (p^2/2\mu - \mu\omega^2 x^2/2)/2\hbar\omega \\ t_3 &\rightarrow T_3 = (p^2/2\mu + \mu\omega^2 x^2/2)/2\hbar\omega. \end{aligned} \tag{56}$$

Another useful realization obtains in terms of the oscillator field operator

$$a = \left(ip/\sqrt{\mu\omega\hbar} + \sqrt{\frac{\mu\omega}{\hbar}}x \right)/\sqrt{2} \tag{57}$$

and its adjoint. We write

$$\begin{aligned} T_+ &= -T_2 + iT_1 = \frac{1}{2}a^\dagger a^\dagger \\ T_- &= -T_2 - iT_1 = \frac{1}{2}aa \\ T_0 &= T_3 = \frac{1}{4}(a^\dagger a + aa^\dagger). \end{aligned} \tag{58}$$

A common parameterization of the $Sp(2, R)$ group is in terms of Euler angles such that a group element would be

$$g(\alpha, \beta, \gamma) = e^{i\alpha t_3} e^{i\beta t_1} e^{i\gamma t_3}. \tag{59}$$

One can readily show that

$$e^{i\beta t_1} = \begin{bmatrix} e^{\beta/2} & 0 \\ 0 & e^{\beta/2} \end{bmatrix} \tag{60}$$

and

$$e^{i\gamma t_3} = \begin{bmatrix} \cos\gamma/2 & -\sin\gamma/2 \\ \sin\gamma/2 & \cos\gamma/2 \end{bmatrix}. \tag{61}$$

The operator $\exp[i\beta T_1]$ is a scale transformation as can be seen from the relation

$$e^{i\beta T_1} f(x) = \exp\left[-\frac{\beta}{2} x \frac{\partial}{\partial x} - \frac{\beta}{4} \right] f(x) = e^{-\beta/4} f(e^{-\beta/2}x). \tag{62}$$

This result is readily shown by considering the transformation of powers of the coordinate. For instance, by using the power series representation

$$f(x)=\sum_{n=0}^{\infty}\frac{f^{(n)}(0)}{n!}x^n \tag{63}$$

and the defining expansion of an exponential operator

$$\exp\left[-\frac{\beta}{2}\left(x\frac{\partial}{\partial x}+\frac{1}{2}\right)\right]=\sum_{k=0}^{\infty}\frac{\beta^k}{k!}\left[-\frac{\beta}{2}\left(x\frac{\partial}{\partial x}+\frac{1}{2}\right)\right]^k, \tag{64}$$

noting that

$$\left[-\frac{1}{2}\left(x\frac{\partial}{\partial x}+\frac{1}{2}\right)\right]^k x^n=(-1)^k\left(\frac{2n+1}{4}\right)^k x^n \tag{65}$$

and

$$\begin{aligned}
&\left[\sum_{k=0}^{\infty}\left(\frac{-\beta}{4}\right)^k\frac{1}{k!}\right]\times\left[\sum_{l=0}^{\infty}\left(\frac{-\beta}{2}\right)^l\frac{1}{l!}x\right]^n\\
&=\left[\sum_{k=0}^{\infty}\left(\frac{-\beta}{4}\right)^k\frac{1}{k!}\right]\times\left[\sum_{l=0}^{\infty}\left(\frac{-n\beta}{2}\right)^l\frac{1}{l!}\right]x^n\\
&=x^n\sum_{j=0}^{\infty}\left(\frac{-\beta}{4}\right)^j\frac{1}{j!}\sum_{k=0}^{j}\binom{j}{k}(2n)^{j-k}\\
&=\sum_{j=0}^{\infty}\left(\frac{-(2n+1)\beta}{4}\right)^j\frac{x^n}{j!}
\end{aligned} \tag{66}$$

the scaling property is shown .

As discussed above, defining a coherent state involves choosing a Lie group G, a unitary irreducible representation, the lowest (or highest) weight state of such a representation, a stability subgroup, and corresponding cosets. This will yield a useful parameter space in which to describe the coherent state. In analogy with the compact Lie group $SO(3)$, the irreducible representations of $Sp(2, R)$ are labeled by the eigenvalues of T_3 corresponding to the lowest weight state. The harmonic oscillator ground state

$$|0\rangle\propto\exp\left[-\frac{1}{2}\left(\frac{x}{b}\right)^2\right] \tag{67}$$

with $b^2=\hbar/\mu\omega$ is the lowest weight state of an irreducible representation labeled by $k=1/8$, where

$$T_3|0\rangle = 2k|0\rangle. \tag{68}$$

We then identify the stability group H as the set

$$H = h|T(h)|0\rangle = e^{i\sigma_h}|0\rangle. \tag{69}$$

Each element of the coset space G/H then corresponds to a coherent state. The decomposition of the group into cosets, taking advantage of the stability group properties, reduces the parameter space of the coherent state to a nonredundant set. In our case the stability group is $SO(2)$ and we can write

$$e^{i\omega T_3}|0\rangle = e^{i\omega\frac{1}{4}(a^\dagger a + aa^\dagger)}|0\rangle e^{i\omega 2k}|0\rangle = e^{i\omega/4}|0\rangle. \tag{70}$$

A new choice of parameters is r, s, ω, ie,

$$\begin{aligned} &g(\alpha, \beta, \gamma) \rightarrow \\ &g(r, s, \omega) = \begin{bmatrix} \sqrt{r} & 0 \\ 0 & 1/\sqrt{r} \end{bmatrix} \begin{bmatrix} 1 & 0 \\ s & 1 \end{bmatrix} \begin{bmatrix} \cos\omega/2 & -\sin\omega/2 \\ \sin\omega/2 & \cos\omega/2 \end{bmatrix} \end{aligned} \tag{71}$$

where the new parameters are identified as

$$\begin{aligned} r &= \cosh\beta + \sinh\beta\cos\alpha \\ s &= \sinh\beta\sin\alpha \end{aligned} \tag{72}$$

$$\omega = \alpha + \gamma - \arctan\left(\frac{\sin\alpha\sinh\beta/2}{\cosh\beta/2 + \cos\alpha\sinh\beta/2}\right) \tag{73}$$

An element of G can now be expressed as

$$g(r, s, \omega) - e^{it_1 \ln r} e^{is(t_3 - t_2)} e^{i\omega t_3}, \tag{74}$$

and going to the unitary irreducible representation carried by even functions the coherent state becomes

$$\begin{aligned} |r, s\rangle &= T(r, s, 0)|0\rangle \\ &= e^{iT_1 \ln r} e^{is(T_3 - T_2)} e^{-\frac{1}{2}\left(\frac{x}{b}\right)^2} \\ &= \exp\left[(is - 1)\frac{x^2}{r2b^2}\right] \end{aligned} \tag{75}$$

where Eq. (56) is used and $b = 1/\sqrt{\mu\omega}$. The parameters are related to average values of the generators such that

$$\begin{aligned}\langle r,s|T_3 - T_2|r,s\rangle &= \frac{1}{b\sqrt{\pi r}}\int_{-\infty}^{\infty}\frac{x^2}{2b^2}\exp\left(-x^2/rb^2\right)dx \\ &= \frac{b^3 r\sqrt{\pi r}}{2b^2 2b\sqrt{\pi r}} = \frac{r}{4}\end{aligned} \tag{76}$$

and

$$\begin{aligned}\langle r,s|T_3|r,s\rangle &= \langle\left(\frac{i}{2}x\frac{\partial}{\partial x} + \frac{i}{4}\right)\rangle \\ &= \frac{1}{b\sqrt{\pi r}}\int_{-\infty}^{\infty}\left[i\frac{x^2}{2b^2 r}(is-1) + \frac{i}{4}\right]\exp\left(-x^2/rb^2\right)dx \\ &= -\frac{s}{4},\end{aligned} \tag{77}$$

where s is real and r is real and positive (see Ref. 10).

A convenient reparameterization of the wave packet in terms of u and w can be accomplished as

$$b^2 r = 2w^2, \quad s = 2uw, \tag{78}$$

so that the Gaussian wave packet becomes

$$\psi(x) \propto \exp\left[-\left(\frac{1-2iuw}{4w^2}\right)(x-q)^2 + ipx\right] = |p,q,u,w\rangle. \tag{79}$$

The Hamiltonian

$$H = -\frac{\hbar^2}{2\mu}\frac{\partial^2}{\partial x^2} + V(x) \tag{80}$$

yields the wave packet average energy

$$E(p,q,u,w) = \frac{p^2}{2\mu} + \frac{u^2}{2\mu} + U(q,w), \tag{81}$$

with $\hbar = 1$ and

$$U(q,w) = \frac{1}{8\mu w^2} + (2\pi)^{-1/2}\int_{-\infty}^{\infty} e^{-\frac{1}{2}y^2} V(wy+q)dy \tag{82}$$

We can now apply the TDVP equations to study the propagation of this wave packet. The elements of the dynamical metric are

$$\eta_{rs} = i\left[\frac{\partial^2}{\partial r'\partial s} - \frac{\partial^2}{\partial r\partial s'}\right] \ln S, \tag{83}$$

with

$$\begin{aligned} S &= \langle p', q', u', w' | p, q, u, w \rangle \\ &= \int_{-\infty}^{\infty} \exp\left[-ax^2 + bx + c\right] dx \\ &= e^c e^{b^2/4a} \int_{-\infty}^{\infty} \exp\left[-(\sqrt{a}x - b/2\sqrt{a})^2\right] dx \\ &= e^c e^{b^2/4a} \sqrt{\pi/a} \end{aligned} \tag{84}$$

and where

$$\begin{aligned} a &= \left[\frac{1+2iu'w'}{4w'^2} + \frac{1-2iuw}{4w^2}\right] \\ b &= 2\left[\frac{1+2iu'w'}{4w'^2}q' + \frac{1-2iuw}{4w^2}q - ip' + ip\right] \\ c &= -\left[\frac{1+2iu'w'}{4w'^2}q'2 + \frac{1-2iuw}{4w^2}q^2\right]. \end{aligned} \tag{85}$$

Differentiation of

$$\ln S = c + b^2/4b - \frac{1}{2}\ln a + \frac{1}{2}\ln \pi \tag{86}$$

yields the elements of the upper triangle

$$\eta_{pq} = 1, \quad \eta_{pu} = 0, \quad \eta_{pw} = 0, \quad \eta_{qu} = 0, \quad \eta_{qw} = 0, \quad \eta_{uw} = 1, \tag{87}$$

in the antisymmetric metric matrix $\{\eta_{rs}\}$. The TDVP equations then become

$$\begin{bmatrix} 0 & 1 & 0 & 0 \\ -1 & 0 & 0 & 0 \\ 0 & 0 & 0 & 1 \\ 0 & 0 & -1 & 0 \end{bmatrix} \begin{bmatrix} \dot{p} \\ \dot{q} \\ \dot{u} \\ \dot{w} \end{bmatrix} = \begin{bmatrix} \partial E/\partial p \\ \partial E/\partial q \\ \partial E/\partial u \\ \partial E/\partial w \end{bmatrix}, \tag{88}$$

or in more detail

$$\begin{aligned}\dot{q} &= \frac{p}{\mu},\\ \dot{p} &= -\frac{\partial U}{\partial q},\\ \dot{w} &= \frac{u}{\mu},\\ \dot{u} &= -\frac{\partial U}{\partial w},\end{aligned} \tag{89}$$

which look very much like the classical Hamilton's equations.

2.2 The Determinant Coherent State for *N* Electrons

For an N-electron system we choose a set of N spin orbitals $\mathbf{u}^{\bullet} = \{u_1, u_2, \ldots, u_N\}$ and form a determinantal wave function

$$\det\{u_1(x_1)u_2(x_2)\ldots u_N(x_N)\}, \tag{90}$$

or in second quantization the state vector

$$|0\rangle = \prod_{i=1}^{N} a_i^{\dagger}|vac\rangle, \tag{91}$$

with $|vac\rangle$ the true vacuum state. We call this the reference state. The basis (of rank K) and the associated field operators are divided into two sets, those that refer to the reference state denoted by $a\bullet$ and the rest denoted by a°, ie,

$$(\mathbf{a}\bullet, \mathbf{a}^{\circ}), \tag{92}$$

and of course a similar partition of the creators so that the reference state is

$$|0\rangle = \prod_{k=1}^{N} a_k^{\bullet\dagger}|vac\rangle. \tag{93}$$

The creation operators and the basis transform in the same manner, so when we apply a general unitary transformation to the basis, the creation operators suffer the same transformation. We can write in matrix form

$$(\mathbf{b}^{\bullet\dagger}, \mathbf{b}^{\circ\dagger}) = (\mathbf{a}^{\bullet\dagger}, \mathbf{a}^{\circ\dagger})\begin{pmatrix}\mathbf{U}^{\bullet} & \mathbf{U}'\\ \mathbf{U}'' & \mathbf{U}^{\circ}\end{pmatrix}, \tag{94}$$

and conclude that the reference state becomes

$$
\begin{aligned}
\prod_{i=1}^{N} & b_i^{\bullet\dagger}|vac\rangle \\
&= \prod_{i=1}^{N}\left[\sum_{l=1}^{N} a_l^{\bullet\dagger} U_{li}^{\bullet} + \sum_{j=N+1}^{K} a_j^{\circ\dagger} U_{ji}''\right]|vac\rangle \\
&= \prod_{i=1}^{N}\left[\sum_{l=1}^{N}\left\{a_l^{\bullet\dagger} + \sum_{j=N+1}^{K}\sum_{k=1}^{N} a_j^{\circ\dagger} U_{jk}''\left(\mathbf{U}^{\bullet-1}\right)_{kl}\right\} U_{li}^{\bullet}\right]|vac\rangle \\
&= \alpha\prod_{i=1}^{N}\left[a_i^{\bullet\dagger} + \sum_{j=N+1}^{K}\sum_{k=1}^{N} a_j^{\circ\dagger} U_{jk}''\left(\mathbf{U}^{\bullet-1}\right)_{ki}\right]|vac\rangle \\
&= \alpha\prod_{i=1}^{N}\left[1 + \sum_{j=N+1}^{K}\sum_{k=1}^{N} a_j^{\circ\dagger} U_{jk}''\left(\mathbf{U}^{\bullet-1}\right)_{ki} a_i^{\bullet}\right] a_i^{\bullet\dagger}\,|\,vac\rangle \\
&= \alpha\prod_{i=1}^{N}\left[1 + \sum_{j=N+1}^{K}\sum_{k=1}^{N} a_j^{\circ\dagger} U_{jk}''\left(\mathbf{U}^{\bullet-1}\right)_{ki} a_i^{\bullet}\right]\prod_{l=1}^{N} a_l^{\bullet\dagger}|vac\rangle.
\end{aligned}
\tag{95}
$$

If we introduce the complex parameters

$$
z_{ji} = \sum_{k=1}^{N} U_{jk}''\left(\mathbf{U}^{\bullet-1}\right)_{ki}, \tag{96}
$$

and write the unnormalized state vector

$$
\prod_{i=1}^{N} b_i^{\bullet\dagger}|vac\rangle = |\mathbf{z}\rangle \tag{97}
$$

with the parameters being time dependent. In terms of the orthonormal spin orbital basis we write

$$
\begin{aligned}
|\mathbf{z}\rangle &= \prod_{i=1}^{N}\left[1 + \sum_{j=N+1}^{K} z_{ji} a_j^{\circ\dagger} a_i^{\bullet}\right]|0\rangle \\
&= \prod_{i=1}^{N}\prod_{j=N+1}^{K}\left[1 + z_{ji} a_j^{\circ\dagger} a_i^{\bullet}\right]|0\rangle \\
&= \prod_{i=1}^{N}\prod_{j=N+1}^{K}\exp\left[z_{ji} a_j^{\circ\dagger} a_i^{\bullet}\right]|0\rangle \\
&= \exp\left[\sum_{i=1}^{N}\sum_{j=N+1}^{K} z_{ji} a_j^{\circ\dagger} a_i^{\bullet}\right]|0\rangle.
\end{aligned}
\tag{98}
$$

In going from the second to the third line in the above equation we have used the fact that the electron field operators are nilpotent and this also is the reason that the exponentiation in the fourth line works. The end result is true because all the operators $a_j^{\circ\dagger} a_i^{\bullet}$ commute.

From these equations it follows straightforwardly that the wave function representative of this state is

$$\det\left[\chi_i(x_j)\right], \tag{99}$$

with the "dynamical spin orbitals"

$$\chi_i = u_i + \sum_{j=N+1}^{K} u_j z_{ji}, \tag{100}$$

where the parameters z_{ji} are complex and are considered to be functions of the time parameter t. As they change during a process involving the electronic system the determinantal state vector can in principle become any determinantal wave function possible to express in the spin orbital basis. The spin orbitals χ_i are not orthonormal even if the basis $\{u_k\}$ is, and in actual application one will often use a raw basis of atomic spin orbitals (often built from Gaussian-type orbitals), which are not orthonormal. In such a case the exponential form of the determinantal state is not applicable and one has to deal with the full complications of the nonunit metric of the basis (see Ref. 3). This is indeed possible and has been coded into the ENDyne program system that uses narrow wave packet nuclei and single determinantal electron states in an explicitly time-dependent, nonadiabatic treatment of molecular processes.

A determinantal wave function expressed in this form is a coherent state. The associated Lie group is the unitary group $U(K)$ and the reference state (Eq. (93)) is the lowest weight state of the irreducible representation $[1^N 0^{K-N}]$ of $U(K)$. The stability group is $U(N) \times U(K-N)$. The norm in an orthonormal basis of spin orbitals is

$$\langle \mathbf{z} | \mathbf{z} \rangle = \det\left[\mathbf{1} + \mathbf{z}^{\dagger}\mathbf{z}\right], \tag{101}$$

which means that we can define an appropriate measure $d\mathbf{z}$ in the parameter space such that the resolution of the identity

$$\int |\mathbf{z}\rangle\langle \mathbf{z}| d\mathbf{z} = 1 \tag{102}$$

holds. The derivation of the form of the measure $d\mathbf{z}$ is nontrivial.[11]

REFERENCES

1. Öhrn, Y.; Deumens, E. A Dynamical Time-Dependent View of Molecular Theory. In: *Theory and Applications of Computational Chemistry: The First 40 Years*; Dykstra, C. E., Frenking, G., Kim, K. S., Scuseria, G. E., Eds.; Elsevier, 2005.
2. Löwdin, P. O.; Mukherjee, P. K. Some Comments on the Time-Dependent Variational Principle. *Chem. Phys. Lett.* **1972**, *14*, 1.
3. Deumens, E.; Diz, A.; Longo, R.; Öhrn, Y. Time-Dependent Theoretical Treatments of the Dynamics of Electrons and Nuclei in Molecular Systems. *Rev. Mod. Phys.* **1994**, *66*(3), 917–983.
4. Broeckhove, J.; Lathouwers, L.; Kesteloot, E.; Van Leuven, P. On the Equivalence of Time Dependent Variational Principles. *Chem. Phys. Lett.* **1988**, *149*, 547.
5. Dirac, P. A. M.. The Principles of Quantum Mechanics. The International Series of Monographs on Physics, 27; Oxford University Press, 1930.
6. Kramer, P.; Saraceno, M. *Geometry of the Time-Dependent Variational Principle in Quantum Mechanics*; Springer: New York, 1981.
7. Öhrn, Y.; Deumens, E. Electron Nuclear Dynamics with Coherent States. In: *Proceedings of the International Symposium on Coherent States: Past, Present, and Future*; Feng, D. H., Klauder, J. R., Eds.; Oak Ridge Associated Universities, World Scientific: Singapore, 1993.
8. Klauder, J. R.; Skagerstam, B. S. *Coherent States, Applications in Physics and Mathematical Physics*; World Scientific: Singapore, 1985.
9. Perelomov, A. M. Coherent States for Arbitrary Lie Group. *Commun. Math. Phys.* **1972**, *26*, 222–236.
10. Sugiara, M. *Unitary Representations and Harmonic Analysis*; John Wiley & Sons: Hoboken, NJ, 1975.
11. Blaizot, J. P.; Orland, H. Path Integrals for the Nuclear Many-Body Problem. *Phys. Rev. C* **1981**, *24*, 1740.

CHAPTER FOUR

Specifics on the Scientific Legacy of Per-Olov Löwdin

Carlos F. Bunge[1]
Instituto de Física, Universidad Nacional Autónoma de México, Mexico City, Mexico
[1]Corresponding author: e-mail address: bunge@fisica.unam.mx

Contents

Advances in Quantum Chemistry, Volume 74
ISSN 0065-3276
http://dx.doi.org/10.1016/bs.aiq.2016.04.004

Abstract

Per-Olov Löwdin was an inspiring and compelling teacher. His most prominent papers were written more than 50 years ago, whereas since then quantum chemistry, its software, and its computers have changed almost beyond recognition. In accurate calculations with truncation energy errors, an important part of Per-Olov's themes and thoughts appears highly relevant today in applications to atoms and small molecules. My purpose here is to place projection operators, natural orbitals, error bounds, and the variational theorem for finite Hermitian matrices, in the light of current challenges in the field.

1. INTRODUCTION

It is an honor and a great pleasure to contribute to the celebration of the centennial anniversary of Per-Olov Löwdin. Many hundreds of people made seminal contributions to quantum chemistry; however, making possible the actual birth of Quantum Chemistry required extra commitments and a strong sense of collaboration within academics and beyond. Per first founded the Quantum Chemistry Group in Sweden at Uppsala University. Technical Note No. 1, forcefully announcing the correlation problem for the first time, came out on June 21, 1957.[1]

In 1960, at the suggestion of several American scientists, specially Harry Shull, he founded the Quantum Theory Project (QTP) at the University of Florida in the United States. Concurrently, Per led a successful funding and pricing bargain that materialized in June 1962 when the existing IBM 650 computer at University of Florida (acquired in 1956, installed in 1957) was replaced by a model 709. This one, being 50 times faster than the old machine, soon spread its influence in science and engineering departments.

1.1 Two Educational Systems: Summer and Winter Institutes

Almost at once, Per invented two original educational systems that made a lasting influence: the International Summer Institutes (1958–1987) held in Sweden–Norway, and the International Winter Institutes (1960–1988) taking place in Florida, decisively affecting the careers of many of today's leading quantum chemists.

Exceptional attractors made them unforgettable. First, there were several weeks of tightly scheduled lectures and discussions on current issues, unavailable elsewhere, and where communication across language barriers was largely untroubled. Socialization was further stimulated by steady calls to participate in various sports and outdoor activities.

Finally, after all the above mentioned lectures, the institutes were capped with an International Symposium lasting over a week. There, graduate students chatted and discussed science on an informal basis with dozens of world class scientists, including an important number of famous quantum chemists. Even if one did not succeed to establish a good communication with a person one intended to, one could readily identify who they were, and who thought highly or poorly about trends and works of one's interest.

That atmosphere was intensified by the awe surrounding the interventions of leading figures like Clemens Roothaan, the rising expectations every time Oktay Sinanoglu took the floor in 1964 after being made full professor at Yale, the hilarious and timely outbursts of the late Howard Taylor, the innovative and perplexing calculations of Enrico Clementi, the witty and sparkling comments of Ernie Davidson, and so on.

Unofficial discussions were even better: configuration interaction is nonsense! blasted Jan Linderberg, to what most agreed, even Per-Olov, although he knew how to remain far from these gatherings. Projection operators are useless! uttered Jürgen Hinze, causing visibly happy faces. Racah algebra is very convenient, asserted Paul Bagus, baffling QTP students. Do not waste your time writing programs in assembler language, whispered a distinguished member of the Chicago group to the relief of many. Why not considering LISP? interceded Harold McIntosh, which prompted me to write my first program to evaluate Hamiltonian matrix elements in such a language, after which I felt enabled to program anything in Fortran.

It is said that it was Aristotle who first proposed a criterion to identify an educated person: it would be some one that would not easily be led to be cheated by experts. The international Summer and Winter Institutes and their associated Symposia were eye openers to the notion that knowledge and cheating always come together. One time, a brash young professor gave a formidable talk on quantum chemical calculations in which the words cheat and cheating were slammed over and over at an astonished and not very congenial audience. Many graduate students were fascinated.

1.2 Quantum Chemistry Would Come of Age Within Ten Years' Notice

Few disciplines have been so controversial at their beginnings. Opposition to early quantum chemistry was widespread. First, there was an intellectual distrust from chemistry faculties unwilling to listen to mathematics and computer jargon with little or no chemical content. On top of it, quantum chemists were as expensive as the computers they thought to deserve, namely, computers that had not even been built.

In connection with the birth of Quantum Chemistry, important work was going on in various groups: at Cambridge, UK under Sir John Lennard-Jones[2,3] and Frank Boys,[4,5] in Oxford under Charles Coulson,[6,7] a pioneer of the concept of Summer Schools, in Chicago under Robert Mulliken[8,9] and Clemens Roothaan,[10,11] and in MIT under John Slater.[12,13] At Indiana University, Harry Shull founded the Quantum Chemistry Program Exchange (QCPE) in 1962–63. QCPE was to serve as a living repository of programs useful for theoretical chemistry, beyond quantum chemistry itself so that programs or part of programs would not have to be rewritten from start.[14]

With these antecedents, in an informal 1963 summer seminar attended by several graduate students at the Chemistry Department of the University of Florida, Joe Hirschfelder eventually announced: "I predict!, I predict than in ten years from now!—Joe searched around after the sympathy of the audience while his tie was vanishing under tons of chalk—in 1973!, Quantum Chemistry will come of age," and that means that "there will be a striking demand for quantum chemists at Chemistry Departments" all over the world.

Joe's prediction came to be remarkably accurate. Many more years later, density functional theory and its applications happened to dominate the field, and the coupled-cluster method took control of the prediction of ground-state geometries of small molecules. Former "Indians"[15] and new-comers at QTP developed a lot of chemically and physically motivated theory, methodology, and software at the frontier of chemistry and physics. Most of what people learned in the old Winter Institutes (1960–1988) has been ebbing and disappearing altogether from quantum chemistry Summer Schools, giving way to new realities. What is the present status of projection operators,[16-18] natural orbitals (NOs),[1,19,20] lower bounds,[21,22] projected Hartree–Fock,[1,16] and the variational theorem?[23,24]

1.3 Preview of Present Status of Some of Löwdin's Themes and Thoughts

Distinctly different than in molecular calculations, configuration interaction (CI) survived in the atomic domain by reasons given elsewhere.[25] We shall use CI-nx for n-tuple excited CI. Likewise, MRCI-nx denotes multireference CI[26] in which each reference carries up to n-excitations.

CI has been and continues to be associated with limitations that never existed (use of nonoptimized orbitals, inefficient evaluation of matrix elements), or that have already been overcome (CI-2x), or that are subject to controversy, such as lack of size-consistency for CI-nx with $n \geq 4$ (why should anyone care about that?).

After Andy Weiss introduced optimized atomic orbitals[27] more than 50 years ago, all atomic CI calculations reported in the literature involved some kind of orbital exponent optimization with possibly no more than one exception.[28] Fully automatic optimization of orbital bases competing with numerical self-consistent field methods[29,30] for atomic calculations has recently been described.[25]

The use of symbolic formulae for efficient evaluation of matrix elements in atomic work[31,32] actually predates the use of the graphical unitary group approach (GUGA) in molecular work.[33] There are some advantages in the use of symbolic formulae for atomic[29,30,34] and molecular[35] work over the GUGA approach. One of them is that the formulas can be used in a priori selected CI (SCI) with truncation energy error,[36,37] whereas the GUGA is oriented toward incorporation of full classes of configurations whose number scales up wildly with orbital basis size.

Size-consistency arguments took strength beginning in 1978.[38] This powerful idea gave rise to many useful methods for molecular structure calculations,[39-41] but it was also used as a valid justification for the demise of *all CI* other than full CI (FCI), when the size-consistency argument was appropriate only regarding CI-2x. For H_2O ground state at the equilibrium geometry, CI-4x has been reported to be comparable to coupled-cluster up to triples,[42] while CI-6x is tantamount to coupled-cluster up to quadruples.[42]

Size consistency is not a synonym nor a requirement for exactness, and this is strictly what is at stake in ab initio work. If exactness is accompanied by a reliable truncation energy error, so much the better The purpose of a priori selected CI (SCI)[36,37] is precisely to approximate full CI with truncation energy errors (TEE) which can quickly be evaluated with an efficient serial program such as ours[25] and before indulging in large-scale CI.

Of course, coupled-cluster has already been used for molecules with 100 electrons, whereas our computer programs have not yet produced highly accurate CI calculations beyond the 10 electron barrier. Nevertheless, for Ne and H_2O, we have published selected CI-6x with large basis sets amounting to 338 radial functions until $\ell = 20$, and 412 C_{2v} symmetry orbitals, respectively, yielding by far the best corresponding energies in the literature.[25,43,44] Our programs can presently approximate FCI with truncation energy errors for any system of 10 or less electrons in any symmetry, including Dirac-type relativistic atomic calculations (we can consider nonabelian groups after having access to corresponding pertinent lists of one- and two-electron integrals). A reasonably parallel program is expected to be operational soon, and that will permit to better evaluate the intrinsic limitations of accurate CI.

Section 2 deals with my student days at QTP, when both lower bounds and projected Hartree–Fock probably exhausted their potentials, and people so involved went through hard times. Here I describe how I survived the debacle by reorienting my Ph.D. work toward CI. Sections 3 and 4 take up projection operators and natural orbitals, respectively.

Because of unfavorable scaling of traditional CI with orbital size, the only hope to approximate FCI with large orbital bases, and to within useful truncation energy errors, appears to be through some sort of a priori SCI, as outlined in Section 5. The basic requirement for our version of selected CI with TEE (SCI-TEE)[36,37] is the development and programming of a priori energy estimates for each configuration K embracing all possible degeneracies. These energy estimates need good CI-3x expansion coefficients and good natural orbitals, which in turn require accurate multireference wave functions from which to extract them.

In Section 6 we deal with an alternative to the traditional relativistic theory of atomic structure.[30] Ours is a variational theory that has been used to evaluate forbidden transition probabilities where negative-energy orbitals make significant contributions, and where natural orbitals play a fundamental role. Our theory can also be used to provide best positive-energy orbitals for Dirac-type relativistic calculations.

2. AT THE EARLY QTP

Following the steps of my dear friend Osvaldo Goscinski, I arrived to the University of Florida in June 1962 with a teaching assistantship that would let me to pursue graduate studies at the QTP under the mythical Per-Olov. I was received by Darwin Smith, very excited about the imminent inauguration of the IBM 709 computer. The following August I briefly met Per on his way to Uppsala after two weeks lecturing on Quantum Foundations of Solid State Theory[45-49] at the 3rd Latin American School of Physics in Mexico City. The real Per-Olov far exceeded expectations: an ebullient European kind of character, radiating endless charm, impeccable manners, and lots of optimism. A true force of nature in a friendly and humane version.

Next September, Per arrived from Sweden earlier than usual. His semi-weekly seminars included studies in perturbation theory,[50-53] the variational theorem for finite Hermitian matrices,[23] natural orbitals,[1,19,20] projection operators,[16-18] and the scaling problem.[54] One year later, in the fall of 1963, Per brought disturbing news: Uppsala was to be dedicated to quantum biology,[55-58] while Florida would work on lower bounds to expectation values.[21,22] I felt discouraged.

2.1 Lost with Lower Bounds

Per assigned me to work on bounds to the mass-polarization operator, for which he filled a couple of calligraphy quality lines in a piece of paper the size of a doctor's prescription with the QTP logo in the upper left. As it turned out, this operator was neither positive- nor negative-definite, thus unsuitable for the existing theory. Happily, right or wrong, I got the idea that the whole subject was heading toward a dead end, and it was therefore unsuitable for a Ph.D. thesis. But that was a silly thought to present to Per since, in retrospective, he probably never thought that all of us were going to produce a Ph.D. thesis out of lower bounds. Osvaldo and Tim Wilson eventually succeeded, but they could not get much further. Nevertheless, this experience strongly motivated me to consider electronic structure calculations with estimates of some kind of error bounds.[36,59]

Meanwhile, my wife Annik had joined QTP and had been assigned to a team dedicated to upper and lower bounds on one-dimensional problems. I was rescued by these people and eventually Annik and I formed a separate group. Our target system was an ice model with hydrogen bonds in which the hydrogen atom was moving in a double-minimum potential expressed in terms of quadratic, cubic, and quartic powers. From it one could extract the usual finite-matrix Hamiltonian representation yielding upper bounds and, as Annik had already proved, a finite-matrix representation of another operator whose eigenvalues gave lower bounds to the exact ones. We both learned how to program in Fortran IV and wrote our first Technical Note, duly published a year later.[60]

2.2 Further Lost with Projected Hartree–Fock

In the fall of 1964, John Slater arrived from MIT bringing a much needed reinforcement to QTP, especially when Per was away. When Per arrived from Uppsala later that year, he "assigned" me to projected Hartree–Fock whereupon I was left adrift once more. Annik had a stronger character than I and thus started work under Ludwig Hofacker.

I began writing various programs. John Connolly, a Slater student who had just joined QTP and with whom I was now sharing my office, keenly asked me what I was really doing. I probably answered "I am working on atoms, projection operators, natural orbitals, etc." But what is your project? No project. John laughed in disbelief.

I soon realized that projected Hartree–Fock was a trap; my problem was how to explain that to Per. When he was safely away in Sweden, in the summer of 1965, I presented my results in a seminar that, according to Darwin Smith, "nobody understood except for Slater." Actually, Darwin had understood quite well what I meant, he was just too perturbed about my rude conclusions. I told him not to worry, that as far as I was concerned projected Hartree–Fock was unworthy of further consideration. In secret, I wrote a paper entitled "Unrestricted Projected Hartree–Fock Solutions for Two-Electron Systems" and put it in a manila envelope, ready to be sent to *Physical Review* after passing my Ph.D. examination, which materialized one year later and was unrelated to that subject. Despite stating in the abstract[61] that "the unrestricted projected Hartree–Fock scheme at the present moment does not represent a reasonable approach to the two-electron and many-electron problems," and of having penned an audacious introduction, Per kindly cited my paper at the next Sanibel Symposia in 1967. Not for a moment I had expected otherwise.

2.3 CI at QTP

Going more than one year back in time, Per was too busy after his arrival in the fall of 1965. Sometime, I called on Per and aimed at telling him about my last summer seminar on projected Hartree–Fock, but he looked distant or so I interpreted him. In late March 1966, around my 25th birthday, while walking in Indian row toward the cafeteria, I took courage to tell him: "I will work on atomic CI." Without turning his head he answered me "CI is pedestrian." I felt relieved: in a flash I realized that Per was letting me free. Next June my CI program was working. Slater agreed to form part of my committee in replacement of Per-Olov, who would not be back until next fall while I wanted to graduate in July. Darwin stepped in as my Ph.D. advisor.

In the meantime I tested the program against invariance in the order of the projected determinants and with respect to linear transformations of the correlation orbitals. Darwin was kept updated on a daily basis. He exulted even more than I when I presented him the best Be energy to date: $E = -14.66250$ a.u., slightly below $E = -16.66179$ a.u. yet unpublished.[62] That same day, on Wednesday, July 20th, I showed my results to Slater. He politely asked: "how do you know your program is right?" I carefully explained to him the invariance tests, while he followed very attentively. After I was over, he smiled leisurely. He raised from his chair and sat down again still smiling. I had it.

On that afternoon I rented an electric typewriter. With Annik at the typewriter I started from page one nonstop for the next five days. In early Sunday hours, before going to bed, Darwin visited by surprise. I handed him about 40 pages; he was pleased, except that Annik had not left blank spaces after commas, which he finally felt forced to approve. On my instructions, Annik was correcting my english without waiting for my approval, something that had her upset for a while until she also realized that otherwise I might not finish on time. On Monday I turned over the five copies at the Dean's office, just ten minutes before the 5 pm deadline. The following day, first thing in the morning, with the seals from the Dean's office still fresh, I gave Slater a copy of my thesis. Is it OK for you if I defend my thesis next Friday? Sure! and he smiled once more. I quickly handed the remaining copies to the other four committee members without further explanations, just telling them that it was OK with Slater. On Friday morning things went well.

From here on, Darwin and Slater were going to be my "guardian angels." I became Assistant Professor without teaching duties. Upon arriving in late November, Per was more kind and effusive with me than ever before. I was able to tell him all about my program in full detail: projection operators all along with *full degeneracies* included (first and probably only atomic CI program to have that), *turn-over rule* for Hamiltonian matrix elements, accurate natural orbitals through a Jacobi diagonalization of the reduced first-order density matrix, with option to keep Hartree–Fock orbitals by zeroing CI coefficients of single excitations. Sufficient core memory was provided by seven chain loadings. Once I incorporated a better configuration selection scheme, I found $E = -14.66419$ a.u. (Be).[63] In Sanibel 1967, Per proudly announced the first "hot" numerical results from QTP. He was also very pleased that I was going to move to Indiana with Harry Shull in the coming months. Our honeymoon lasted forever.

Starting May 1, 1967, I spent one of the best years of my life at Harry Shull's lab.

3. PROJECTION OPERATORS

In September 1966, Annik was technically once more under Per-Olov, after Ludwig settled in Northwestern University, and the situation inclined her to join my CI efforts. She decided to write a program to evaluate orthonormal configuration-state-functions (CSF) F_{gK}:

$$F_{gK} = \mathcal{O}(\Gamma,\gamma)\sum_{i=1}^{g} D_{iK} b_i^g = \sum_{i=1}^{n_K} D_{iK} c_i^g, \quad g = 1,\ldots,g_K \tag{1}$$

in terms of which the CI expansion reads:

$$\Psi^{FCI} = \sum_{K=1}^{K_x}\sum_{g=1}^{g_K} F_{gK} C_{gK}, \tag{2}$$

where K and g label configurations and degenerate elements, respectively. F_{gK} is a linear combination of n_K Slater determinants D_{iK}, and $\mathcal{O}(\Gamma,\gamma)$ is a symmetric projection operator[64] for all pertinent symmetry operators Γ and a given (N-electron) irreducible representation γ.[65-68]

The first running version of Annik's program PROJEC[69] was operational the following month, after which hand calculation of determinant projections was over. Her projection operator $\mathcal{O}(\Gamma,\gamma)$ was the product of angular orbital momentum and spin orbital momentum operators $\mathcal{O}(\mathbf{L}^2;L,M_L)$ and $\mathcal{O}(\mathbf{S}^2;S,M_S)$, respectively, where $\mathcal{O}(\mathbf{M}^2;k,m)$ was taken as[18]:

$$\mathcal{O}(\mathbf{M}^2;k,m) = (2k+1)!\frac{(k+m)!}{(k-m)!}\sum_{\nu=0}^{k_{max}-k}\frac{(-1)^{\nu}\mathbf{M}_-^{(k-m+\nu)}\mathbf{M}_+^{(k-m+\nu)}}{\nu(2k+\nu+1)!}. \tag{3}$$

Earlier work by Rotenberg[70,71] used instead the traditional projection operator of Gray and Wills.[19,72]

Program PROJEC featured the first implementation of Löwdin's projection operator for angular momentum,[18] but that was not realized for almost 20 years.[65] Then, it was found that Löwdin's method was many times faster than the one of Gray and Wills; however, it became numerically unstable, even in quadruple precision arithmetic, for configurations with $\ell = 30$ starting with B ground state, and there was no cure to that. On the other hand, Gray and Wills' projection operator performed considerably slower than Per's, although it proved to be numerically stable with respect to high ℓ values at least up to $\ell = 30$.

After it became fully automatic,[65-68] PROJEC was renamed AUTOCL. At this point Annik left the project to fully dedicate to her own group on atmospheric chemistry. Starting with 2300 lines of Fortran code in 1966 (slightly more than one full box of cards), and more than 100,000 lines in 2010,[43] AUTOCL now consists of about 200,000 lines, and it can produce selected symmetry-configuration lists with truncation energy errors for atoms and molecules[25] before taking up large-scale CI.

The use of projection operators provides a very convenient modular framework for nonrelativistic and relativistic, atomic and molecular calculations, in a single computer program. Thus, each time our program ATMOL for wave function calculations was tested, first for relativistic calculations and later for molecular calculations, correct results were quickly obtained in a matter of days.

3.1 Partitioning of Degenerate Spaces

Degenerate spaces arise in open-shell configurations. In MRCI the occurrence of open-shell configurations is almost unavoidable, even for closed-shell ground states. The idea is to partition the degenerate spaces of singly and doubly excited configurations into two parts: one with the maximum possible dimension which has null matrix elements with the Hartree–Fock configurations, called the maximum projected noninteracting space (MAXPNI) and the other, its orthogonal complement,[73] called the minimum projected interacting space (MINPI). MAXPNI, MINPI, and the best calculation on the ground state of atomic carbon at the time were the subject of the Ph.D. thesis[73-75] of Annik, the first woman to earn a Ph.D. from QTP.

Interacting spaces of a smaller size are also possible.[35,76] The quotient between the dimension of MINPI and that one of the full space may exceed the value of 10 for a single reference, which turns out to be decisive to optimize basis orbitals and to evaluate optimally decoupled positive-energy orbitals. In an MRCI, that quotient may even be greater than that. For singly excited configurations, MINPI is expanded entirely in terms of singly and doubly excited determinants,[76] whereas for doubly excited configurations, MINPI is exclusively expanded by doubly excited determinants.[76]

Good NOs for selected CI-6x require the use of multireference CI-3x in the interacting space. MRCI in the interacting space was first advanced by Bernie Brooks and Fritz Schaefer in the late 1970s.[77,78]

4. FAST-CONVERGING CI EXPANSIONS

4.1 Natural Orbitals

In 1955 Löwdin introduced natural (spin)orbitals[19] χ_i and occupation numbers n_i:

$$\gamma(1,1') = \sum n_i \chi_i^*(1)\chi_i(1'), \tag{4}$$

as the eigenfunctions and eigenvalues of the reduced first-order density matrix $\gamma(1,1')$:

$$\gamma(1,1') = N \int \Psi(1,2,\ldots,N)^* \Psi(1',2,\ldots,N) d(2,3,\ldots,N). \tag{5}$$

In Eq. (4) the n_i are in nonincreasing order. Let $\mathcal{G} \equiv \{\phi_i(1), i = 1,2,\ldots,p\}$ be a given spin orbital basis $\mathcal{G}$ of size p, and let Ψ be an FCI expansion expressed in these spin orbitals:

$$\Psi = \sum D_K(\mathcal{G}) c_K. \tag{6}$$

It is not hard to see that the diagonal elements γ_{ii} of $\gamma(1,1')$ are given by:

$$\gamma_{ii} = \sum_K^{(i)} |c_K|^2, \tag{7}$$

where the summation goes over all K holding index i. The corresponding natural spin orbitals $\mathcal{S} \equiv \{\chi_i(1), i = 1,2,\ldots,p\}$ afford an alternative representation $\mathcal{S}$ of the original $\mathcal{G}$.

Being an FCI, Ψ is invariant upon nonsingular linear transformations of the spin orbital basis. Thus Ψ can also be expressed in terms of its natural spin orbitals as:

$$\Psi = \sum D_K(\mathcal{S}) a_K, \tag{8}$$

whereby,

$$n_i = \sum_K^{(i)} |a_K|^2. \tag{9}$$

After a nontrivial application of the variational theorem for finite Hermitian matrices[79,80] one arrives to:

$$\sum_{i=1}^{r} n_i \geq \sum_{i=(1)}^{(r)} \gamma_{ii}, \quad r < p, \tag{10}$$

where the sum in the right hand side goes over *any* r indices, whence it follows:

$$\sum_{i=1}^{r} \sum_{K}^{(i)} |a_K|^2 \geq \sum_{i=(1)}^{(r)} \sum_{K}^{(i)} |c_K|^2, \quad r < p, \tag{11}$$

expressing the optimum convergence property of the natural (spin) orbitals.[1,81]

As mentioned earlier, SCI-TEE gets nowhere without NOs. Calculation of suitable NOs must proceed by parts, first time by MRCI-2x, after which MRCI-3x becomes feasible, all these MRCI-nx taken in the interacting space. One may still find final NOs obtained from a preliminary SCI-TEE wave function.

4.2 Ergonals

Ergonals[82] (or energy-optimized spin orbitals[83]) ζ_i and ergonal energies ϵ_i were introduced in 1973:

$$h^{(p)}(1,1') = \sum \epsilon_i \zeta_i^*(1)\zeta_i(1'), \tag{12}$$

as the eigenfunctions and eigenvalues of the reduced, averaged, and projected first-order energy density matrix $h^{(p)}(1,1')$:

$$h^{(p)}(1,1') = N \int \Psi(1,2,\ldots,N)^* \mathcal{O}^{(p)} \frac{H(1,2,\ldots,N) + H(1',2,\ldots,N)}{2} \mathcal{O}^{(p)} \Psi(1',2,\ldots,N) d(2,3,\ldots,N), \tag{13}$$

where $\mathcal{O}^{(p)}$ projects any manifold into the space $\mathcal{G}$ of the previous section, and H is the usual Hamiltonian. Ψ is still invariant thanks to the use of the projection operator $\mathcal{O}^{(p)}$ in Eq. (13); therefore:

$$\Psi = \sum D_K(\mathcal{G})c_K = \sum D_K(\mathcal{S})a_K = \sum D_K(\mathcal{E})e_K. \tag{14}$$

Similarly as with NOs, one gets[83]:

$$\sum_{i=1}^{r}\sum_{K}^{(i)}\sum_{L} e_K e_L H_{KL}(\mathcal{E}) \le \sum_{i=(1)}^{(r)}\sum_{K}^{(i)}\sum_{L} a_K a_L H_{KL}(\mathcal{S}), \tag{15}$$

expressing the optimum convergence property of ergonals[83] relative to NOs themselves. For atoms, orbital bases are rather good, thus ergonals should be almost identical to NOs. For molecules, however, current basis sets carry rather large CBS errors, thus time is ripe to verify whether ergonals afford a significant energy convergence advantage over NOs.

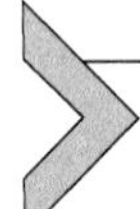

5. NEARING FULL CI WITH TRUNCATION ENERGY ERRORS

Given Schrödinger's equation for stationary states and an orbital basis, FCI is obtained from a matrix-eigenvalue equation,

$$\mathbf{H}\ \mathbf{C}_\mu = E_\mu^{\mathrm{FCI}}\mathbf{C}_\mu, \tag{16}$$

and the full CI expansion reads:

$$\Psi^{\mathrm{FCI}} = D_0 c_0 + \sum_i\sum_a D_i^a c_i^a + \sum_{i<j}\sum_{a<b} D_{ij}^{ab} c_{ij}^{ab} + \sum_{i<j<k}\sum_{a<b<c} D_{ijk}^{abc} c_{ijk}^{abc} + \cdots, \tag{17}$$

where i, j, k, refer to occupied spin orbitals or Dirac bispinors, and a, b, c denote corresponding unoccupied one-electron functions. Distinct from the variational coefficients C_{gK} in Eq. (2), the variational coefficients $c_{ijk\ldots}^{abc\ldots}$ of Eq. (17) satisfy approximate relationships among themselves, useful to predict coefficients of a given excitation order in terms of products of coefficients of lower excitation order.[84,85] This is closely related to our method[36] to carry out a priori selection of configurations with TEE.

The exact eigenvalues E_μ of Schrödinger's equation can be written as:

$$E_\mu = E_\mu^{\mathrm{FCI}} + \Delta E_\mu^{\mathrm{CBS}}, \tag{18}$$

where $\Delta E_\mu^{\mathrm{CBS}}$ is a complete basis set (CBS) extrapolated energy error.[59,86] The subscript μ will now be dropped in the understanding that it applies to any stationary state.

The FCI energy E^{FCI} can be made more explicit:

$$E^{\mathrm{FCI}} = E_{\mathrm{up}}^{\mathrm{FCI}} + \Delta E_{\mathrm{up}}^{\mathrm{FCI}} \pm \delta E_{\mathrm{up}}^{\mathrm{FCI}} + \Delta E_{\mathrm{d}}^{\mathrm{FCI}} \pm \delta E_{\mathrm{d}}^{\mathrm{FCI}}, \tag{19}$$

where $E_{\mathrm{up}}^{\mathrm{FCI}}$ is a variational (upper bound) energy, and $\Delta E_{\mathrm{up}}^{\mathrm{FCI}}$ is the error in the variational upper bound due to our use of CI by parts.[37] $\Delta E_{\mathrm{d}}^{\mathrm{FCI}}$ is the error due to discarded configurations in SCI.[25,36,37] The δ quantities are corresponding uncertainties.

The CBS energy error is approximated by:

$$\Delta E^{\mathrm{CBS}} = \Delta E_{\mathrm{patt}}^{\mathrm{CBS}} \pm \delta E_{\mathrm{patt}}^{\mathrm{CBS}}, \tag{20}$$

where $\Delta E_{\mathrm{patt}}^{\mathrm{CBS}}$ is an estimate of the CBS energy error based on empirical formulas aiming to simulate energy patterns of convergence.[34,87] For Ne and H_2O, $\delta E_{\mathrm{patt}}^{\mathrm{CBS}}$ is, by far, the leading uncertainty in calculated energies.

5.1 Dick Brown's Formula for Configurational Energy Contributions

Just for the purpose of selecting configurations K, rather than CSFs F_{gK}, let us sum over all degenerate elements g in Eq. (2), getting:

$$\Psi^{\mathrm{FCI}} = \sum_{\mathrm{K}=1}^{\mathrm{K}_x} G_{\mathrm{K}} B_{\mathrm{K}}; \quad G_{\mathrm{K}} = N_{\mathrm{K}} \sum_{\mathrm{g}=1}^{\mathrm{g_K}} F_{\mathrm{gK}} C_{\mathrm{gK}}, \tag{21}$$

in terms of normalized symmetry-configurations G_{K} and coefficients B_{K}. The reason for this manipulation is that distinct from the C_{gK} coefficients of Eq. (2), the B_{K} coefficients of three and higher excited configurations can be predicted from B_{K} coefficients of lower excited ones.[36] We will use Eq. (21) just for preselecting and selecting configurations K rather than CSFs; for all other purposes we will be using Eq. (2).

Selection of configurations K is achieved by use of the Brown formula[36,88] for individual energy contributions ΔE_{K} of K:

$$\Delta E_{\mathrm{K}} = (E - H_{\mathrm{KK}}) B_{\mathrm{K}}^2 / (1 - B_{\mathrm{K}}^2), \tag{22}$$

where B_{K} is the corresponding CI coefficient, and the diagonal matrix element H_{KK} is well approximated by any diagonal element between Slater determinants $D_{i\mathrm{K}}$ of K:

$$H_{\mathrm{KK}} \approx \langle D_{i\mathrm{K}} | H | D_{i\mathrm{K}} \rangle, \tag{23}$$

since $(E - H_{KK})$ is of the order of a few Hartree for those ΔE_K of the order of one μHartree or smaller, whereupon ΔE_K can be evaluated for *any* values of n_K and g_K in Eq. (1).

Eq. (22) is applicable to disconnected Ks, viz., those n-excited Ks that can be written as products of excitations of order lower than n. All other configurations are called connected ones. The latter occur because of the use of symmetry orbitals.

For connected Ks, it is unknown how to predict B_K coefficients a priori. For selection purposes, a rough approximation is given by[43]:

$$\Delta E_K = (E - H_{KK}) \prod_{i=1}^{q} n_{K_i}, \tag{24}$$

where the n_{K_i}s are NO occupation numbers of K running up to excitation order q. Since estimates (24) have large uncertainties, they must be accompanied by a sensitivity analysis.[25]

The a priori truncation energy error ΔE_d^{FCI} of discarded configurations is obtained by:

$$\Delta E_d^{FCI} = \sum_{\text{discarded K}} \Delta E_K. \tag{25}$$

Different selection thresholds, $T_{sd}(q)$ and $T_{sc}(q)$, are used for (22) and (24), for disconnected or connected configurations, respectively, and for each order of excitation q.

5.2 SCI with Truncation Energy Error

The latest update of SCI-TEE is given in Ref. 25, to which we refer for numerical results and for detailed explanations. SCI-TEE aims to approximate full CI with TEE. At present, success is (optimistically) guaranteed for atomic and molecular systems with no more than 10 electrons.

5.3 Role of Projectors and NOs in SCI with Truncation Energy Error

Good NOs require multireference CI-2x and CI-3x wave functions, with the excitations of the reference configurations limited to their respective *interactive spaces*. The sole purpose of interacting spaces is to obtain good NOs, both for nonrelativistic as well as for relativistic calculations.

Best NOs are obtained from CI-nx with highest possible n. The number of NOs needed to converge in CI-2x is much larger than that needed for convergence of CI-3x, and this number keeps becoming smaller and smaller for increasing values of n in CI-nx.[25] Whenever the error in CI-2x becomes easy to evaluate, a molecular calculation like in Ref. 25 would be recovering 99.9% of the correlation energy.[25]

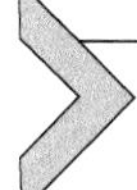

6. UNAMBIGUOUS DIRAC-TYPE CALCULATIONS

6.1 Ambiguity of Perturbative Quantum Electrodynamics for Bound States

The relativistic theory of atomic structure (RTAS)[30] is understood to be an adaptation of perturbative quantum electrodynamics (PQED)[89,90] from which an effective Hamiltonian is formed.[30] PQED works with ad hoc positive Hamiltonians.[90] The adaptation of PQED to RTAS revolves around justifying and choosing positive-energy orbitals, together with the exclusion of the negative-energy ones.[30] The problem with relying in the use of ad hoc positive-energy orbitals is that they are not unique, therefore expectation values become ambiguous, viz., they are dependent upon the (ambiguous) choice of positive-energy orbitals. Variational QED is still at works.

Alternatively, relativistic quantum mechanics for atoms and molecules (RQMAM)[25] starts from a Dirac-type Hamiltonian H_{Dt}, such as discussed in Ref. 25, and attempts to correct it in various ways not necessarily invoking PQED. RQMAM seeks *variational* approximations to bound-state solutions of H_{Dt}:

$$H_{Dt}\Psi_{\mu} = E_{\mu}\Psi_{\mu}, \tag{26}$$

which eliminate the ambiguities of RTAS. In this connection, RQMAM has been employed to produce atomic radiative transition probabilities using fully variational wave functions[91] where the effect of negative-energy orbitals upon some forbidden transitions was predominant, in fair agreement with perturbation theory results.[92]

In RQMAM, H_{Dt} is represented by a complete orbital basis incorporating positive-energy and negative-energy orbitals on an equal footing,[93,94] viz., the orbital basis is invariant upon separate nonsingular linear transformations of its upper and lower components. Using the variational theorem of RQMAM,[93,94] *unique* stationary state solutions of Eq. (26) are obtained, thus removing the ambiguities in the traditional treatment, as it should fit a bona fide physical theory.

As an extra bonus, the "disease" of the complete Hamiltonian is clarified[95] and a less arbitrary and more exact method to obtain an effective Hamiltonian is obtained. An effective Hamiltonian is still needed for larger than two-electron states in order to avoid calculation of *very* highly excited states of the complete Hamiltonian,[25] beyond the scope of current methods, but this may be subject to change in the future.

We are interested in bound-state solutions of Eq. (26). Nevertheless, in 1951, while using $H_{Dt} \equiv H_{DC}$, namely, the Dirac–Coulomb Hamiltonian, Gerry Brown and David Ravenhall[95] asserted that with a basis of four-component Dirac bispinors,[30]

$$\Psi^{(n)}_{n\ell jm_j} = \frac{1}{r}\begin{pmatrix} P_{n\ell j}(r)\mathcal{Y}_{\kappa m_j} \\ i\,Q_{n\ell' j}(r)\mathcal{Y}_{-\kappa m_j} \end{pmatrix}, \quad n = 1, 2, \ldots, n_x(\ell, j), \tag{27}$$

H_{DC} could have negative-energy expectation values, which is true. Furthermore, they concluded that H_{DC} is unbounded from below, which is also true. However, they came to believe that H_{DC} could not have bona fide bound-state solutions and therefore was declared "sick": thus was created the "Brown–Ravenhall disease." The "disease" does exist *only* if the variational theorem for relativistic states is ignored, which was not their fault, since that was only made public in 1997.[93]

As a "cure" for the "Brown–Ravenhall disease," a no-pair Hamiltonian was advanced, $\mathbf{H}^{+}_{DC}$[95,96]:

$$\mathbf{H}^{+}_{DC} = \Lambda^{++} H_{DC} \Lambda^{++}, \tag{28}$$

$$\mathbf{H}^{+}_{DC}\Psi^{+}_{i} = E^{+}_{i}\Psi^{+}_{i}, \tag{29}$$

where Λ^{++} is a product of one-particle projection operators,

$$\Lambda^{++} = \prod_{i=1}^{N} \lambda^{+}(i), \tag{30}$$

$$\lambda^{+}(1) = \sum_{n(\epsilon_n > 0)} |u^{+}_{n}(1)\rangle\langle u^{+}_{n}(1)|, \tag{31}$$

and the u^{+}_{n}'s are the positive-energy (orthonormal) eigenfunctions of a one-electron operator $h_0(1)$ yet to be specified:

$$h_0(1)u^{+}_{n}(1) = \epsilon_n u^{+}_{n}(1), \quad \epsilon_n > 0. \tag{32}$$

The negative-energy eigenfunctions, u_n^-,

$$h_0(1)u_n^-(1) = \epsilon_n u_n^-(1), \quad \epsilon_n < 0, \tag{33}$$

are defined likewise.

The choice of h_0 completely specifies the Hamiltonian $\mathbf{H}_{DC}^+$ and defines a set of orbitals, viz., a one-electron basis (not *states*!) over which the quantized QED fields may act, while allowing for the consistent incorporation of QED effects using perturbation theory.[90] Notice that distinct from H_{DC}, the no-pair Hamiltonian $\mathbf{H}_{DC}^+$ is a finite matrix, and therefore it is bounded below.

There is great interest[97] in removing the QED restriction of using a "positive" Hamiltonian,[90] particularly in connection with the calculation of transition probabilities,[92] which are gauge dependent unless negative-energy orbitals are incorporated.

Notwithstanding, and quite in general, the concept of a no-pair Hamiltonian is a useful one. However, in order to understand which one is the *best possible no-pair Hamiltonian*, it is necessary to appreciate and to use the complete Hamiltonian, as we shall see in the following.

6.2 Variational Theorem for Nonrelativistic Bound States

Let A be a Hermitian operator bounded from below, let $a_i^{(n)}$ be the ordered eigenvalues of a matrix representation $A^{(n)}$ of A, of order n,

$$a_1^{(n)} \le a_2^{(n)} \le \cdots \le a_n^{(n)}, \tag{34}$$

and let $a_i^{(n+1)}$ be the ordered eigenvalues (in nonincreasing order) of the Hermitian matrix $A^{(n+1)}$ derived by bordering $A^{(n)}$ by one column and one row. According to the bracketing theorem, it holds[79,80,98]:

$$a_i^{(n+1)} \le a_i^{(n)} \le a_{i+1}^{(n+1)}. \tag{35}$$

This is the variational theorem for Hermitian operators bounded from below.

6.3 Variational Theorem for Relativistic Bound States

A pertinent corollary of Eq. (35) is obtained by noticing that its last inequality can be continued indefinitely, while new bordering columns and rows keep being added[93]:

$$a_i^{(n)} \le a_{i+1}^{(n+1)} \le a_{i+2}^{(n+2)} \le \cdots \le a_{i+q}^{(n+q)}, \tag{36}$$

for any positive value of q. Eq. (36) relates the eigenvalues of an n-dimensional representation of a Hermitian operator with an $(n + q)$-dimensional representation of the same operator which includes the original n-dimensional representation. Since in Eq. (36) we are now dealing with *finite matrices*, the result (36) is *independent* on whether or not the original operator A is bounded below. Hereon, A will be taken as any four-component Dirac-type Hamiltonian H_{Dt}.

The key to uniqueness is the use of invariant spaces, leading to invariance of eigenvalues and invariance of full CI wave functions, a concept any of us at QTP heard from Per-Olov well over a hundred times. We shall spend some time to define an orbital basis which is invariant with respect to separate nonsingular linear transformations of its upper and lower components. In order to be specific we use the notation for atomic jj-JM wave functions[30]; translation to the molecular realm is straightforward.

For any neutral atom (all neutral atoms are known to have bound states), one can assume, without arbitrariness, that a proper one-body operator h_0 can be found, and that the set $\{\Psi^{(n)}_{n\ell jm_j}, n=1,2,\ldots,n_x(\ell,j)\}$ is identical with a *suitable* set of positive-energy orbitals $\{u_n^+; n=1,2,\ldots,w\}$, where $w=\sum_{\ell,j} n_x(\ell,j)$, just to confirm that there is a one-to-one identification between the elements of both sets. The radial functions $P_{n_k\ell j}$ have been discussed elsewhere.[25]

In principle, one can use the basis $\{\Psi^{(n)}_{n\ell jm_j}, n=1,2,\ldots,n_x(\ell,j)\}$ to set up an FCI of dimension $\mathcal{N}^+$, and provided the calculation is numerically stable, the eigenvalues E_i^+ of Eq. (29) will give reasonable results; however, these are clearly not unique because the positive-energy orbitals do depend on the definition of h_0. Moreover, in general, the positive-energy orbitals are not invariant upon separate nonsingular linear transformations of their own upper and lower components. So, in relativistic calculations we have the paradoxical result that the full CI is not an invariant unless it is taken with respect to a *complete basis* in the sense of being invariant with respect to separate nonsingular linear transformations of its upper and lower components.

In order to insure uniqueness, a complementary basis $\{\Psi^{(n_x+k)}_{k\ell jm_j}\}$ needs to be introduced:

$$\Psi^{(n_x+k)}_{k\ell jm_j} = \frac{1}{r}\begin{pmatrix} P_{k\ell j}(r)\mathcal{Y}_{\kappa m_j} \\ -i\, Q_{k\ell' j}(r)\mathcal{Y}_{-\kappa m_j} \end{pmatrix}, \quad n=n_x(\ell,j),\ k=1,2,\ldots,n_x(\ell,j), \quad (37)$$

where the lower component has an explicit negative sign. The new set, also of dimension w, is linearly independent of the first set, and thus it can always by orthogonalized to it. The union of both sets is a $2w$-dimensional set, called the double-primitive set, which is *invariant* upon separate nonsingular linear transformations of the upper and lower components, and that property, together with hermiticity, guarantees that the corresponding full CI matrix $\mathbf{H}_{DC}$, of dimension $\mathcal{N}$, will have a unique set of eigenvalues:

$$\mathbf{H}_{DC}\mathbf{C}_j = E_j^{\mathrm{FCI}}\mathbf{C}_j. \tag{38}$$

Recalling that $\mathcal{N}^+$ is the dimension of the no-pair Hamiltonian $\mathbf{H}_{DC}^+$, and that $\mathcal{N}$ is the dimension of the complete matrix $\mathbf{H}_{DC}$, we define $\mathcal{N}^-$ by,

$$\mathcal{N}^- = \mathcal{N} - \mathcal{N}^+. \tag{39}$$

Since $\mathbf{H}_{DC}^+$ is contained in $\mathbf{H}_{DC}$, from Eq. (36) it follows:

$$E_i^{+\mathrm{FCI}} \leq E_{i+\mathcal{N}^-}^{\mathrm{FCI}}, \quad i = 1, 2, \ldots, \mathcal{N}^+. \tag{40}$$

Thus, the eigenvalues $E_{i+\mathcal{N}^-}^{\mathrm{FCI}}$ of a full and invariant representation of $\mathbf{H}_{DC}$ are upper bounds to the eigenvalues of the corresponding "no-pair" or "positive" Hamiltonian $\mathbf{H}_{DC}^+$.[93] An analogous result is obtained when H_{DC} is replaced by the Breit–Dirac Hamiltonian.[99]

The eigenvalues $E_i^{\mathrm{FCI}}, i = 1, 2, \ldots, \mathcal{N}^-$ provide approximations to the negative continuum. Let us now take the ionization threshold of all electrons as equal to zero. In such a case, it is observed that, for small orbital bases, these eigenvalues accumulate around $-mc^2, -2mc^2, \ldots -Nmc^2$, where N is the number of electrons, m is the electron mass, and c is the speed of light. As the size of the orbital basis continues to be enlarged, the eigenvalues begin to spread with increased uniformity below $E_{\mathcal{N}^-+1}$.

Furthermore, the value of the electron densities obtained from the eigenfunctions $\mathbf{C}_j, j \leq \mathcal{N}^-$ oscillates along the radial coordinate, while those corresponding to the right hand side of Eq. (40) exhibit the monotonic decrease proper of the nonrelativistic regime.

In a thought experiment in which numerical stability could be kept for unlimitedly large orbital bases, (approximations to) the negative continuum always remains below the ground state $E_{\mathcal{N}^-+1}$ as predicted by Eq. (40). Above the ground state, up to 12 accurate excited states of the same symmetry have been calculated for three-electron states[100] of both O^{+5} and Ne^{+7}.

6.4 What Is Wrong with the Brown–Ravenhall Paper?

Brown and Ravenhall[95] declared meaningless the Dirac–Coulomb Hamiltonian H_{DC}, and by extension all other Dirac-type Hamiltonians, only because orbitals may have negative-energy components. Since their argument involves a finite number of matrix elements, and not the operators themselves, their reasoning is refuted by Eq. (40). Eq. (40) is valid for all systems for which one can construct a no-pair Hamiltonian, such as all neutral atoms and all positive atomic ions, for states sufficiently below from the ionization threshold.

Other people may argue by first choosing a complete continuum basis set with a continuum spectra in $[-\infty, +\infty]$ and then proceed to immerse a bound state approximation into that continuum. But that leads to a well-known example of overcompleteness[81] which becomes worse and worse as more positive-energy orbitals are introduced. Thus, there is no way for an approximate bound state to dissolve into a bona fide complete continuum in $[-\infty, +\infty]$ without falling into overcompleteness.

Atomic resonances constitute a totally different case. Resonances are *not* bound states; however, they can be *approximated* by a superposition of an approximate bound state and continuum orbitals which may or may not be approximated in Hilbert space, viz., they dissolve or do not dissolve into a continuum by construction. In both cases, useful physical consequences can be extracted therefrom.

6.5 Joe Sucher's Folk Theorem Is Wrong

In a letter entitled "Continuum dissolution and the relativistic many body problem. A solvable problem," Sucher[101] addresses a two-electron problem for a Dirac-type Hamiltonian represented by two (actually four) one-electron Dirac bispinors: $\phi_{+}^{-1/2}$, $\phi_{+}^{+1/2}$, $\phi_{-}^{-1/2}$, and $\phi_{-}^{+1/2}$, where the superscripts represent the quantum number m_j with possible values $-1/2$ and $+1/2$. These span a four-dimensional one-electron space which is invariant upon nonsingular linear transformations upon them. We will use Slater determinants instead of the Hartree products used by Sucher, and $\mathbf{J}^2$ two-electron eigenfunctions, since the proper counting of these is of paramount importance.

There will be a total of $4 \cdot 3/2 = 6$ determinants yielding three J=1 components (with $M_J = -1$, 0, and +1) and three $J = 0$ components.

The latter are given by the two ordered determinants $\det|\phi_+^{-1/2}(1)\phi_+^{+1/2}(2)|$ and $\det|\phi_-^{-1/2}(1)\phi_-^{+1/2}(2)|$, and by $\det|\phi_+^{-1/2}(1)\phi_-^{+1/2}(2)| + \det|\phi_-^{-1/2}(1)\phi_+^{+1/2}(2)|$. The full CI symmetry-adapted $\mathbf{H}_{DC}$ matrix splits into three one-dimensional blocks for the three eigenfunctions with J=1 plus one three-dimensional block corresponding to J=0 two-electron bases, thus for J=0 $\mathcal{N}=3$, while $\mathcal{N}^+=1$ since $\mathbf{H}_{DC}^+$ is fully represented in $\det|\phi_+^{-1/2}(1)\phi_+^{+1/2}(2)|$. Thus $\mathcal{N}^-=2$.

Sucher makes his argument by looking at the lowest eigenvalue of its representation of a Dirac-type Hamiltonian, whereas the correct way to solve his proposed problem is by looking at the lowest two-electron *stationary state*, namely, the eigenvalue $E_{i+\mathcal{N}^-}^{\text{FCI}}$ of order $1+\mathcal{N}^-=3$, Eq. (40). The corresponding eigenfunction contains both positive- and negative-energy one-electron functions, while its eigenvalue is an upper bound to the expectation value of the no-pair Hamiltonian with respect to $\det|\phi_+^{-1/2}(1)\phi_+^{+1/2}(2)|$.[93] Thus concludes our explanation of why Sucher's "Folk's Theorem"[101] is wrong.

6.6 Optimal Decoupling of Positive- and Negative-Energy Orbitals

Except for Sucher's example above, all other relativistic calculations outside our group deal just with positive-energy orbitals, viz., with ad hoc bases that are *not invariant* upon separate nonsingular linear transformations of its upper and lower components.

Let us define optimally decoupled positive-energy orbitals as those that minimize the energy difference between the full (invariant) result $E_{i+\mathcal{N}^-}^{\text{FCI}}$ and the one obtained with just positive-energy orbitals, $E_i^{+\text{FCI}}$, for a given value of the subindex *i*.

As shown in a 1998 paper,[94] when the Sucher problem is solved (with 18 instead of 2 Dirac bispinors) for the ground state of U^{+90}, $\mathcal{N}^+=9\cdot 10/2=45$, $\mathcal{N}^+ + \mathcal{N}^- = 18\cdot 19/2 = 171$, and $\mathcal{N}^- = 171-45=126$. The positive-energy orbitals yielding an energy closest (from below) to the (invariant) eigenvalue of order $1+\mathcal{N}^-=127$ (corresponding to the one of order 3 in Sucher's example) turn out to be nothing less than the corresponding *natural orbitals*. Further details are given in Ref. 94. The present status of best positive-energy orbitals has been recently reviewed.[25]

7. FINAL REMARKS

Per-Olov Löwdin was a remarkable man, and those of us who were lucky to feel, to enjoy, and to learn from his "better side" are grateful to Sweden in particular, and to the world in general, for having produced such a man.

If no biographer already started to write about his life, she (he) must make a quick decision to start working at once, since Per's life contains many stories worth knowing, and his close friends, associates, and beneficiaries will pass away before not too long. Of course, the same may be said in connection with other distinguished quantum chemists.

One important aspect to recover from Per's vast legacy is his disarmingly plural approach to science, when most other successful scientists tend to be single minded.

From my own personal viewpoint, I have this to say:

(i) Atomic structure calculations may probably be of interest for many more centuries. At present, numerical multiconfigurational self-consistent field[29,30] and SCI[25] are the leading approaches, both anchored in the variational theorem. Also, SCI is unthinkable without making use of projection operators and natural orbitals.

(ii) Projection operators and the "turn-over rule" presently lack better replacements for efficient wholesale evaluation of matrix elements such as required in SCI. The latter is presently, by far, the most accurate method for both atomic and molecular 6–10 electron systems. Projectors also provide for a modular framework for nonrelativistic and relativistic, atomic and molecular calculations, in a single computer program.

(iii) The variational theorem will remain in the headlines while a truthful relativistic quantum mechanics for atoms and molecules is sought, namely, a complete atomic and molecular theory incorporating positive-energy and negative-energy orbitals.[25] This theory is necessary for unambiguous expectation values, and mostly for transition values such as in atomic forbidden transitions.[91] The variational theorem may also show the path to develop variational QED. Both are necessary to understand stationary states, excited states in particular. The variational theorem and its consequences are at the root of explaining bound states at the Hartree–Fock level,[102] and at the full CI level.[102]

(iv) Natural orbitals dearly accompanied Per-Olov slightly more than the second half of his life. Presently, selected CI-nx and selected MRCI-nx offer a very good method to evaluate potential-energy surfaces.[25] Whether this method will ever become the best one will strongly depend on code efficiency and its corresponding ease of use. Natural orbitals for these applications will survive, while MRCI-nx does not become obsolete.

(v) Natural orbitals will likely be present through centuries in an unexpected application: the determination of best positive-energy orbitals in relativistic calculations.[25,94]

Per-Olov lives and will live in the hearts and minds of those who shared or will eventually share some of his themes, of his thoughts, and of his lively optimism.

ACKNOWLEDGMENTS

I am grateful to the Instituto de Física and to the Dirección General de Tecnologías de la Información (DGTIC) of Universidad Nacional Autónoma de México for excellent and free computer services. As always, I am deeply indebted to all my coauthors and colleagues who shared their knowledge with me through the years.

REFERENCES

1. Löwdin, P. O. Correlation Problem in Many-Electron Quantum Mechanics. I. Review of Different Approaches and Discussion of Some Current Ideas. *Adv. Chem. Phys.* **1959**, *2*, 207. [First published as Tech. Note No. 1, Uppsala Quantum Chemistry Group, June 21, 1957].
2. Lennard-Jones, J. E. The Electronic Structure of Some Diatomic Molecules. *Trans. Faraday Soc.* **1929**, *25*, 668.
3. Hall, G. G. The Molecular Orbital Theory of Chemical Valency. VI. Properties of Equivalent Orbitals. *Proc. R. Soc. Lond. A* **1950**, *202*, 336.
4. Boys, S. F. Electronic Wave Functions. I. A General Method of Calculation for the Stationary States of Any Molecular System. *Proc. R. Soc. Lond. A* **1950**, *200*, 542.
5. Coulson, C. A. Samuel Francis Boys 1911-1972. *Biogr. Mem. Fellows R. Soc.* **1973**, *19*, 94.
6. Coulson, C. A. Self-Consistent Field for Molecular Hydrogen. *Proc. Camb. Philos. Soc.* **1938**, *34*, 204.
7. Coulson, C. A. Present State of Molecular Structure Calculations. *Rev. Mod. Phys.* **1960**, *32*, 170.
8. Mulliken, R. S. The Interpretation of Band Spectra. III. Electron Quantum Numbers and States of Molecules and Their Atoms. *Rev. Mod. Phys.* **1932**, *4*, 1.
9. Mulliken, R. S.; Roothaan, C. C. J. Broken Bottlenecks and the Future of Molecular Quantum Mechanics. *Proc. Natl. Acad. Sci. U.S.A.* **1959**, *45*, 394.
10. Roothaan, C. C. J. New Developments in Molecular Orbital Theory. *Rev. Mod. Phys.* **1951**, *23*, 69.
11. Roothaan, C. C. J. Self-Consistent Field Theory for Open Shells of Electronic Systems. *Rev. Mod. Phys.* **1960**, *32*, 179.

12. Slater, J. C. The Theory of Complex Spectra. *Phys. Rev.* **1929**, *34*, 1293.
13. Barnett, M. P. Mechanized Molecular Calculations—The POLYATOM System. *Rev. Mod. Phys.* **1963**, *35*, 571.
14. Boyd, D. B. Quantum Chemistry Program Exchange, Facilitator of Theoretical and Computational Chemistry in Pre-Internet History. In: Pioneers of Quantum Chemistry, Strom, E. T.; Wilson, A. K. (Eds.), Vol. 1122. ACS Symposium Series. pp. 221–273. Chapter 8.
15. Linderberg, J. Per-Olov Löwdin 1916-2000. *Proc. Am. Philos. Soc.* **2003**, *147*, 175.
16. Löwdin, P. O. Quantum Theory of Many-Particle Systems. III. Extension of the Hartree-Fock Scheme to Include Degenerate Systems and Correlation Effects. *Phys. Rev.* **1955**, *97*, 1509.
17. Löwdin, P. O. The Normal Constants of Motion in Quantum Mechanics Treated by Projection Technique. *Rev. Mod. Phys.* **1962**, *34*, 520. [First published as Tech. Note No. 68, Uppsala Quantum Chemistry Group, November 15, 1961].
18. Löwdin, P. O. Angular Momentum Wavefunctions Constructed by Projection Operators. *Rev. Mod. Phys.* **1964**, *36*, 966. [First published as Tech. Note No. 12, Uppsala Quantum Chemistry Group, May 10, 1958].
19. Löwdin, P. O. Quantum Theory of Many-Particle Systems. I. Physical Interpretations by Means of Density Matrices, Natural Spin-Orbitals, and Convergence Problems in the Method of Configurational Interaction. *Phys. Rev.* **1955**, *97*, 1474.
20. Löwdin, P. O.; Shull, H. Natural Orbitals in the Quantum Theory of 2-Electron Systems. *Phys. Rev.* **1956**, *101*, 1730.
21. Löwdin, P. O. Studies in Perturbation Theory. X. Lower Bounds to Energy Eigenvalues in Perturbation-Theory Ground State. *Phys. Rev.* **1965**, *139*, A357.
22. Löwdin, P. O. Studies in Perturbation Theory. XI. Lower Bounds to Energy Eigenvalues, Ground State, and Excited States. *J. Chem. Phys.* **1965**, *43*, S175.
23. Shull, H.; Löwdin, P. O. Variation Theorem for Excited States. *Phys. Rev.* **1958**, *110*, 1466. [First published as Tech. Note No. 9, Uppsala Quantum Chemistry Group, April 1, 1958].
24. Löwdin, P. O. *Linear Algebra for Quantum Theory*. Wiley: New York, 1998.
25. Almora-Díaz, C. X.; Rivera-Arrieta, H. I.; Bunge, C. F. Recent Progress in the Variational Orbital Approach to Atomic and Molecular Electronic Structure. *Adv. Quant. Chem.* **2016**, *72*, 129.
26. Schmidt, M. W.; Gordon, M. S. The Construction and Interpretation of MCSCF Wavefunctions. *Annu. Rev. Phys. Chem.* **1998**, *49*, 233.
27. Weiss, A. W. Configuration Interaction in Simple Atomic Systems. *Phys. Rev.* **1961**, *122*, 1826.
28. Eissner, W.; Jones, M.; Nussbaumer, H. Techniques for the Calculation of Atomic Structures and Radiative Data Including Relativistic Corrections. *Comput. Phys. Commun.* **1974**, *8*, 270.
29. Fischer, C. F. Atomic Structure: Multiconfiguration Hartree-Fock Theories. In *Springer Handbook of Atomic, Molecular and Optical Physics*; Drake, G. W. F. Ed.; Springer Science+Business Media, Inc.: New York, 2006; pp. 307–324.
30. Grant, I. P. Relativistic Atomic Structure. In *Springer Handbook of Atomic, Molecular and Optical Physics*; Drake, G. W. F. Ed.; Springer Science+Business Media, Inc.: New York, 2006; pp. 325–358.
31. Hibbert, A. A general Program for Calculating Angular Momentum Integrals in Atomic Structure. *Comput. Phys. Commun.* **1970**, *1*, 359.
32. Glass, R.; Hibbert, A. Relativistic Effects in Many Electron Atoms. *Comput. Phys. Commun.* **1978**, *16*, 19.
33. Shavitt, I. Matrix Element Evaluation in the Unitary Group Approach to the Electron Correlation Problem. *Int. J. Quant. Chem.* **1978**, *14*, 5.

34. Jitrik, O.; Bunge, C. F. Atomic Configuration Interaction and Studies of He, Li, Be and Ne Ground States. *Phys. Rev. A* **1997**, *56*, 2614.
35. Liu, B.; Yoshimine, M. The Alchemy Configuration Interaction Method. I. The Symbolic Matrix Method for Determining Elements of Matrix Operators. *J. Chem. Phys.* **1981**, *74*, 612.
36. Bunge, C. F. Selected Configuration Interaction with Truncation Energy Error and Application to the Ne Atom. *J. Chem. Phys.* **2006**, *125*, 014107.
37. Bunge, C. F.; Carbó-Dorca, R. Select-Divide-and-Conquer Method for Large-Scale Configuration Interaction. *J. Chem. Phys.* **2006**, *125*.
38. Pople, J. A.; Krishnan, R.; Schlegel, H. B.; Binkley, J. S. Many-Body Perturbation Theory, Coupled-Pair Many-Electron Theory, and the Importance of Quadruple Excitations for the Correlation Problem. *Int. J. Quant. Chem.* **1978**, *14*, 545.
39. Gdanitz, R. J.; Ahlrichs, R. The Averaged Coupled-Pair Functional (ACPF): A Size-Extensive Modification of MR CI(SD). *Chem. Phys. Lett.* **1988**, *143*, 413.
40. Taylor, P. R. Coupled-Cluster Methods in Quantum Chemistry. Lecture Notes in Quantum Chemistry: European Summer School; Springer-Verlag: Berlin, 1994; pp. 125–202.
41. Szalay, P. Configuration Interaction: Corrections for Size-Consistency. In *Encyclopedia of Computational Chemistry*; von Ragué Schleyer, P. Ed.; Chichester: Wiley Interscience, 2005.
42. Olsen, J.; Jorgensen, P.; Koch, H.; Balkova, A.; Bartlett, R. J. Full Configuration-Interaction and State of the Art Correlation Calculations on Water in a Valence Double-Zeta Basis with Polarization Functions. *J. Chem. Phys.* **1996**, *104*, 8007.
43. Bunge, C. F. A Priori Selected Configuration Interaction with Truncation Energy Error, General Sensitivity Analysis and Application to the Ne Atom. *Mol. Phys.* **2010**, *108*, 3279.
44. Almora-Díaz, C. X. Highly Correlated Configuration Interaction Calculations on Water with Large Orbital Bases. *J. Chem. Phys.* **2014**, *140*, ((184302)).
45. Löwdin, P. O. On the Non-Orthogonality Problem Connected with the Use of Atomic Wave Functions in the Theory of Molecules and Crystals. *J. Chem. Phys.* **1950**, *18*, 365.
46. Löwdin, P. O. A Note on the Quantum-Mechanical Perturbation Theory. *J. Chem. Phys.* **1951**, *19*, 1396.
47. Löwdin, P. O. Quantum Theory of Cohesive Properties of Solids. *Adv. Phys.* **1956**, *5*, 1.
48. Löwdin, P. O. Band Theory, Valence Bond, and Tight-Binding Calculations. *J. Appl. Phys.* **1962**, *33*, 251.
49. Löwdin, P. O. Exchange, Correlation, and Spin Effects in Molecular and Solid-State Theory. *Rev. Mod. Phys.* **1962**, *34*, 80.
50. Löwdin, P. O. Studies in Perturbation Theory. IV. Solution of Eigenvalue Problem by Projection Operator Formalism. *J. Math. Phys.* **1962**, *3*, 969. [First published as Tech. Note No. 47, Uppsala Quantum Chemistry Group, August 1, 1960].
51. Löwdin, P. O. Studies in Perturbation Theory. V. Some Aspects on the Exact Self-Consistent Field Theory. *J. Math. Phys.* **1962**, *3*, 1171. [First published as Tech. Note No. 48, Uppsala Quantum Chemistry Group, August 1, 1960].
52. Löwdin, P. O. Wave and Reaction Operators in the Quantum Theory of Many-Particle Systems. *Rev. Mod. Phys.* **1963**, *35*, 702.
53. Löwdin, P. O. Studies in Perturbation Theory. IX. Connection Between Various Approaches in the Recent Development. Evaluation of Upper Bounds to Energy Eigenvalues in Schrödinger's Perturbation Theory. *J. Math. Phys.* **1965**, *6*, 1341.

54. Löwdin, P. O. Scaling Problem, Virial Theorem, and Connected Relations in Quantum Mechanics. *J. Mol. Spectrosc.* **1959**, *3*, 46.
55. Löwdin, P. O. Quantum Genetics and the Aperiodic Solid. *Adv. Quant. Chem.* **1966**, *2*, 213. [First published as Tech. Note No. 85, Uppsala Quantum Chemistry Group, November 11, 1962].
56. Löwdin, P. O. Proton Tunnelling in DNA and Its Biological Implications. *Rev. Mod. Phys.* **1963**, *35*, 724.
57. Löwdin, P. O. Effect of Proton Tunnelling in DNA on Genetic Information and Problems of Mutations, Aging and Tumors. *Biopolym. Symp.* **1964**, *13*, 161.
58. Löwdin, P. O. Some Aspects of Quantum Biology. *Biopolym. Symp.* **1964**, *13*, 293.
59. Bunge, C. F. Electronic Wave Functions for Atoms. II. Some Aspects of the Convergence of the CI Expansion for the He Isoelectronic Series. *Theor. Chim. Acta* **1970**, *16*, 126.
60. Bunge, C. F.; Bunge, A. Upper and Lower Bounds to the Eigenvalues of Double Minimum Potentials. *J. Chem. Phys.* **1965**, *43*, S194. [First published as Tech. Rep. No. 60, Quantum Theory Project, July 1964].
61. Bunge, C. F. Unrestricted Projected Hartree-Fock Solutions for Two-Electron Systems. *Phys. Rev.* **1967**, *154*, 70.
62. Miller, K. J.; Ruedenberg, K. Electron Correlation and Augmented Separated-Pair Expansion in Berylliumlike Atomic Systems. *J. Chem. Phys.* **1968**, *48*, 3450.
63. Bunge, C. F. Electronic Wave Functional For Atoms. I. Ground State of Be. *Phys. Rev.* **1968**, *168*, 92.
64. Wigner, E. P. *Group Theory (English Translation)*. Academic: New York, 1959.
65. Bunge, A. V.; Bunge, C. F.; Jáuregui, R.; Cisneros, G. Symmetry Eigenfunctions for Many-Electron Atoms and Molecules: A Unified and Friendly Approach for Frontier Research and Student Training. *Comput. Chem.* **1989**, *13*, 201.
66. Jáuregui, R.; Bunge, C. F.; Bunge, A. V.; Cisneros, G. Angular Momentum Eigenfunctions for Many-Electron Calculations. *Comput. Chem.* **1989**, *13*, 223.
67. Bunge, A. V.; Bunge, C. F.; Jáuregui, R.; Cisneros, G. Spin Eigenfunctions for Many-Electron Calculations. *Comput. Chem.* **1989**, *13*, 239.
68. Cisneros, G.; Jáuregui, R.; Bunge, C. F.; Bunge, A. V. Molecular Symmetry Eigenfunctions for Many-Electron Calculations. *Comput. Chem.* **1989**, *13*, 255.
69. Bunge, C. F.; Bunge, A. Eigenfunctions of Spin and Orbital Angular Momentum by the Projection Operator Technique. *J. Comput. Phys.* **1971**, *8*, 409.
70. Rotenberg, A. Calculation of Exact Eigenfunctions of Spin and Orbital Angular Momentum Using the Projection Operator Method. *J. Chem. Phys.* **1963**, *39*, 512.
71. Rotenberg, A. Program QCPE 37 from the Quantum Chemistry Program Exchange, Distribution Presently Discontinued.
72. Gray, N. M.; Wills, L. A. Note on the Calculation of Zero Order Eigenfunctions. *Phys. Rev.* **1931**, *38*, 248.
73. Bunge, A. Electronic Wave Functions for Atoms. III. Optimal Partition of Degenerate Spaces and Ground State of Carbon. *J. Chem. Phys.* **1970**, *53*, 20.
74. Bunge, A. A Configuration Interaction Study of the Ground State of the Carbon Atom. [First published as Tech. Rep. No. 135, Quantum Theory Project, April 1968].
75. Bunge, A.; Bunge, C. F. Correlation Energy of the Ground State of C. *Phys. Rev. A* **1970**, *1*, 1599.
76. Bunge, C. F.; Bunge, A. Symmetry Eigenfunctions Suitable for Many-Electron Theories and Calculations. I. Mainly Atoms. *Int. J. Quant. Chem.* **1973**, *7*, 927.

77. Brooks, B. R.; Schaefer, H. F. The Graphical Unitary Approach to the Electron Correlation Problem. Methods and Preliminary Applications. *J. Chem. Phys.* **1979**, *70*, 5092.
78. Brooks, B. R.; Laidig, W. D.; Saxe, P.; Handy, N. C.; Schaefer, H. F. The Loop-Driven Graphical Unitary Group Approach: A Powerful Method for the Variational Description of Electron Correlation. *Phys. Scr.* **1980**, *21*, 312.
79. Hylleraas, E. A.; Undheim, B. Numerische berechnung der 2S -terme von ortho- und par-helium. *Z. Phys.* **1930**, *65*, 759.
80. MacDonald, J. K. L. Successive Approximations by the Rayleigh-Ritz Variation Method. *Phys. Rev.* **1933**, *43*, 830.
81. Löwdin, P. O. Present Situation of Quantum Chemistry. *J. Phys. Chem.* **1957**, *61*, 56.
82. Bunge, C. F. Ergonals: Optimum Spinorbitals in Accurate Electronic Calculations? In *Energy Structure and Reactivity*; Smith, D. W., McRae, W., Eds.; Wiley, 1973; pp. 68–74.
83. Bunge, C. F. Energy Optimized Spinorbitals in Accurate Electronic Calculations. *Phys. Rev. A* **1973**, 7, 15.
84. Sinanoğlu, O. Many-Electron Theory of Atoms and Molecules. I. Shells, Electron Pairs vs. Many-Electron Correlations. *J. Chem. Phys.* **1962**, *36*, 706.
85. Harris, F. E.; Monkhorst, H. J.; Freeman, D. L. *Algebraic and Diagrammatic Methods in Many-Fermion Theory*. Oxford University Press: New York, 1992.
86. Dunning, T. H., Jr.; Peterson, K. A.; Woon, D. E. Correlation Consistent Basis Sets for Molecular Calculations. In *Encyclopedia of Computational Chemistry*; Schleyer, P. v. R. Ed.; Wiley: New York, 1998; pp. 88–115.
87. Klopper, W.; Bak, K. L.; Jorgensen, P.; Olsen, J.; Helgaker, T. Highly Accurate Calculations of Molecular Electronic Structure. *J. Phys. B* **1999**, *32*, R103.
88. Brown, R. E. A Configuration Interaction Study of the $^1\Sigma^+$ States of the Lithium Hydride Molecule 1967. Ph.D. Thesis.
89. Sapirstein, J. R. Quantum Electrodynamics. In *Springer Handbook of Atomic, Molecular and Optical Physics*; Drake, G. W. F. Ed.; Springer Science+Business Media, Inc.: New York, 2006; pp. 413–428.
90. Weinberg, S. *The Quantum Theory of Fields*. Cambridge University Press: Cambridge, UK, 1998.
91. Jitrik, O.; Bunge, C. F. Atomic Radiative Transition Probabilities using Negative-Energy Orbitals in Fully Variational Wave Functions. *Nucl. Instrum. Methods B* **2005**, *235*, 105.
92. Johnson, W. R.; Plate, D. R.; Sapirstein, J. Transition Amplitudes in the Helium Isoelectronic Sequence. *Adv. At. Mol. Opt. Phys.* **1995**, *35*, 255.
93. Jáuregui, R.; Bunge, C. F.; Ley-Koo, E. Upper Bounds to the Eigenvalues of the No-Pair Hamiltonian. *Phys. Rev. A* **1997**, *55*, 1781.
94. Bunge, C. F.; Jáuregui, R.; Ley-Koo, E. Optimal Decoupling of Positive- and Negative-Energy Orbitals in Relativistic Electronic Structure Calculations Beyond Hartree-Fock. *Int. J. Quant. Chem.* **1998**, *70*, 805.
95. Brown, G. E.; Ravenhall, D. G. On the Interaction of Two Electrons. *Proc. R. Soc. Lond. A* **1951**, *208*, 552.
96. Sucher, J. Foundations of the Relativistic Theory of Many-Electron Atoms. *Phys. Rev. A* **1980**, *22*, 348.
97. Sapirstein, J. Theory of Many-Electron Atoms. *Phys. Scr.* **1993**, *T46*, 52.
98. Bateman, H. On a Set of Kernels Whose Determinants Form a Sturmian Sequence. *Bull. Am. Math. Soc.* **1912**, *18*, 179.

99. Bunge, C. F.; Ley-Koo, E.; Jáuregui, R. Variational Incorporation of Negative-Energy Orbitals in Relativistic Electronic Structure Calculations. *Int. J. Quant. Chem.* **2000**, *80*, 461.
100. Rivera-Arrieta, H. I. *Highly Accurate Method for Atomic Excited States with Application to Spectroscopy of* O^{5+} *and* Ne^{7+} *(in Spanish), M.Sc. Thesis.* Facultad de Química, Universidad Nacional Autónoma de México, 2013; December 5.
101. Sucher, J. Continuum Dissolution and the Relativistic Many Body Problem. A Solvable Model. *Phys. Rev. Lett.* **1985**, *55*, 1033.
102. Bunge, C. F.; Ley-Koo, E.; Jáuregui, R. Justification of Relativistic Dirac-Hartree-Fock and Configuration Interaction. *Mol. Phys.* **2000**, *98*, 1067.

CHAPTER FIVE

Time-Dependent Perturbation Theory with Application to Atomic Systems

Ingvar Lindgren[1]
University of Gothenburg, Gothenburg, Sweden
[1]Corresponding author: e-mail address: ingvar.lindgren@physics.gu.se

Contents

Abstract

A new form of time-dependent perturbation theory has been developed based upon the covariant-evolution operator (CEO), previously introduced by us. This has made it possible to combine time-dependent perturbations, like the quantum-electrodynamical (QED) perturbations, with time-independent interactions, like the Coulomb interaction (electron correlation) in a single perturbation expansion. For the first time quantum-electrodynamical

Advances in Quantum Chemistry, Volume 74
ISSN 0065-3276
http://dx.doi.org/10.1016/bs.aiq.2016.06.002

perturbations can then be combined with electron correlation beyond second order. The experimental accuracy is in many cases so high that these effects have become significant. A numerical scheme has been developed where first-order QED effects are combined with the electron correlation and applied to highly charged helium-like ions. This scheme contains the dominating part of the higher-order QED effects and has been applied to highly charged helium-like ions, for which effects beyond second order (two-photon effects) have for the first time been evaluated. The calculations have been performed using Feynman as well as Coulomb gauge. In evaluating effects beyond second order involving radiative QED it was necessary to employ the Coulomb gauge.

1. INTRODUCTION

Various forms of perturbation theory were developed already in the 18th and the 19th centuries, particularly in connection with astronomical calculations. With the advent of quantum mechanics in the 20th century a wide new field for perturbation theory emerged. The most frequently used form, the *Rayleigh–Schrödinger* perturbation theory, was developed by Erwin Schrödinger,[1] based upon early work by Lord Rayleigh, and another form, the *Brillouin–Wigner* perturbation theory, by Léon Brillouin and Eugine Wigner.

The application of perturbation theory in the quantum field was further developed and systemized particularly by Per-Olov Löwdin, during the 1950s and 1960s, as reported in a long sequence of seminal papers.[2–12]

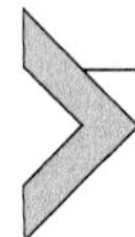

2. STANDARD TIME-INDEPENDENT PERTURBATION THEORY

2.1 Nondegenerate and Degenerate Theory

We shall start by reviewing the standard nondegenerate Rayleigh–Schrödinger perturbation theory, largely following the work of Löwdin, see, eg, Ref. 8.

We want to solve the eigenvalue equation (Schrödinger equation)

$$H\Psi = E\Psi \tag{1}$$

by successive approximations,

$$\Psi = \Psi_0 + \Psi^{(1)} + \Psi^{(2)} + \cdots . \tag{2}$$

We partition the Hamiltonian (H) into a zeroth-order or *model Hamiltonian* (H_0) and a *perturbation* (V)

$$H = H_0 + V. \tag{3}$$

Nonrelativistically

$$H_0 = \sum_{n=1}^{N} h_S(n) \tag{4}$$

is the sum of single-electron Schrödinger Hamiltonians and V is primarily the electron–electron interaction, using conventional notations,

$$V = \sum_{n<m}^{N} \frac{e^2}{4\pi\epsilon_0\, r_{nm}}. \tag{5}$$

The zeroth-order or *model function* (Ψ_0) is then an eigenfunction of H_0,

$$H_0\Psi_0 = E_0\Psi_0. \tag{6}$$

We assume for the moment that the model energy, E_0, is *nondegenerate*, which implies that no other eigenfunction of H has the same model energy. The model function defines a (one-dimensional) *model space* with the *projection operator*

$$P = |\Psi_0\rangle\langle\Psi_0|. \tag{7}$$

Following Löwdin, we introduce a *wave operator,* Ω,[a] which transforms the model function to the full wave function,

$$\Psi = \Omega\Psi_0. \tag{8}$$

We employ the *intermediate normalization* (IN), implying that the inner product of the full wave function and the model function is unity

$$\langle\Psi_0|\Psi\rangle = 1. \tag{9}$$

It then follows that *the model function is the projection of the full wave function onto the model space*

$$P|\Psi\rangle = |\Psi_0\rangle\langle\Psi_0|\Psi\rangle = \Psi_0. \tag{10}$$

[a] By Löwdin denoted by W.

The wave operator satisfies in IN the *Bloch equation*, derived by Claude Bloch in 1955,[13]

$$\boxed{(E_0 - H_0)P = (V\Omega - \Omega PV\Omega)P,} \tag{11}$$

which can also be written as

$$\boxed{\Omega P = \Gamma_Q(V\Omega - \Omega PV\Omega)P,} \tag{12}$$

introducing the Löwdin *reduced resolvent*[8]

$$\Gamma_Q = \frac{Q}{E_0 - H_0}. \tag{13}$$

The Bloch equation can be used to generate the Rayleigh–Schrödinger perturbation expansion to all orders.

An *effective Hamiltonian* can be defined so that it generates the exact energy by operating on the model function

$$H_{\text{eff}}\Psi_0 = E\Psi_0. \tag{14}$$

It then follows that this operator has the form

$$H_{\text{eff}} = PH\Omega P = P(H_0 + V)\Omega P. \tag{15}$$

The second part we refer to as the *effective interaction*

$$W = PV\Omega P. \tag{16}$$

2.2 Quasi-Degenerate Theory

The presentation above is still valid, if there are several eigenfunctions with the same model energy, ie, a degenerate model space. The situation is different if we have several closely lying states, *quasi-degeneracy*, within the model space, such as a fine-structure multiplet. Then the standard procedure might lead to severe convergence problems. It is then often more efficient to consider an *extended model space*, containing all model states of the multiplet.

The problem with quasi-degeneracy can be illustrated by the $1s2p$ fine structure of He-like ions. In the relativistic treatment we have here the unperturbed states, $1s2p_{1/2}$ and $1s2p_{3/2}$, which can form two closely lying $J = 1$ states. Using the standard procedure, based on the Bloch equation with a one-dimensional model space (11), there has been problem of getting convergence for these states.[14] By means of an "extended model space," on the other hand, all states could easily be calculated.[15]

We consider now a set of eigenfunctions of the Schrödinger equation, *target states*,

$$H|\Psi^{\alpha}\rangle = E^{\alpha}|\Psi^{\alpha}\rangle \quad (\alpha = 1\cdots d). \tag{17}$$

For each target state there exists a *model state*, $|\Psi_0^{\alpha}\rangle$ $(\alpha = 1\cdots d)$, forming a *model space*. The projection operator for the entire model space is P and for the complementary space $Q = 1 - P$. With intermediate normalization (18) we have

$$\langle\Psi_0^{\alpha}|\Psi^{\alpha}\rangle = 1\,; \quad |\Psi_0^{\alpha}\rangle = P|\Psi^{\alpha}\rangle \quad (\alpha = 1\cdots d). \tag{18}$$

A single *wave operator, Ω*, transforms all model states to the corresponding target states,

$$\Omega|\Psi_0^{\alpha}\rangle = |\Psi^{\alpha}\rangle \quad (\alpha = 1\cdots d). \tag{19}$$

The *effective Hamiltonian* is defined as before (14)

$$H_{\text{eff}}\Psi_0^{\alpha} = E^{\alpha}\Psi_0^{\alpha}, \tag{20}$$

which leads to the same expressions as in the nondegenerate case

$$H_{\text{eff}} = PH\Omega P; \qquad W = PV\Omega. \tag{21}$$

The wave operator now satisfies a "generalized" Bloch equation with a commutator on the left-hand side[16,17]

$$\boxed{[\Omega, H_0]P = (V\Omega - \Omega PV\Omega)P = (V\Omega - \Omega W)P.} \tag{22}$$

2.3 Linked-Diagram Theorem

The perturbation expansion can conveniently be transformed into graphical form in terms of (Goldstone) diagrams by means of second quantization, as explained in many text books, eg, Ref. 17. The electron orbitals are separated into core orbitals, occupied in all states of the model space, valence orbitals, occupied in some, and virtual orbitals, not occupied in any of the model-space states. This leads to so-called linked and unlinked diagrams, where a diagram is said to be unlinked, if it contains a disconnected, closed part. A diagram is said to be closed if the initial and final states lie in the model space. It was conjectured by Brueckner in 1955[18] that in the RS expansion with a one-dimensional model space the unlinked diagrams are essentially cancelled out, and this was somewhat later rigorously proved to all orders by Goldstone.[19] This is the *linked-diagram or linked-cluster theorem*, which normally simplifies the evaluation of the expansion considerably. The linked-diagram theorem holds also for a degenerate and quasi-degenerate model space, provided that it is complete in the sense that all electron configurations that can be formed by the open shells are included in the model space.[20,21]

3. RELATIVISTIC AND QED EFFECTS

The time-independent perturbation theory sketched above works in principle also relativistically, if we replace the nonrelativistic Hamiltonian (3) by the relativistic one. Then the model Hamiltonian is the sum of single-electron Dirac Hamiltonians

$$H_0 = \sum_{n=1}^{N} h_{\rm D}(n), \tag{23}$$

and the perturbation is the nonrelativistic electron–electron interaction (38), representing the electrostatic interaction, and the instantaneous Breit interaction

$$V_{\rm B} = -\frac{e^2}{8\pi\epsilon_0}\sum_{m<n}\left[\frac{\boldsymbol{\alpha}_m\cdot\boldsymbol{\alpha}_n}{r_{mn}} + \frac{(\boldsymbol{\alpha}_m\cdot\boldsymbol{r}_{mn})(\boldsymbol{\alpha}_m\cdot\boldsymbol{r}_{mn})}{r_{mn}^3}\right], \tag{24}$$

representing the magnetic interaction and the leading retardation effect. This Hamiltonian can lead to singularities, known as the Brown–Ravenhall effect,[22] when the electrons are "excited" into the negative continuum. For that reason Sucher[23] has introduced projection operators, which prevent that from happening,

$$H_{\rm NVPA} = \Lambda_+\left[\sum_{n=1}^{N} h_{\rm D}(i) + V_{\rm C} + V_{\rm B}\right]\Lambda_+, \tag{25}$$

which is known as the *No-Virtual-Pair Approximation* (NVPA).

Effects beyond the NVPA are conventionally referred to as quantum-electrodynamic (QED) effects, although there is no sharp distinction. Some low-order Feynman diagrams are exhibited in Fig. 1.

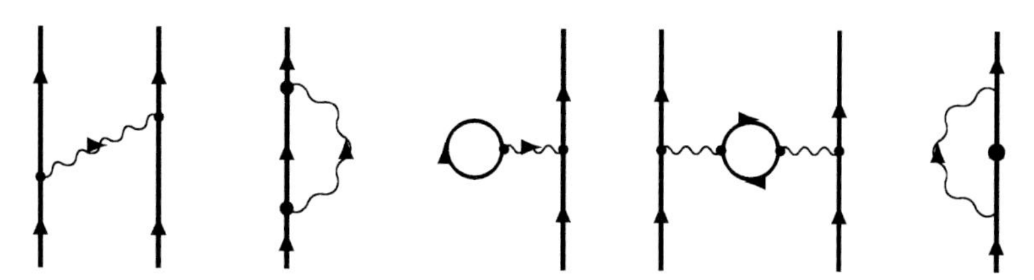

Fig. 1 Some low-order Feynman diagrams of QED effects, as defined in the text, representing (in order) the photon retardation, the electron self-energy, vacuum polarization (two diagrams), and vertex correction.

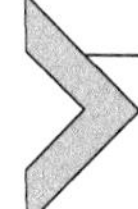

4. TIME-DEPENDENT PERTURBATION THEORY

4.1 The Time-Evolution Operator

A basic tool in time-dependent perturbation theory is the time-evolution operator, which describes how, for instance, the wave function evolves in time ($t > t_0$)

$$\Psi(t) = U(t, t_0)\Psi(t_0). \tag{26}$$

Here, the interaction picture (IP) is used, related to the standard Schrödinger picture (S) by

$$|\Psi(t)\rangle = \mathrm{e}^{\mathrm{i}H_0 t/\hbar}\,|\Psi_\mathrm{S}(t)\rangle; \quad \mathcal{O}(t) = \mathrm{e}^{\mathrm{i}H_0 t/\hbar}\,\mathcal{O}_\mathrm{S}\,\mathrm{e}^{-\mathrm{i}H_0 t/\hbar}. \tag{27}$$

The evolution operator satisfies the equation ($\hbar = 1$)[b]

$$\mathrm{i}\frac{\partial U(t, t_0)}{\partial t} = VU(t, t_0), \tag{28}$$

where V is the perturbation (3). This gives

$$\begin{aligned} U(t, t_0) &= 1 - \mathrm{i}\int_{t_0}^{t} \mathrm{d}t_1\; V(t_1)U(t_1, t_0) \\ &= 1 - \mathrm{i}\int_{t_0}^{t} \mathrm{d}t_1\; V(t_1) + \mathrm{i}^2\int_{t_0}^{t} \mathrm{d}t_1\, V(t_1)\int_{t_0}^{t_1} \mathrm{d}t_2\; V(t_2) + \cdots \\ &= 1 - \mathrm{i}\int_{t_0}^{t} \mathrm{d}t_1\; V(t_1) + \frac{\mathrm{i}^2}{2!}\int_{t_0}^{t} \mathrm{d}t_1\int_{t_0}^{t} \mathrm{d}t_2\; T[V(t_1)V(t_2)] + \cdots \\ &= \sum_{n=0}^{\infty}\frac{(-\mathrm{i})^n}{n!}\int_{t_0}^{t} \mathrm{d}t_1 \ldots \int_{t_0}^{t} \mathrm{d}t_n\; T[V(t_1)\ldots V(t_n)], \end{aligned} \tag{29}$$

[b] We maintain throughout all remaining fundamental constants.

where T is the time-ordering operator, which orders the operators in decreasing time order.

We introduce the perturbation density $\mathcal{H}(x)$

$$V(t)=\int \mathrm{d}^3\boldsymbol{x}\,\mathcal{H}(x), \tag{30}$$

giving

$$U(t,t_0)=\sum_{n=0}^{\infty}\frac{1}{n!}\left(\frac{-\mathrm{i}}{c}\right)^n\int_{t_0}^{t}\mathrm{d}x_1^4\ldots\int_{t_0}^{t}\mathrm{d}x_n^4\;T[\mathcal{H}(x_1)\ldots\mathcal{H}(x_n)]. \tag{31}$$

We shall mostly assume that the perturbation density is

$$\mathcal{H}(x)=-\hat{\psi}^{\dagger}(x)ec\alpha^{\mu}A_{\mu}(x)\hat{\psi}(x), \tag{32}$$

which is the interaction between an electron and the electromagnetic field. This represents the emission or the absorption of a photon, and two such perturbations are needed to represent the exchange of a virtual photon between the electrons. Then we shall denote the perturbation (30) by $v(t)$ to indicate that it represents only "half" a photon. We assume that a *damping factor* $\mathrm{e}^{-\gamma|t|}$ is applied to the perturbation, where γ is a small, positive number, which eventually $\to 0$.

4.2 The Electron Propagator

Second-quantized operators can be contracted, and the contraction between two electron-field operators is defined as *the difference between the time and normal orderings* (see, for instance, Ref. 17, Ch. 11)

$$\overbrace{\hat{\psi}(x_1)\hat{\psi}^{\dagger}}(x_2)=T\left[\hat{\psi}(x_1)\hat{\psi}^{\dagger}(x_2)\right]-N\left[\hat{\psi}(x_1)\hat{\psi}^{\dagger}(x_2)\right]=\left\langle 0\middle|T\left[\hat{\psi}(x_1)\hat{\psi}^{\dagger}(x_2)\right]\middle|0\right\rangle. \tag{33}$$

The vacuum expectation value vanishes for every normal-ordered product, and hence the contraction is equal to the vacuum expectation value of the time-ordered product. We define the electron propagator by

$$\overbrace{\hat{\psi}(x_1)\hat{\psi}^{\dagger}}(x_2)=\left\langle 0\middle|T\left[\hat{\psi}(x_1)\hat{\psi}^{\dagger}(x_2)\right]\middle|0\right\rangle=\mathrm{i}\,S_{\mathrm{F}}(x_1,x_2). \tag{34}$$

The electron propagator can be expressed as a complex integral

$$S_{\mathrm{F}}(x_1,x_2)=\int\frac{\mathrm{d}\omega}{2\pi}\frac{\phi_j(\boldsymbol{x}_1)\,\phi_j^{\dagger}(\boldsymbol{x}_2)}{\omega-\varepsilon_j+\mathrm{i}\eta\,\mathrm{sgn}(\varepsilon_j)}\,\mathrm{e}^{-\mathrm{i}\omega(t_1-t_2)} \tag{35}$$

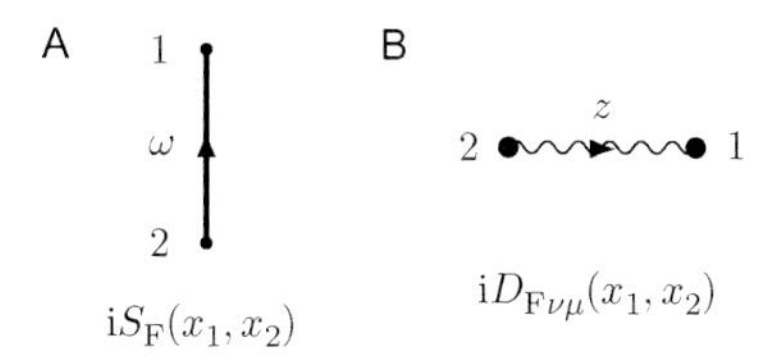

Fig. 2 Graphical representation of the (bound-state) electron propagator (34) (A) and photon propagator (50) (B).

with the fourier transform (in operator form)

$$S_F(\omega) = \frac{|j\rangle\langle j|}{\omega - \varepsilon_j(1 - i\eta)} = \frac{1}{\omega - h_D(1 - i\eta)}, \tag{36}$$

where h_D is the single-electron Dirac Hamiltonian. This is identical to the single-electron resolvent (c.f. (13)) (Fig. 2)

$$S_F(\mathcal{E}) = \Gamma(\mathcal{E}) = \frac{1}{\mathcal{E} - H_0}. \tag{37}$$

4.3 The S-Matrix

The *S-matrix* is defined

$$S = U(\infty, -\infty) \tag{38}$$

or with (31)

$$S = \sum_{n=0}^{\infty} \frac{1}{n!}\left(\frac{-i}{c}\right)^n \int_{-\infty}^{\infty} dx_1^4 \ldots \int_{-\infty}^{\infty} dx_n^4\, T[\mathcal{H}(x_1)\ldots\mathcal{H}(x_n)]. \tag{39}$$

4.4 The Green's Function

The one-body Green's function can in the nondegenerate case (closed-shell case) be expressed as

$$G(x, x_0) = \frac{\langle 0|T[\hat{\psi}(x)\, S\hat{\psi}^\dagger(x_0)]|0\rangle}{\langle 0|S|0\rangle} = \frac{1}{\langle 0|S|0\rangle}\sum_{n=0}^{\infty}\frac{1}{n!}\left(\frac{-i}{c}\right)^n \times \int_{-\infty}^{\infty} dx_1^4 \ldots \int_{-\infty}^{\infty} dx_n^4 \langle 0|T[\hat{\psi}(x)\mathcal{H}(x_1)\ldots\mathcal{H}(x_n)\hat{\psi}^\dagger(x_0)]|0\rangle, \tag{40}$$

using (39).

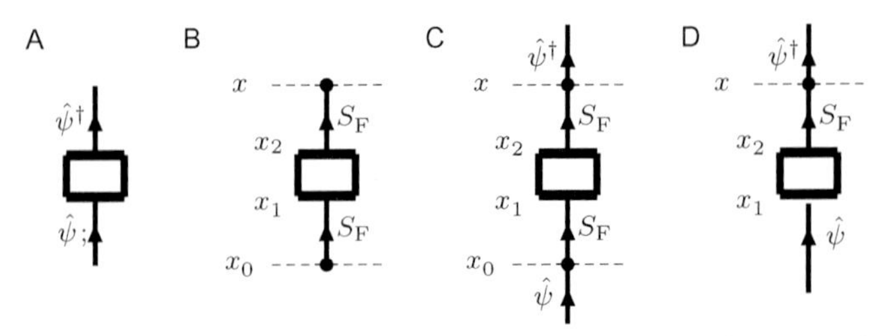

Fig. 3 One-particle evolution operator or S-matrix (A), Green's function (B), and the covariant evolution operator (CEO) (C). (D) is the CEO with initial time $t_0 = -\infty$.

The extra electron-field operators are contracted to the free ends of the S-matrix, forming electron propagators (see Fig. 3B). Hence, time can flow in both directions, and the Green's function is relativistically covariant. The denominator in (40) eliminates the disconnected parts that are singular. Therefore, the Green's function is completely connected and regular.

In second order the one-body Green's function becomes

$$G(x,x_0) = \frac{1}{\langle 0|S|0\rangle}\left(\frac{-\mathrm{i}}{c}\right)^2 \iint \mathrm{d}^4x_1\mathrm{d}^4x_2 \langle 0|T\left[\hat{\psi}(x)\mathcal{H}(x_1)\,\mathcal{H}(x_2)\,\hat{\psi}^\dagger(x_0)\right]|0\rangle. \tag{41}$$

With a perturbation $V(x_1, x_2)$

$$\overbrace{\mathcal{H}(x_1)\mathcal{H}_2} = \hat{\psi}^\dagger(x_1)\,\mathrm{i}V(x_1,x_2)\hat{\psi}(x_2) \tag{42}$$

this gives, after contracting the electron-field operators forming the electron propagators (see Fig. 3B),

$$G(x,x_0) = \frac{1}{\langle 0|S|0\rangle}\frac{1}{c^2}\left\{\iint \mathrm{d}^4x_1\mathrm{d}^4x_2\,\mathrm{i}S_\mathrm{F}(x,x_2)\,(-\mathrm{i})\,V(x_2,x_1)\,\mathrm{i}S_\mathrm{F}(x_1,x_0)\right\}. \tag{43}$$

(Assuming that $x_2 > x_1$, the factor of 1/2 can be left out.)

4.5 The Covariant Evolution Operator

The covariant evolution operator (CEO) is obtained by attaching additional electron-field operators to the free ends of the (numerator of the) Green's function (Fig. 3C), making it into an *operator*. The corresponding analytical expression is

$$U_{\rm Cov}(t,t_0) = \iint {\rm d}^3\boldsymbol{x}\,{\rm d}^3\boldsymbol{x}_0\,\langle 0|T\big[\hat{\psi}^\dagger(x)\hat{\psi}(x)\;S\,\hat{\psi}^\dagger(x_0)\hat{\psi}(x_0)\big]|0\rangle. \tag{44}$$

In second order this becomes

$$\begin{aligned} U^1_{\rm Cov}(t,t_0) = &\frac{1}{c^2}\iint {\rm d}^3\boldsymbol{x}\,{\rm d}^3\boldsymbol{x}_0\,\hat{\psi}^\dagger(x)\Big\{\iint {\rm d}^4x_1{\rm d}^4x_2\,{\rm i}S_{\rm F}(x,x_1)\\ &(-{\rm i})\,V(x_1,x_2)\times{\rm i}S_{\rm F}(x_2,x_0)\Big\}\,\hat{\psi}(x_0),\end{aligned} \tag{45}$$

where the expression in the curly brackets is the numerator of the corresponding Green's function (43).

When the initial time $t_0 \to -\infty$, we can make the replacement

$$\int {\rm d}^3\boldsymbol{x}_0\,{\rm i}S_{\rm F}(x_1,x_0)\,\hat{\psi}(x_0) \Rightarrow \hat{\psi}(x_1), \tag{46}$$

and (45) becomes (Fig. 3D)

$$\begin{aligned} U^1_{\rm Cov}(t,-\infty) = &\frac{1}{c^2}\iint {\rm d}^3\boldsymbol{x}\,\hat{\psi}^\dagger(x)\Big\{\iint {\rm d}^4x_1{\rm d}^4x_2\,{\rm i}S_{\rm F}(x,x_1)\,(-{\rm i})\,V(x_1,x_2)\Big\}\\ &\times\hat{\psi}(x_2).\end{aligned} \tag{47}$$

This can be expressed in operator form as

$$U_{\rm Cov}(t,-\infty)P_{\mathcal{E}} = {\rm e}^{-{\rm i}t(\mathcal{E}-H_0)}\,{\rm i}S_{\rm F}(\mathcal{E})(-{\rm i})\,V(\mathcal{E})P_{\mathcal{E}} = {\rm e}^{-{\rm i}t(\mathcal{E}-H_0)}\Gamma(\mathcal{E})\,V(\mathcal{E})P_{\mathcal{E}}, \tag{48}$$

using the identity (37). This can be iterated, forming a "ladder diagram"

$$U_{\rm Cov}(t,-\infty)P_{\mathcal{E}} = {\rm e}^{-{\rm i}t(\mathcal{E}-H_0)}\,[\Gamma(\mathcal{E})\,V(\mathcal{E})\,\Gamma(\mathcal{E})\,V(\mathcal{E})\cdots]P_{\mathcal{E}}. \tag{49}$$

This holds also in the many-body case.

4.6 Single-Photon Exchange

The photon propagator is defined by means of the contraction of two radiation-field operators (see Eq. 32)

$$\overbrace{A_\mu(x_1)A_\nu}(x_2) = \langle 0|T\big[A_\mu(x_1)A_\nu(x_2)\big]|0\rangle = {\rm i}\,D_{{\rm F}\nu\mu}(x_1,x_2). \tag{50}$$

The corresponding contraction of the interactions (32) is

$$\overbrace{\mathcal{H}(x_1)\mathcal{H}_2} = e^2c^2\alpha_1^\mu\alpha_2^\nu\mathrm{i}\, D_{\mathrm{F}\mu\nu}(x_1,x_2) = e^2c^2\mathrm{i}D_\mathrm{F}(x_1,x_2) \tag{51}$$

with

$$D_\mathrm{F}(x_1,x_2) = \alpha_1^\mu\alpha_2^\nu\, D_{\mathrm{F}\mu\nu}(x_1,x_2), \tag{52}$$

using the summation convention.

We introduce a gauge-dependent function $f(\kappa)$ by means of the Fourier transform of (52) with respect to time

$$D_\mathrm{F}(q,\boldsymbol{x}_1,\boldsymbol{x}_2) = \frac{1}{ce^2}\int_0^\infty \frac{2\kappa\,\mathrm{d}\kappa\, f(\kappa)}{q^2-\kappa^2+\mathrm{i}\eta}, \tag{53}$$

where $\kappa = |\boldsymbol{k}|$, $\boldsymbol{k}$ being the photon momentum vector, and $q = k_0$ the zeroth component, now as a free parameter.

In the Feynman gauge we have

$$f^\mathrm{F}(\kappa) = -\frac{e^2}{4\pi^2\epsilon_0}\,\alpha_1^\mu\alpha_{2\mu}\,\frac{\sin\kappa r_{12}}{r_{12}} \tag{54}$$

(with $r_{12} = |\boldsymbol{r}_1 - \boldsymbol{r}_2|$) and for the transverse part in the Coulomb gauge

$$f_\mathrm{T}^\mathrm{C}(\kappa) = \frac{e^2}{4\pi^2\epsilon_0}\frac{\sin(\kappa r_{12})}{r_{12}}\left[\alpha_1\cdot\alpha_2 - \frac{(\boldsymbol{\alpha}_1\cdot\nabla_1)(\boldsymbol{\alpha}_2\cdot\nabla_2)}{\kappa^2}\right], \tag{55}$$

where $\boldsymbol{\alpha}$ is the three-dimensional vector part of the Dirac α symbol and $\boldsymbol{\nabla}$ is the nabla vector (see Ref. 24, App. A).

The CEO for the exchange of a single photon between the electrons is, using (48) (see Fig. 4),

$$U_\mathrm{sp}(t,-\infty)|ab\rangle = \frac{\mathrm{e}^{-\mathrm{i}t(E_0-H_0)}}{E_0-H_0}\,V_\mathrm{sp}|ab\rangle, \tag{56}$$

where

$$V_\mathrm{sp} = e^2c^2\alpha_1^\mu\alpha_2^\nu D_{\mathrm{F}\mu\nu}(z;\boldsymbol{x}_1,\boldsymbol{x}_2) = \int\frac{\mathrm{d}z}{2\pi}\int\frac{2c^2\kappa\,\mathrm{d}\kappa\, f(\kappa,\boldsymbol{x}_1,\boldsymbol{x}_2)}{z^2-c^2\kappa^2+\mathrm{i}\eta} \tag{57}$$

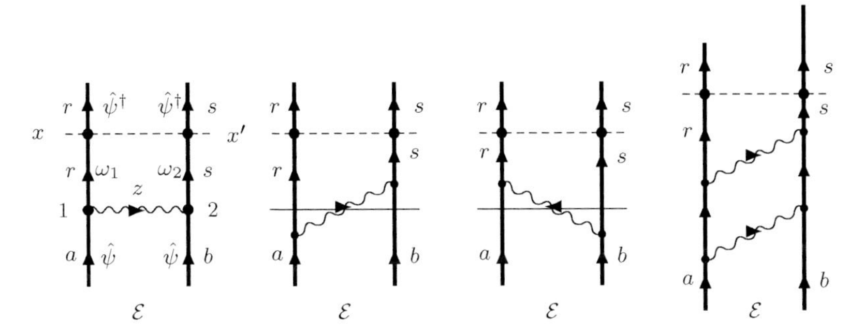

Fig. 4 The evolution-operator diagram for the retarded one- and two-photon exchange.

is the single-photon interaction. Integration over $z = cq$ yields

$$\langle rs|V_{\rm sp}|tu\rangle = \left\langle rs\left|\int_0^\infty c\,{\rm d}\kappa f(\kappa)\left[\frac{1}{\varepsilon_a-\varepsilon_r-(c\kappa-{\rm i}\gamma)_r}+\frac{1}{\varepsilon_b-\varepsilon_s-(c\kappa-{\rm i}\gamma)_s}\right]\right|tu\right\rangle. \tag{58}$$

The two terms correspond here to the two time orderings shown in the figure. The denominators can formally be obtained by the standard rules for the denominators of a many-body Goldstone diagram, as the sum of the orbital energies of the incoming lines minus those of the outgoing lines that cross a horizontal line,[17,25] if we include a term $-c\kappa$ for the crossing of a photon line.

The interaction $V_{\rm sp}$ is energy dependent, ie, it depends on the energy of the state it is operating on. This can be demonstrated more clearly by writing

$$V_{\rm sp}(\mathcal{E})P_{\mathcal{E}} = \int_0^\infty c\,{\rm d}\kappa f(\kappa)\left[\frac{1}{\mathcal{E}-\varepsilon_r-\varepsilon_u-(c\kappa-{\rm i}\gamma)_r}+\frac{1}{\mathcal{E}-\varepsilon_t-\varepsilon_s-(c\kappa-{\rm i}\gamma)_s}\right]P_{\mathcal{E}}. \tag{59}$$

The photon interaction can be iterated, and the evolution operator for an iterated interaction becomes (49)

$$U_{\rm Cov}(t,-\infty)P_{\mathcal{E}} = {\rm e}^{-{\rm i}t(\mathcal{E}-H_0)}\,\Gamma_Q(\mathcal{E})\,V_{\rm sp}(\mathcal{E})\Gamma_Q(\mathcal{E})\,V_{\rm sp}|\mathcal{E}\rangle\cdots P_{\mathcal{E}}, \tag{60}$$

as long as there are no intermediate model-space states (see last diagram in Fig. 4).

4.7 Gell-Mann–Low Theorem

Gell-Mann–Low theorem[26] states that for a nondegenerate model space

$$\Psi = \lim_{\gamma \to 0} \frac{U_\gamma(0, -\infty)|\Psi_0\rangle}{\langle \Psi_0|U_\gamma(0, -\infty)|\Psi_0\rangle} \tag{61}$$

is a solution of the time-independent Schrödinger equation

$$(H_0 + v)\Psi = E\Psi, \tag{62}$$

where Ψ_0 is the model function (2) and v is the perturbation with the density (32). This equation does not preserve the number of photons, and the solutions lie generally in the *photonic Fock space*. This perturbation is time independent in the Schrödinger picture in the limit $\gamma \to 0$.

With an extended model space the Gell-Mann–Low theorem becomes[27]

$$\Psi^\alpha = \lim_{\gamma \to 0} \frac{U_\gamma(0, -\infty)|\Phi^\alpha\rangle}{\langle \Psi_0^\alpha|U_\gamma(0, -\infty)|\Phi^\alpha\rangle}, \tag{63}$$

where Φ^α is the *parent state*, which is the limit of the target state as the damping is adiabatically switched off. (Note that this is for a multidimensional model space generally different from the model state.)

4.8 Green's Operator

Like the numerator of the Green's operator expression (40), the CEO (44) can be singular. In the closed-shell Green's function case this could be remedied by dividing by the expectation value of the S-matrix. In the present case we shall use a more general technique, valid also in the general open-shell case.

We define a *Green's operator*, $\mathcal{G}$, by[24,27]

$$U(t, t_0)P = \mathcal{G}(t, t_0) \cdot PU(0, t_0)P, \tag{64}$$

where $\mathcal{G}$ does not operate beyond the dot. This operator is a time-dependent wave operator for the relativistic wave function

$$\Psi^\alpha(t) = \mathcal{G}(t, -\infty)\Psi_0^\alpha. \tag{65}$$

For $t = 0$ this goes over into the standard wave operator (8). (In the following we shall leave out the initial time and assume it to be $t_0 = -\infty$. Instead, we shall insert the energy parameter.)

From the definition (64) it follows that the lowest-order Green's operators become

$$
\begin{aligned}
\mathcal{G}^{(0)}(t,\mathcal{E})P_{\mathcal{E}} &= U^{(0)}(t,\mathcal{E})P_{\mathcal{E}} \\
\mathcal{G}^{(1)}(t,\mathcal{E})P_{\mathcal{E}} &= U^{(1)}(t,\mathcal{E})P_{\mathcal{E}} - \mathcal{G}^{(0)}(t,\mathcal{E}')\,P_{\mathcal{E}'}U^{(1)}(0,\mathcal{E})P_{\mathcal{E}} \\
\mathcal{G}^{(2)}(t,\mathcal{E})P_{\mathcal{E}} &= U^{(2)}(t,\mathcal{E})P_{\mathcal{E}} - \mathcal{G}^{(0)}(t,\mathcal{E}')\,P_{\mathcal{E}'}U^{(2)}(0,\mathcal{E}) \\
&\qquad P_{\mathcal{E}} - \mathcal{G}^{(1)}(t,\mathcal{E}')\,P_{\mathcal{E}'}U^{(1)}(0,\mathcal{E})P_{\mathcal{E}}.
\end{aligned}
\tag{66}
$$

The superscript (n) indicates the nth-order approximation. The zeroth-order operator is

$$U^{(0)}(t,\mathcal{E}) = \mathcal{G}^{(0)}(t,\mathcal{E}) = \mathrm{e}^{-\mathrm{i}t(\mathcal{E}-H_0)}. \tag{67}$$

The first-order evolution operator is (48)

$$U^{(1)}(t,\mathcal{E}) = \mathrm{e}^{-\mathrm{i}t(\mathcal{E}-H_0)}\Gamma(\mathcal{E})\,V(\mathcal{E}), \tag{68}$$

which becomes singular, when the final state lies in the model space. Then we insert a counterterm

$$\mathrm{e}^{-\mathrm{i}t(\mathcal{E}'-H_0)}\,P_{\mathcal{E}'}\frac{1}{\mathcal{E}-H_0}V_{\mathrm{sp}}(\mathcal{E}). \tag{69}$$

(Note that the energy parameter of the first factor is $\mathcal{E}'$.) We now have

$$P_{\mathcal{E}'}\frac{1}{\mathcal{E}-H_0} = P_{\mathcal{E}'}\frac{1}{\mathcal{E}-\mathcal{E}'}, \tag{70}$$

and this makes the difference between (68) and (69) into a difference ratio

$$\frac{\mathcal{G}^{(0)}(t,\mathcal{E}) - \mathcal{G}^{(0)}(t,\mathcal{E}')}{\mathcal{E}-\mathcal{E}'} = \frac{\delta\mathcal{G}^{(0)}(t,\mathcal{E})}{\delta\mathcal{E}}, \tag{71}$$

which in the limit of complete degeneracy leads to a partial derivative. The first-order Green's operator then becomes[c]

$$\mathcal{G}^{(1)}(t,\mathcal{E})P_{\mathcal{E}} = \mathcal{G}^{(0)}(t,\mathcal{E})\Gamma_Q(\mathcal{E})V(\mathcal{E})P_{\mathcal{E}} + \frac{\delta\mathcal{G}^{(0)}(t,\mathcal{E})}{\delta\mathcal{E}}P_{\mathcal{E}'}V(\mathcal{E})P_{\mathcal{E}}. \tag{72}$$

Similarly, one finds that the second-order Green's operator becomes[24]

$$\mathcal{G}^{(2)}(t,\mathcal{E})P_{\mathcal{E}} = \mathcal{G}_0^{(2)}(t,\mathcal{E})P_{\mathcal{E}} + \frac{\delta\mathcal{G}^{(0)}(t,\mathcal{E})}{\delta\mathcal{E}}W_0^{(2)} + \frac{\delta\mathcal{G}^{(1)}(t,\mathcal{E})}{\delta\mathcal{E}}W^{(1)}. \tag{73}$$

[c] We shall normally keep the difference ratios in our expressions, which leads to different rules than using the derivatives (see Ref. 28).

Here

$$W^{(1)} = PV(\mathcal{E})P_{\mathcal{E}} \tag{74}$$

is the first-order effective interaction and

$$W_0^{(2)} = P_{\mathcal{E}} V(\mathcal{E}) \Gamma_Q(\mathcal{E}) V(\mathcal{E}) P_{\mathcal{E}} \tag{75}$$

is the second-order effective interaction without any intermediate model-space state. The second-order Green's operator without any intermediate model-space states is

$$\mathcal{G}_0^{(2)}(t,\mathcal{E})P_{\mathcal{E}} = U_0^{(2)}(t,\mathcal{E}) = \mathcal{G}^{(0)}(t,\mathcal{E})\,\Gamma_Q(\mathcal{E})V(\mathcal{E})\Gamma_Q(\mathcal{E})V(\mathcal{E})P_{\mathcal{E}}. \tag{76}$$

In the present formalism the effective interaction (16) is given by

$$W = P\mathcal{R}P = PV_{\mathrm{F}}\Omega_{\mathrm{Cov}}P = P\left(\mathrm{i}\frac{\partial}{\partial t}\mathcal{G}(t,\mathcal{E})\right)_{t=0} P. \tag{77}$$

Applying this relation to (73) yields

$$W^{(2)} = W_0^{(2)} + \frac{\delta W^{(1)}}{\delta \mathcal{E}} W^{(1)}. \tag{78}$$

This shows that the second-order effective interaction can in the time-dependent formalism have an internal model-space state, which is not the case in the time-independent formalism.

4.9 Time Dependence of the Green's Operator

We shall conclude the presentation of the formalism by considering the question that may seem mysterious to some readers. We have seen previously that in the expansion of the Green's operator the terms have the time dependence (49)

$$\mathrm{e}^{-\mathrm{i}t(\mathcal{E}-H_0)},$$

where $\mathcal{E}$ is the unperturbed model-space energy the operator is acting on. This is in the interaction picture, and transforming to the standard Schrödinger picture (27) the time dependence becomes

$$\mathrm{e}^{-\mathrm{i}t\mathcal{E}}.$$

But this differs from that of the standard quantum-mechanical picture, where the time dependence of the wave function is

$$\mathrm{e}^{-\mathrm{i}tE},$$

E being the exact energy of the state. This can be explained in the following way.

It can be shown that the general Green's operator can be expanded in terms of the Green's operator without intermediate states

$$\mathcal{G}(t,\mathcal{E})P_{\mathcal{E}} = \mathcal{G}_0(t,\mathcal{E})P_{\mathcal{E}} + \sum_{n=1} \frac{\delta^n \mathcal{G}_0(t,\mathcal{E})}{\delta \mathcal{E}^n} W^n P_{\mathcal{E}}. \tag{79}$$

The sum represents here all model-space contributions (MSCs). If we let this expression operate on a model function $|\Psi_0\rangle$ of energy E_0, we will form a Taylor series with W being the expansion parameter.[d] But

$$W|\Psi_0\rangle = \Delta E|\Psi_0\rangle, \tag{80}$$

where ΔE is the difference between the exact and model energies. The relation (79) can then be reformulated as

$$\mathcal{G}(t,E_0)|\Psi_0\rangle = \mathcal{G}_0(t,E)|\Psi_0\rangle, \tag{81}$$

ie, the MSC to all orders have the effect of shifting the energy parameter from the model energy to the exact energy. This is true also for the time dependence of the time factor.

The CEO (or Green's operator) with no model-space states has according to (49) the time dependence $\mathrm{e}^{-\mathrm{i}t(E_0-H_0)}$ when operating on a state of energy E_0. If we now apply the MSCs, the time dependence will shift to $\mathrm{e}^{-\mathrm{i}t(E-H_0)}$ in the interaction picture. But this is exactly the time dependence of a wave function or state vector in the quantum-mechanical picture. Therefore, we see that the time dependence of the Green's operator agrees with that picture, when we include the MSCs.

[d] This differs from a standard Taylor series, since it is based upon the difference ratio rather than the partial derivative. For more details, see Ref. 28 and Ref. 24, Ch. 6.

5. QED AND ELECTRON CORRELATION

We shall now see how we can combine QED and electron correlation beyond second order, which has been performed for the first time only recently. Such effects cannot be evaluated by the standard methods for QED calculations.

5.1 Nonradiative QED in Combination with Electron Correlation

We consider first the combination of electron correlation and the non-radiative QED effects, ie, photon retardation and virtual electron–positron pairs, which is illustrated in Fig. 5. The CEO expression (wave function) for the first diagram in the figure, with no intermediate model-space state, is according to (49)

$$U_{\mathrm{Cov}}(t,\mathcal{E})P_{\mathcal{E}} = \mathrm{e}^{-it(\mathcal{E}-H_0)}\Gamma(\mathcal{E})\,V_{\mathrm{sp}}(\mathcal{E})\,\Gamma_Q(\mathcal{E})\,V_{\mathrm{C}}\Gamma_Q(\mathcal{E})\,V_{\mathrm{C}}P_{\mathcal{E}}, \tag{82}$$

where V_{sp} is the general single-photon interaction (58) and

$$V_{\mathrm{C}} = \frac{e^2}{4\pi\epsilon_0\, r_{12}} \tag{83}$$

is the instantaneous Coulomb interaction. The corresponding effective interaction (energy contribution) is obtained by (77)

$$W = P_{\mathcal{E}}\,V_{\mathrm{sp}}(\mathcal{E})\,\Gamma_Q(\mathcal{E})\,V_{\mathrm{C}}\Gamma_Q(\mathcal{E})\,V_{\mathrm{C}}P_{\mathcal{E}}. \tag{84}$$

If there is an intermediate model-space state as indicated in the second diagram in the figure, then there will also be a MSC to the Green's operator (66, 73),

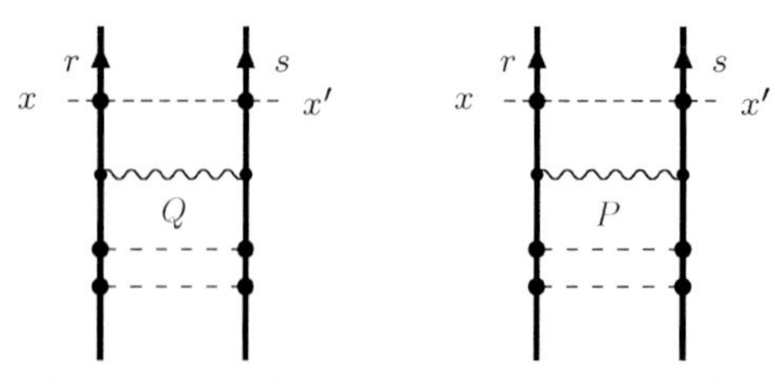

Fig. 5 The retarded photon interaction combined with Coulomb interactions (electron correlation). We show explicitly the *outgoing arrows* to indicate that this represent a wave-function contribution.

$$\mathcal{G}(t,\mathcal{E}) = \frac{\delta \mathcal{G}^{(1)}}{\delta \mathcal{E}} W_0^{(2)}, \tag{85}$$

where $\mathcal{G}^{(1)} = \mathcal{G}^{(0)} \Gamma_Q V$ is the first-order Green's operator and $W_0^{(2)}$ is the second-order effective interaction (75) without intermediate model-space states. (There is also an imaginary component of the Green's operator that lies in the model-space which we have left out.) The corresponding effective interaction becomes (c.f. (72))

$$W = \frac{\delta W^{(1)}}{\delta \mathcal{E}} W_0^{(2)}. \tag{86}$$

If there is a model-space state between the Coulomb interactions, there will be an additional MSC, and $W_0^{(2)}$ is replaced by the last term in (78).

The effects of electron correlation on retardation/virtual pairs were calculated for the first time by Daniel Hedendahl[29] a few years ago for the ground states of helium-like ions and compared with previous experimental and theoretical results. These higher-order results supplemented extensive second-order (two-photon) calculations performed by Artemyev et al.[30] using the two-time Green's function as well as by Plante et al.[14] using many-body perturbation theory with first-order QED energy corrections. The effect of QED correlations is small but improves the numerical accuracy compared to the two-photon results.

5.2 Radiative QED Effects in Combination with Electron Correlation

In numerical QED calculations involving radiative QED effects the Feynman gauge has previously been applied almost exclusively. The Coulomb gauge is considerably more complicated to use but might be advantageous in combination with electron correlation, due to the dominating Coulomb interaction. The basic formalism for the dimensional regularization in Coulomb gauge was developed in the 1980s by Adkins,[31,32] but the first numerical calculation of that kind was not performed until 2011 by Holmberg and Hedendahl on hydrogen-like ions.[33]

More recently Holmberg et al.[34] have evaluated the effect of electron correlation on the radiative QED (electron self-energy and vertex correction) for helium-like ions, and the results are somewhat surprising.

We consider first the first-order Coulomb-screened self-energy and vertex corrections for He-like systems, illustrated in Fig. 6. Diagram (A) with an

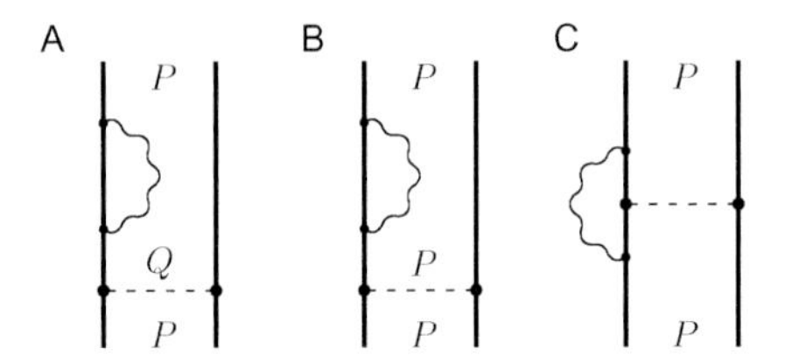

Fig. 6 Second-order Coulomb-screened self-energy and vertex correction for He-like systems. Diagrams (B) and (C) are divergent.

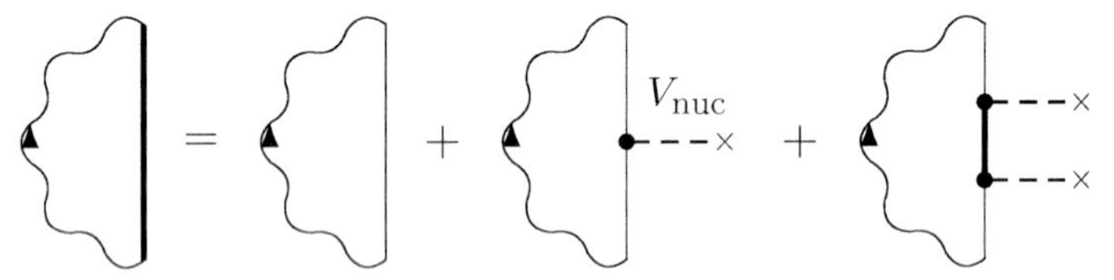

Fig. 7 Expanding the bound-state self-energy into zero-, one-, and many-potential terms. The *heavy vertical lines* represent the bound-state orbitals, generated in an extern (nuclear) potential, and the *thin lines* represent free-electron states.

intermediate Q-state is irreducible, ie, cannot be separated into two legitimate energy diagrams. This is regular after renormalization. Diagram (B) with an intermediate model-space state is singular, and the singularity is cancelled by diagram (C), the vertex correction (due to the Ward identity).

For the dimensional regularization the bound states have to be expanded in terms of free states and (nuclear) potential interactions, *zero-, one-, and many-potential parts* (see Fig. 7),

$$\Sigma^{\rm bou} = \Sigma^{\rm free}_{\rm zp} + \Sigma^{\rm free}_{\rm op} + \Sigma^{\rm free}_{\rm mp}. \tag{87}$$

The numerical results of all second-order diagrams are shown in Table 1, and it is found that this sum is gauge independent, as expected. It is remarkable, though, that the individual terms behave quite differently in the two gauges. In the Feynman there are enormous cancellations between the various contributions. Some terms are here hundred times larger than the corresponding ones in the Coulomb gauge. This phenomenon has drastic consequences in higher orders, where all contributions cannot be evaluated.

In Table 2 we show the corresponding contributions beyond second order. Here, it is only possible to evaluate the leading zero-potential (free-electron) contribution of the MSC and vertex parts. This then shows that it is possible to achieve sensible results beyond second order in the Coulomb

Table 1 Two-Photon Exchange for the Electron Self-Energy and Vertex Correction for the Ground State of He-Like Argon Ion, Using the Coulomb and Feynman Gauges (in meV)

Diagram	Coulomb Gauge	Feynman Gauge
Irreducible (A)	− 104.2	− 87.0
MSC (B), zero-pot.	− 24.8	3890
Vertex (C), zero-pot.	16.2	− 3653
MSC/vertex, multi-pot.	− 1.1	− 192
Total second order	113.8(8)	113(1)

From Holmberg, J.; Salomonson, S.; Lindgren, I. Coulomb-Gauge Calculation of the Combined Effect of the Correlation and QED for Heliumlike Highly Charged Ions. *Phys. Rev. A* **2015,** *92*, 012509.

Table 2 Correlation Effect Beyond Two-Photon Exchange for the Electron Self-Energy and Vertex Correction for the Ground State of He-Like Argon Ion, Using the Coulomb and Feynman Gauges (in meV).

Diagram	Coulomb Gauge	Feynman Gauge
Irreducible (A)	4.24(3)	− 71.1
MSC (B), zero-pot.	1.16	− 24.3
Vertex (C), zero-pot.	− 0.73	54.2
Total	4.67(3)	− 41

From Holmberg, J.; Salomonson, S.; Lindgren, I. Coulomb-Gauge Calculation of the Combined Effect of the Correlation and QED for Heliumlike Highly Charged Ions. *Phys. Rev. A* **2015,** *92*, 012509.

gauge, which was quite an unexpected result. Our interpretation of this is that the Coulomb gauge is a more "physical gauge."

6. CONCLUDING REMARKS

For the helium-like ions the higher-order effects, beyond second order, are just barely detectable with the present experimental accuracy. However, there are other systems for which laser spectroscopy can be performed and where the experimental accuracy is considerably higher. This can be illustrated by the $^3P_2$–$^3P_1$ transition in helium-like fluorine. This transition has been measured by Myers et al.[35] to be 4364.517(6) μH, while the best theoretical value is 4362(5) μH. Another example is the transition $1s2s\ ^1S_0$–$1s2p\ ^3P_1$ in silicon, which has also been measured by Myers et al.[36] to be 7230.585(6) cm^{-1}, while the best theoretical value is 7229(2) cm^{-1}.

It would be a great challenge to test the new accurate computational procedure on this type of results in order to see to what degree it is capable of reproducing these accurate results. Unfortunately, we have no possibility to do so at our laboratory due to lack of manpower. But it is our hope that some other group will have the capacity in the future.

ACKNOWLEDGMENTS

I am grateful to Per-Olov Löwdin for introducing me to the field of perturbation theory some 40–50 years ago. I have benefitted much from that education. I also want to express my gratitude to my coworkers Sten Salomonson, Daniel Hedendahl, Björn Åsén, and Johan Holmberg together with whom the work described here has been performed.

REFERENCES

1. Schrödinger, E. Quantisierung als Eigenwertproblem. *Ann. Phys.* **1926**, *80*, 437.
2. Löwdin, P. O. Quantum Theory of Many-Particle Systems. III. Extension of the Hartree-Fock Scheme to Include Degenerate Systems and Correlation Effects. *Phys. Rev.* **1958**, *97*, 1509–1520.
3. Löwdin, P. O. Quantum Theory of Many-Particle Systems. III. Extension of the Hartree-Fock Scheme to Include Degenerate Systems and Correlation Effects. *Phys. Rev.* **1955**, *97*, 1509–1529.
4. Löwdin, P. O. Quantum Theory of Many-Particle Systems. II. Study of the Ordinary Hartree-Fock Approximation. *Phys. Rev.* **1955**, *97*, 1490–1508.
5. Löwdin, P. O. Quantum Theory of Many-Particle Systems. I. Physical Interpretations by Means of Density Matrices, Natural Spin-Orbitals, and Convergence Problems in the Method of Configurational Interaction. *Phys. Rev.* **1955**, *97*, 1474–1489.
6. Löwdin, P. O. Correlation Problem in Many-Electron Quantum Mechanics. I. Review of Different Approaches and Discussion of Some Current Ideas. *Adv. Chem. Phys.* **1959**, *3*, 207–322.
7. Löwdin, P. O. Expansion Theorems for the Total Wave Function and Extended the Hartree-Fock Schemes. *Rev. Mod. Phys.* **1960**, *32*, 328–334.
8. Löwdin, P. O. Studies in Perturbation Theory. IV. Solution of Eigenvalue Problem by Projection Operators. *J. Math. Phys.* **1962**, *3*, 969–982.
9. Löwdin, P. O. Studies in Perturbation Theory. IV. Solution of Eigenvalue Problem by Projection Operators. V. Some Aspects on the Exact Self-Consistent Field Theory. *J. Math. Phys.* **1962**, *3*, 969, 1171.
10. Löwdin, P. O. Studies in Perturbation Theory. X. Lower Bounds to Eigenvalues in Perturbation-Theory Ground State. *Phys. Rev.* **1965**, *139*, A357–A372.
11. Löwdin, P. O. Some Aspects on the Bloch-Lindgren Equation and a Comparison with the Partitioning Technique. *Adv. Quantum Chem.* **1998**, *30*, 415–432.
12. Löwdin, P. O. Connection Between Semi-empirical and Ab Initio Methods in Quantum Theory of Molecules. *Int. J. Quantum Chem.* **1999**, *72*, 379–391.
13. Bloch, C. Sur la théorie des perurbations des etats liés. *Nucl. Phys.* **1958**, *6*–7, 329–451.
14. Plante, D. R.; Johnson, W. R.; Sapirstein, J. Relativistic All-Order Many-Body Calculations of the n=1 and n=2 States of Heliumlike Ions. *Phys. Rev. A* **1994**, *49*, 3519–3530.
15. Mårtensson-Pendrill, A. M.; Lindgren, I.; Lindroth, E.; Salomonson, S.; Staudte, D. S. Convergence of Relativistic Perturbation Theory for the 1*s*2*p* States in Low-Z Heliumlike Systems. *Phys. Rev. A* **1995**, *51*, 3630–3635.

16. Lindgren, I. The Rayleigh-Schrödinger Perturbation and the Linked-Diagram Theorem for a Multi-configurational Model Space. *J. Phys. B* **1974**, 7, 2441–2470.
17. Lindgren, I.; Morrison, J. *Atomic Many-Body Theory*, 2nd ed.; Springer-Verlag: Berlin, 1986. Reprinted 2009.
18. Brueckner, K. A. Many-Body Problems for Strongly Interacting Particles. II. Linked Cluster Expansion. *Phys. Rev.* **1955**, *100*, 36–45.
19. Goldstone, J. Derivation of the Brueckner Many-Body Theory. *Proc. R. Soc. Lond. Ser. A* **1957**, *239*, 267–279.
20. Brandow, B. H. Linked-Cluster Expansions for the Nuclear Many-Body Problem. *Rev. Mod. Phys.* **1967**, *39*, 771–828.
21. Lindgren, I. A Coupled-Cluster Approach to the Many-Body Perturbation Theory for Open-Shell Systems. *Int. J. Quantum Chem.* **1978**, *S12*, 33–58.
22. Brown, G. E.; Ravenhall, D. G. On the Interaction of Two electrons. *Proc. R. Soc. Lond. Ser. A* **1951**, *208*, 552–559.
23. Sucher, J. Foundations of the Relativistic Theory of Many Electron Atoms. *Phys. Rev. A* **1980**, *22*, 348–362.
24. Lindgren, I. *Relativistic Many-Body Theory: A New Field-Theoretical Approach*, 2nd ed.; Springer-Verlag: New York, 2016.
25. Lindgren, I.; Salomonson, S.; Hedendahl, D. Many-Body Procedure for Energy-Dependent Perturbation: Merging Many-Body Perturbation Theory with QED. *Phys. Rev. A* **2006**, *73*, 062502.
26. Gell-Mann, M.; Low, F. Bound States in Quantum Field Theory. *Phys. Rev.* **1951**, *84*, 350–354.
27. Lindgren, I.; Salomonson, S.; Åsén, B. The Covariant-Evolution-Operator Method in Bound-State QED. *Phys. Rep.* **2004**, *389*, 161–261.
28. Lindgren, I.; Salomonson, S.; Hedendahl, D. Many-Body-QED Perturbation Theory: Connection to the Two-Electron Bethe-Salpeter Equation. Einstein Centennial Review Paper. *Can. J. Phys.* **2005**, *83*, 183–218.
29. Hedendahl, D. Towards a Relativistic Covariant Many-Body Perturbation Theory. Ph. D. Thesis. University of Gothenburg: Gothenburg, Sweden, 2010.
30. Artemyev, A. N.; Shabaev, V. M.; Yerokhin, V. A.; Plunien, G.; Soff, G. QED Calculations of the n=1 and n=2 Energy Levels in He-Like Ions. *Phys. Rev. A* **2005**, *71*, 062104.
31. Adkins, G. One-Loop Renormalization of Coulomb-Gauge QED. *Phys. Rev. D* **1983**, *27*, 1814–1820.
32. Adkins, G. One-Loop Vertex Function in Coulomb-Gauge QED. *Phys. Rev. D* **1986**, *34*, 2489–2492.
33. Holmberg, J.; Hedendahl, D. (to appear in ArXiv:quant.ph).
34. Holmberg, J.; Salomonson, S.; Lindgren, I. Coulomb-Gauge Calculation of the Combined Effect of the Correlation and QED for Heliumlike Highly Charged Ions. *Phys. Rev. A* **2015**, *92*, 012509.
35. DeVore, T. R.; Crosby, D. N., Myers, E. G. Improved Measurement of the $1s2s\ ^1S_0 - 1s2p\ ^3P_1$ Interval in Heliumlike Silicon. *Phys. Rev. Lett.* **2008**, *100*, 243001.
36. Myers, E. G.; Margolis, H. S.; Thompson, J. K.; Farmer, M. A.; Silver, J. D.; Tarbutt, M. R. Precision Measurement of the $1s2p\ ^3P_2 - ^3P_1$ Fine Structure Interval in Heliumlike Fluorine. *Phys. Rev. Lett.* **1999**, *82*, 4200–4203.

CHAPTER SIX

Quantum Partitioning Methods for Few-Atom and Many-Atom Dynamics☆

David A. Micha[1]
University of Florida, Gainesville, FL, United States
[1]Corresponding author: e-mail address: micha@qtp.ufl.edu

Contents

Abstract

The subject of this contribution is how projection operators can be constructed to treat a variety of time-dependent phenomena involving interacting molecules, and to treat the dissipative dynamics of a localized subsystem in a large environment. It develops partitioning methods in a functional space of wavefunctions introduced to construct molecular effective potentials and long-lived states from distortion, adiabatic, and fast motion states. It also gives a treatment starting from the statistical density operator, for partitioning in a many-atom system undergoing dissipative dynamics, and shows how to construct contracted density operators with selected total system states, or reduced density operators for localized phenomena in a primary region. The presentation displays related mathematical procedures useful for partitioning of both wavefunctions and density operators and for derivation of their equations of motion.

☆This work has been partly supported by the NSF of the USA and by the Dreyfus Foundation. It is a pleasure to dedicate this publication to the celebration of the 100th anniversary of Per-Olov Löwdin's birthday.

Advances in Quantum Chemistry, Volume 74
ISSN 0065-3276
http://dx.doi.org/10.1016/bs.aiq.2016.06.001

1. INTRODUCTION

Mathematical partitioning methods of quantum mechanics provide ways to model physical systems and their properties, and ways to systematically improve calculations. Partitioning may refer to separation of states in a functional space into two or more sets to be chosen in the definition of partition operators. Alternatively they may refer to partition in physical space of a many-atom system into fragments of given atomic structure, with a primary region to be treated in atomic detail and a remaining secondary region which can be described in statistical terms. Partitioning methods also can be introduced to organize quantum mechanical calculations by systematically enlarging chosen sets of quantum states or primary regions to be treated in sequence until a desired convergence is attained.

Per-Olov Löwdin wrote a long series of papers on partitioning techniques and their application to stationary states of a quantal system with a given Hamiltonian operator $\widehat{H}$, where one is interested in the eigenvalues E and eigenfunctions Ψ of the associated Schroedinger equation. In his Part 1 he used a partition of a functional basis set, separating one reference state Φ_1, to derive an implicit equation $E=f(E)$ for an eigenvalue of the Schroedinger equation.[1] In a later paper Part IV of the series, he introduced Hermitian complementary projection operators $\widehat{O}$ and $\widehat{P}$, with $\widehat{O}+\widehat{P}=\widehat{I}$, and a nonnormal projector, or wave operator $\widehat{\Omega}(E)=\widehat{O}+\widehat{P}\left[\alpha\widehat{O}+\widehat{P}\left(E-\widehat{H}\right)\widehat{P}\right]^{-1}\widehat{P}\widehat{H}\widehat{O}$, to formally construct in addition eigenfunctions $\Psi=\widehat{\Omega}\Phi$ of the equation.[2] He also elaborated on the usefulness of partitioning to deal with atomic substitution in molecules, atoms in crystal fields, and "dressing" of molecules in an electromagnetic field, and how the resulting equations could be numerically solved by iterative methods more powerful than the usual perturbation expansions.[3,4]

Partitioning methods using projection operators in a functional space of quantum states had also been introduced in a unified theory of nuclear reactions and in particular to treat the formation of nuclear compound states.[5] Projection operators had been found useful in the formal treatments of time-dependent scattering[6] and of decaying states.[7] This is a subject subsequently developed for molecular systems by our research group[8–15] and by several other ones.[16–18] In our own work, we introduced partitioning treatments of molecular long-lived states formed during molecular interactions, and related optical potential models for molecular interactions.

Another wide area where partitioning methods have been very useful is the dynamics of systems with many atoms under thermodynamical constrains, where a region of a system is of primary interest and evolves over time under the influence of the remaining secondary region, leading to fluctuation and dissipation phenomena. These aspects have been covered in recent research monographs on the dynamics of molecular systems and of condensed phases,[19–26] and in our treatments of optical properties in extended many-atom systems.[27–33]

The subject of this contribution is how projection operators can be constructed to treat a variety of time-dependent phenomena involving interacting molecules, and to treat the dissipative dynamics of a localized subsystem in a large environment. Section 2 develops the partitioning methods in a functional space of wavefunctions needed to construct molecular effective potentials and long-lived states. Section 3 gives a treatment starting from the statistical density operator for partitioning in a many-atom system undergoing dissipative dynamics and shows how to construct contracted density operators or reduced density operators (RDOp). The presentations display parallel mathematical procedures useful for partitioning of both wavefunctions and density operators.

2. MOLECULAR EFFECTIVE POTENTIALS AND LONG-LIVED STATES FROM PARTITIONING METHODS

2.1 Partitioning of the Functional Space of Molecular States

Considering two interactiong molecules A and M, a consistent treatment of collisional and long-lived states can best be done within a time-dependent formulation and introducing wavefunction packets $\Psi(t)$ which depend on all electronic and nuclear position and spin variables. The wavefunctions must satisfy over time the equation

$$\widehat{H}\Psi(t) = i\hbar\partial\Psi/\partial t$$

which must be solved with boundary conditions $\langle\Psi|\,\Psi\rangle = 1$ and the initial condition $\Psi(t_0) \approx \Psi_0(t_0) = \Psi^{(0)}(\lambda)$ at an initial time before interactions, where λ signifies a set of parameters with a given initial distribution of values. The total Hamiltonian can be written as $\widehat{H} = \widehat{H}_0 + \widehat{V}$, a sum of a time-independent Hamiltonian describing a reference (unperturbed) system plus a perturbation which in a general case can depend on time.

The total time-dependent wavefunction can be partitioned by means of a Hermitian projection operator $\widehat{P}$ satisfying idempotency $\widehat{P}^2 = \widehat{P}$, and of its complementary projector $\widehat{Q} = \widehat{I} - \widehat{P}$. Writing $\widehat{P}\Psi = \Psi_P$, $\widehat{Q}\Psi = \Psi_Q$ and applying the projection operators to the two sides of the differential equation for the time-dependent wavefunction one finds the two partitioned equations

$$\widehat{P}\widehat{H}(t)\widehat{P}\Psi_P(t) + \widehat{P}\widehat{H}(t)\widehat{Q}\Psi_Q(t) = i\hbar\, \partial\Psi_P/\partial t$$
$$\widehat{Q}\widehat{H}(t)\widehat{P}\Psi_P(t) + \widehat{Q}\widehat{H}(t)\widehat{Q}\Psi_Q(t) = i\hbar\, \partial\Psi_Q/\partial t$$

The second equation can be formally solved for Ψ_Q introducing a time-evolution operator in Q subspace

$$U_Q(t,t') = \exp_T\left[(-i/\hbar)\int_{t'}^{t} ds \widehat{Q}\widehat{H}(s)\widehat{Q}\right]$$

constructed with a time-ordered exponential and giving the solution

$$\Psi_Q(t) = \Psi_Q(t_0) - (i/\hbar)\int_{t_0}^{t} dt' U_Q(t,t')\widehat{Q}\widehat{H}(t')\widehat{P}\Psi_P(t')$$

where the first term to the right is an initial state in the Q subspace, which can be constructed to be null, or such that it averages to zero over a set of initial conditions with a given initial probability distribution. Replacing this in the first of the partitioned equations leads to

$$\begin{aligned} i\hbar\, \partial\Psi_P/\partial t = {} & \widehat{P}\widehat{H}(t)\widehat{P}\Psi_P(t) + \widehat{P}\widehat{H}(t)\widehat{Q}\Psi_Q(t_0) \\ & - (i/\hbar)\int_{t_0}^{t} dt' \widehat{P}\widehat{H}(t)\widehat{Q}U_Q(t,t')\widehat{Q}\widehat{H}(t')\widehat{P}\Psi_P(t') \end{aligned}$$

This is an integrodifferential equation giving a formal general solution for Ψ_P. The right hand terms can be physically interpreted when the projectors have been chosen to correspond to specific phenomena in P and Q subspaces. The first term gives the time evolution in the P subspace when the Q subspace can be ignored, the second term describes, divided by $i\hbar$, a driving rate resulting from phenomena initially present external to the P subspace, and the last term accounts for delayed transient effects due to coupling of the P subspace to the Q subspace and back, and is mathematically given by a memory kernel under its integral. The operators describing the driving rate and the transient rate are found to be closely related. The solution $\Psi_P(t;\lambda)$ depends, as shown, parametrically on the initial conditions and averaging both sides of the equation over an interval of parameter values $\Delta\lambda$, such as molecular rotational orientations, may lead to a simpler equation without the driving term and with a simplified memory kernel.

2.2 Partitioning the Wave Operator

For time-independent interaction potentials and Hamiltonians it is possible to further simplify the treatment. A formal solution to the equation with a time-independent Hamiltonian can be written in terms of propagator (or Green's) functions $G^{(\pm)}(t, t_0)$, equal to zero for $t < t_0$ (the + retarded form) or $t > t_0$ (the − advanced form), respectively, and satisfying

$$\left(i\hbar\,\partial/\partial t - \hat{H}\right)\hat{G}^{(\pm)}(t, t_0) = \delta(t - t_0)$$
$$\hat{G}^{(\pm)}(t, t_0) = \mp(i/\hbar)\theta[\pm(t - t_0)]\hat{U}(t - t_0)$$

in terms of the step function θ and the time-evolution operator $\hat{U}$ generated by $\hat{H}$. A similar expression applies for the unperturbed propagator $\hat{G}_0^{(\pm)}(t, t_0)$, and the evolving wavefunction is given by the solution of the integral equation

$$\Psi^{(+)}(t) = \Psi_0^{(+)}(t) + \int_{-\infty}^{+\infty} dt'\,\hat{G}_0^{(+)}(t,t')\,\hat{V}\,\Psi^{(+)}(t')$$

For time-independent interactions, it is convenient to introduce a Fourier transform between time t and energy u by means of

$$\tilde{f}(u) = \int_{-\infty}^{+\infty} dt \exp(iut/\hbar) f(t),\; f(t) = \int_{-\infty}^{+\infty} du (2\pi\hbar)^{-1} \exp(-iut/\hbar)\tilde{f}(u)$$

which gives in the limit $u \approx E + i\varepsilon = E^{(+)}$, with $\varepsilon \to 0$, new transformed equations containing an energy dependent, instead of the previous time dependent, resolvent function $\tilde{G}^{(+)}(E) - R^{(+)}(E)$. It satisfies

$$\left(E^{(+)} - \hat{H}\right)\hat{R}^{(+)}(E) = \hat{I}$$

and its unperturbed form can be used to construct the solution $\Psi_E^{(+)} = \Psi_{0E} + \hat{R}_0^{(+)}(E)\,\hat{V}\,\Psi_E^{(+)}$, with $(E - H_0)\Psi_{0E} = 0$ giving the unperturbed state, and with $\hat{R}^{(+)}(E) = \hat{R}_0^{(+)}(E) + \hat{R}_0^{(+)}(E)\,\hat{V}\,\hat{R}^{(+)}(E)$. This in turn can be formally solved and gives[12]

$$\Psi_E^{(+)} = \widehat{W}^{(+)}(E)\Psi_{0E}, \quad \widehat{W}^{(+)}(E) = \widehat{I} + \widehat{R}^{(+)}(E)\widehat{V}$$

where the wave operator to the right can be expanded in powers of V. This solution can also be rewritten as $\left(E^{(+)} - \widehat{H}\right)\Psi_E^{(+)} = i\varepsilon\Psi_{0E}$ which gives $\Psi_E^{(+)} = i\varepsilon\widehat{R}^{(+)}(E)\Psi_{0E}$ showing that the wave operator can be defined in terms of the full Hamiltonian alone.

These equations are suitable for partitioning of states and operators. We briefly present here a compact derivation, of a more detailed treatment previously given in Ref. 12. Any operator can be decomposed into diagonal and nondiagonal forms as in $\widehat{A} = \widehat{A}_D + \widehat{A}_N$, $\widehat{A}_D = \widehat{P}\widehat{A}\widehat{P} + \widehat{Q}\widehat{A}\widehat{Q}$, $\widehat{A}_N = \widehat{P}\widehat{A}\widehat{Q} + \widehat{Q}\widehat{A}\widehat{P}$, and products of operators can be expressed in terms of these components. The diagonal (or D-) form allows separation of terms relating to P and Q subspaces with exclusion of their coupling, a feature that greatly simplifies calculations. In particular, one can apply the decomposition to the Hamiltonian and resolvent, to extract the diagonal part of the resolvent function solving for the nondiagonal part in the second line of

$$\left(E^{(+)} - \widehat{H}_D\right)\widehat{R}_D^{(+)} - \widehat{H}_N\widehat{R}_N^{(+)} = \widehat{I}$$
$$\left(E^{(+)} - \widehat{H}_D\right)\widehat{R}_N^{(+)} - \widehat{H}_N\widehat{R}_D^{(+)} = 0$$

and replacing it in the first line to obtain a desired equation with an effective Hamiltonian containing only diagonal parts in P and Q subspaces,

$$\left(E^{(+)} - \widehat{\mathcal{H}}_D\right)\widehat{R}_D^{(+)} = \widehat{I}$$
$$\widehat{\mathcal{H}}_D = \widehat{H}_D + \widehat{H}_N\left(E^{(+)} - \widehat{H}_D\right)^{-1}\widehat{H}_N$$

This effective Hamiltonian separates interactions occurring within each subspace without their coupling, in its first term, from interactions involving the coupling of the two subspaces in the second term, with the latter depending on the energy E of the system. This can be used to construct a compact form of the diagonal resolvent and from it a wave operator in D-form, provided $P\Psi_{0E} = \Psi_{0E}$. In this case $\widehat{P}\Psi_E = i\varepsilon\widehat{P}\widehat{R}_D^{(+)}\widehat{P}\Psi_{0E} = \widehat{P}\widehat{W}_D^{(+)}\widehat{P}\Psi_{0E}$, with the diagonal part of the wave operator given by

$$\widehat{W}_D^{(+)}(E) = \widehat{I} + \widehat{R}_D^{(+)}(E)\left(\widehat{\mathcal{H}}_D - E\right)$$

This wave operator can be compared with the one introduced earlier[2] to solve for stationary bound states. The present formulation allows a treatment for states in the P subspace with discrete as well as continuous eigenenergies, and can therefore be applied to scattering and decaying states as well. Here the diagonal resolvent $\widehat{R}_D^{(+)}(E)$ can be expanded in powers of the recoupling energy operator $\widehat{H}_N\left(E^{(+)}-\widehat{H}_D\right)^{-1}\widehat{H}_N$ when the P–Q interaction is weak.

2.3 Projection Operators for Molecular Effective Potentials

We consider a physical system composed of two molecular species A and M at a relative position $\vec{R}$ and electronic variables and other nuclear position coordinates included in the set $\boldsymbol{R}_{int}$, interacting with a potential energy function V going to zero at large distances R between the colliding molecules. The total Hamiltonian operator for this system, in terms of a relative kinetic energy operator $\widehat{K}$ for a reduced mass m and of internal Hamiltonian operators, is $\widehat{H}=\widehat{H}_0+\widehat{V}$ with $\widehat{H}_0=\widehat{K}+\widehat{H}_A+\widehat{H}_M$ the Hamiltonian in the absence of interactions. Its steady-state eigenfunctions $\Psi_{0\nu}$ satisfy $\widehat{H}_0\Psi_{0\nu}=E\Psi_{0\nu}$ with the index $\nu=\left(\vec{p}, n\right)$ giving the relative momentum and internal states n of the molecules, and with a total energy $E=p^2/(2m)+W_n$ from the reduced mass m of the pair with internal energy W_n. The initial state can be constructed as a superposition of unperturbed steady states, with each given by a product of the internal states of the molecules and their relative unperturbed state in free space, and evolves over time as a wavepacket $\Psi_0(t)$.

We are interested in situations where molecules initially in a P subspace of chosen electronic, vibrational, and rotational molecular states, collide subject to an interaction potential which vanishes at large distances, leading to transitions between P and Q subspaces at finite distances, and we want to concentrate on final states also in the P subspace. It is then convenient to partition states and operators by means of a Hermitian projection operator $\widehat{P}$ satisfying idempotency $\widehat{P}^2=\widehat{P}$ and the asymptotic condition $\lim\left[\widehat{H},\widehat{P}\right]=0$ for large distances $R\to\infty$. The desired states $\widehat{P}\Psi_E$ required only the wave operator $\widehat{P}\widehat{W}_D^{(+)}\widehat{P}$ and this contains only the diagonal effective Hamiltonian

$$\widehat{\mathcal{H}}_P=\widehat{P}\widehat{H}\widehat{P}+\widehat{\mathcal{V}}_P(E),\quad \widehat{\mathcal{V}}_P(E)=\widehat{P}\widehat{V}\widehat{P}+\widehat{P}\widehat{H}\widehat{Q}\left(E^{(+)}-\widehat{Q}\widehat{H}\widehat{Q}\right)^{-1}\widehat{Q}\widehat{H}\widehat{P}$$

with an effective potential $\widehat{\mathcal{V}}_P(E)$ defined in the P subspace and going to zero at large distances. Its second term describes the influence of states outside P and its properties depend on the reduced resolvent $\widehat{Q}\left(E^{(+)} - \widehat{Q}\widehat{H}\widehat{Q}\right)^{-1}\widehat{Q}$ as given by the eigenvalues of $\widehat{Q}\widehat{H}\widehat{Q}$. This second term can be approximated and used to estimate whether states in the Q subspace can be discarded or not.

If its lowest eigenenergy is $\mathcal{E}_Q$ then the second term is a real valued positive operator for $E < \mathcal{E}_Q$, while for $E > \mathcal{E}_Q$ it is complex with a negative valued imaginary part

$$Im\widehat{\mathcal{V}}_P(E) = -\pi\widehat{P}\widehat{H}\widehat{Q}\delta\left(E - \widehat{Q}\widehat{H}\widehat{Q}\right)\widehat{Q}\widehat{H}\widehat{P}$$

as follows from the special functions relation $\lim(u + i\varepsilon)^{-1} = \mathcal{P}u^{-1} - i\pi\delta(u)$. A complex potential with a negative imaginary part leads to a negative divergence of the scattering flux[34] and therefore in our treatment to a loss of flux from the P subspace to the Q subspace. This formalism provides the foundation for optical models of collisions, which we have extensively used to construct complex valued intermolecular potentials.[35]

To proceed, it is necessary to specify the projection operators corresponding to different physical situations. Given the asymptotic Hamiltonian $\widehat{H}_0 = \widehat{K} + \widehat{H}_A + \widehat{H}_M$, it is more general to define a reference Hamiltonian $\widehat{H}^{(0)}$ valid for all intermolecular distances R and such that $\widehat{H}^{(0)} \approx \widehat{H}_0$ for $R \to \infty$. Letting $\widehat{H} = \widehat{H}^{(0)} + \widehat{H}^{(1)}$ we require that the projector P satisfies $\left[\widehat{H}^{(0)}, \widehat{P}\right] = 0$ for all distances. It can be chosen so that effective potentials for molecular interactions correspond at least to: (a) a distortion potential; (b) an adiabatic potential; and (c) a fast motion potential. The related projectors are given in what follows.

(a) Distortion potentials

Using internal molecular states for species A and M and their product to construct $|u_n\rangle = \left|u_j^{(A)} u_k^{(M)}\right\rangle$, which depend on electronic and internal nuclear variables and have internal energy W_n, projectors $\widehat{P}_n = |u_n\rangle\langle u_n|$ over a selected set of states $n_1 \leq n \leq n_2$ can be included in $\widehat{P} = \sum_n \widehat{P}_n$ to form the distortion Hamiltonian and potential operators in

$$\widehat{H}^{(0)}_{dis} = \widehat{H}_0 + \widehat{V}^{(0)}, \quad \widehat{V}^{(0)} = \sum_n V_{nn}\left(\vec{R}\right)\widehat{P}_n$$

where the distortion potential $V_{nn}\left(\vec{R}\right) = \left\langle u_n \right| \widehat{V}\left(\vec{R}\right) \left| u_n \right\rangle$ is yet a function of relative position. To second order in the coupling of P and Q subspaces the effective Hamiltonian looks now like

$$\widehat{\mathcal{H}}_P = \sum_{n,n'} |u_n\rangle \left(\widehat{\mathcal{H}}_{nn'}^{(0)} + \widehat{\mathcal{H}}_{nn'}^{(1)} + \widehat{\mathcal{H}}_{nn'}^{(2)} \right) \langle u_{n'}|$$

$$\widehat{\mathcal{H}}_{nn'}^{(0)} = \delta_{nn'} \left(\widehat{K} + W_n + V_{nn} \right), \quad \widehat{\mathcal{H}}_{nn'}^{(1)} = V_{nn'} - \delta_{nn'} V_{nn}$$

$$\widehat{\mathcal{H}}_{nn'}^{(2)} = \sum_m V_{nm} \widehat{R}_{m,dis}^{(+)} \left(\vec{R}, \widehat{\nabla}_{\vec{R}}; E \right) V_{mn'}$$

where $n_1 > m > n_2$ spans the Q subspace. The third line is a function of relative position and relative gradient $\widehat{\nabla}_{\vec{R}}$, through the distortion Hamiltonian resolvent, as shown. To assess its importance, the resolvent magnitude can be estimated to be about $\Delta E^{-1} \cong (\Delta p.v)^{-1}$ where v is the local relative velocity and Δp is the change in relative momentum, so that the second-order term is of magnitude $(\Delta V)^2 \Delta E^{-1}$ and small for small coupling energies between internal states or for large relative velocities.

(b) Adiabatic potentials

When internal molecular states are strongly coupled as the distance R changes, it is necessary to introduce molecular adiabatic eigenstates $\left|z_\alpha\left(\vec{R}\right)\right\rangle$, which are functions of the internal coordinates of A and M, and eigenenergies $\mathcal{E}_\alpha(R)$ of the adiabatic Hamiltonian $\widehat{H}_{\vec{R}} = \widehat{H}_A + \widehat{H}_M + \widehat{V}$ which excludes the relative kinetic energy $\widehat{K} = -\left(\hbar^2/2m\right)\nabla^2$. In this case, a suitable projector dependent on the relative position is given by

$$\widehat{P}\left(\vec{R}\right) = \sum_\alpha \widehat{P}_\alpha\left(\vec{R}\right), \quad \widehat{P}_\alpha\left(\vec{R}\right) = \left|z_\alpha\left(\vec{R}\right)\right\rangle \left\langle z_\alpha(\vec{R})\right|$$

with a P subspace selected to contain states $\alpha_1 \leq \alpha \leq \alpha_2$ with asymptotic energies $\mathcal{E}_\alpha^{(0)}$ below the total energy E and therefore energetically open. Now the coupling between P and Q subspaces comes from the relative kinetic energy operator with matrix elements

$$\left\langle z_\alpha\left(\vec{R}\right) \middle| \widehat{K} \middle| z_{\alpha'}\left(\vec{R}\right) \right\rangle = -\left(\hbar^2/2m\right)\delta_{\alpha\alpha'}\nabla^2 + C_{\alpha\alpha'}\left(\vec{R}, \vec{\nabla}_{\vec{R}}\right)$$

where the last term is a nonadiabatic coupling involving the gradient and second gradient of the adiabatic state $z_\alpha\left(\vec{R}\right)$ times the state $z_{\alpha'}\left(\vec{R}\right)$, and integration over internal molecular coordinates.[12] The effective Hamiltonian is now

$$\widehat{\mathcal{H}}_P\left(\vec{R},\vec{R}'\right)=\sum_{\alpha,\alpha'}\left|z_\alpha\left(\vec{R}\right)\right\rangle\left(\widehat{\mathcal{H}}^{(0)}_{\alpha\alpha'}+\widehat{\mathcal{H}}^{(1)}_{\alpha\alpha'}+\widehat{\mathcal{H}}^{(2)}_{\alpha\alpha'}\right)\left\langle z_{\alpha'}\left(\vec{R}'\right)\right|$$

$$\widehat{\mathcal{H}}^{(0)}_{\alpha\alpha'}\left(\vec{R},\vec{R}'\right)=\delta_{\alpha\alpha'}\delta\left(\vec{R}-\vec{R}'\right)\left[-\left(\hbar^2/2m\right)\nabla^2+\mathcal{E}_\alpha(R)+C_{\alpha\alpha}\left(\vec{R},\vec{\nabla}_{\vec{R}}\right)\right]$$

$$\widehat{\mathcal{H}}^{(1)}_{\alpha\alpha'}\left(\vec{R},\vec{R}'\right)=\delta\left(\vec{R}-\vec{R}'\right)\left[C_{\alpha\alpha'}\left(\vec{R},\vec{\nabla}_{\vec{R}}\right)-\delta_{\alpha\alpha'}C_{\alpha\alpha}\left(\vec{R},\vec{\nabla}_{\vec{R}}\right)\right]$$

$$\widehat{\mathcal{H}}^{(2)}_{\alpha\alpha'}=\sum_{\beta}C_{\alpha\beta}\left(\vec{R},\vec{\nabla}_{\vec{R}}\right)R^{(+)}_{\beta,ad}\left(\vec{R},\vec{R}';E\right)C_{\beta\alpha'}\left(\vec{R}',\vec{\nabla}'_{\vec{R}}\right)$$

where the last line accounts for coupling of P and Q subspaces due to relative motion and also depends on the energy E through the adiabatic resolvent. Here the summation is over $\alpha>\beta>\alpha'$, which defines the Q subspace. The calculation of adiabatic states in Q subspace can be quite demanding insofar it may involve crossings and avoided crossings of potentials, and the expression above can provide estimates of the magnitude of couplings to states outside the P subspace and whether they can be discarded.

(c) Fast motion potentials

A fast motion description is advantageous when motion along $\vec{R}$ is fast compared with the other coordinates $\boldsymbol{R}_{int}=(\boldsymbol{R}_A,\boldsymbol{R}_M)$, slow by comparison. This may apply for example to electron scattering or emission, when an electron at position $\vec{R}$ interacts with a molecule. Taking the slow variables as fixed and the Hamiltonian to be $\widehat{H}_{\boldsymbol{int}}=\widehat{K}+\widehat{V}_{int}\left(\vec{R}\right)$, its fast scattering eigenfunctions $\varphi^{(+)}_{\vec{p}}\left(\vec{R};\boldsymbol{R}_{int}\right)=\left\langle\vec{R}\mid\vec{p};\boldsymbol{R}_{int}\right\rangle$, with scattering boundary conditions for outgoing waves and relative momentum $\vec{p}$, for eigenenergies $p^2/(2m)$, can be constructed with normalization $\left\langle\vec{p};\boldsymbol{R}_{int}\right|\left.\vec{p}';\boldsymbol{R}_{int}\right\rangle=\delta\left(\vec{p}-\vec{p}'\right)$. The basis set $\left\{\left|\vec{p};\boldsymbol{R}_{int}\right\rangle\right\}$ satisfies the completeness relation $\int\int d^3p\,d\boldsymbol{R}_{int}\left|\vec{p};\boldsymbol{R}_{int}\right\rangle\left\langle\vec{p};\boldsymbol{R}_{int}\right|=\widehat{I}$. The total Hamiltonian in a compact notation with masses of slow motions shown as M, is $\widehat{H}=\widehat{H}_{\boldsymbol{int}}-\left(\hbar^2/2M\right)\nabla^2_X$ with the second term giving the kinetic energy operator for slow motions. A projector $\widehat{P}$ can be introduced to select scattering states with momenta within a range $\Delta\vec{p}$, which form a wavepacket when added, and describe the fast relative motion for fixed internal variables, in the form

$$\widehat{P} = \int d\boldsymbol{R}_{int}\widehat{P}(\boldsymbol{R}_{int}),\quad \widehat{P}(\boldsymbol{R}_{int}) = \int_{\Delta\vec{p}} d^3p|\vec{p}\,;\boldsymbol{R}_{int}\rangle\langle\vec{p}\,;\boldsymbol{R}_{int}|$$

With this, the leading term in the effective Hamiltonian has matrix elements in the basis set $\{|\vec{p}\,;\boldsymbol{R}_{int}\rangle\}$ given by $\mathcal{H}^{(0)}_{\vec{p}\vec{p}'}(\boldsymbol{R}_{int},\boldsymbol{R}'_{int}) = \delta(\boldsymbol{R}_{int}-\boldsymbol{R}'_{int})\delta(\vec{p}-\vec{p}\,')p^2/(2m)$. States in the P subspace are coupled by $\mathcal{H}^{(1)}_{\vec{p}\vec{p}'}(\boldsymbol{R}_{int},\boldsymbol{R}'_{int}) = H_{\vec{p}\vec{p}'}(\boldsymbol{R}_{int},\boldsymbol{R}'_{int}) - \mathcal{H}^{(0)}_{\vec{p}\vec{p}'}(\boldsymbol{R}_{int},\boldsymbol{R}'_{int})$. The contribution from coupling with the Q subspace through the kinetic energy operator of slow motions is derived similarly to the adiabatic case, and becomes to second order in the couplings

$$\widehat{\mathcal{H}}^{(2)}_{\vec{p}\vec{p}'}(\boldsymbol{R}_{int},\boldsymbol{R}'_{int}) = \int d^3q C_{\vec{p}\vec{q}}(\boldsymbol{R}_{int},\vec{\nabla}_{int})R^{(+)}_{\vec{q},fast}(\boldsymbol{R}_{int},\boldsymbol{R}'_{int};E)C_{\vec{q}\vec{p}'}(\boldsymbol{R}'_{int},\vec{\nabla}'_{int})$$

where the integral excludes the range $\Delta\vec{p}$, and with the coupling $C_{\vec{p}\vec{q}}$ involving first- and second-order derivatives of a fast scattering state $\varphi^{(+)}_{\vec{q}}$ with respect to the slow variables, times another fast state $\varphi^{(+)}_{\vec{p}}$ and integration over the $\vec{R}$ variables. This expression can be used to estimate whether the P–Q subspace interactions are small and can be neglected.

2.4 Projection Operators for Long-Lived States

Two interacting molecules A and M may be found in a joint transient state induced by light absorption or by their collision with each other. Some transient states (A, M) may be long lived compared with the duration of the exciting event (a light pulse or a collision duration) and may as a result be detectable in molecular spectroscopy, or may contribute to properties of an assembly of those molecules. It is advantageous to describe the long-lived states starting from some chosen combination of a set of $\kappa = 1$ to K orthonormalized states Φ_κ of the molecular pair. These states are functions of the intermolecular distance R and vanish at long distances. They can be used to construct a projector $\widehat{Q} = \sum_\kappa |\Phi_\kappa\rangle\langle\Phi_\kappa|$ and a Hamiltonian $\widehat{Q}\widehat{H}\widehat{Q}$ with eigenstates $\xi_\kappa = \sum_{\kappa'} c_{\kappa\kappa'}\Phi_{\kappa'}$, shown as linear combinations of the originally chosen states, and eigenenergies $\mathcal{E}_\kappa$. The eigenvalue spectra of Hamiltonians $\widehat{Q}\widehat{H}\widehat{Q}$, $\widehat{P}\widehat{H}\widehat{P}$, and $\widehat{H}$ are shown schematically in the following figure, with an origin of the energy axis such that the $\mathcal{E}_\kappa$ are positive (Fig. 1).

The contribution of long-lived states to the dynamics of the molecular pair induced by excitations or collisions leads in a time-independent description to the appearance in the wavefunction $\widehat{P}\Psi_{\nu}^{(+)} = i\varepsilon\widehat{P}\widehat{R}_{D}^{(+)}\widehat{P}\Psi_{0\nu} = \widehat{P}\Psi_{\nu,dir}^{(+)} + \widehat{P}\Psi_{\nu,res}^{(+)}$ of the shown second term, a scattering resonance state given by[12]

$$\widehat{P}\Psi_{\nu,res}^{(+)} = \left(E^{(+)} - \widehat{P}\widehat{H}\widehat{P}\right)^{-1}\widehat{P}\widehat{H}\widehat{Q}\left(E^{(+)} - \widehat{\mathcal{H}}_{Q}\right)^{-1}\widehat{Q}\widehat{H}\widehat{P}\Psi_{\nu,dir}^{(+)}$$

where the effective Hamiltonian $\mathcal{H}_Q(E)$ is in general a normal non-Hermitian operator for energies E in the energy continuum of $\widehat{P}\widehat{H}\widehat{P}$, and is defined in a finite functional space of dimension K. Its complex eigenvalues $f_k(E)$ and eigenfunctions $|\zeta_k(E)\rangle$, or resonance states, satisfy the equation

$$\left[f_k(E) - \widehat{\mathcal{H}}_Q(E)\right]|\zeta_k(E)\rangle = 0$$

which can be used to construct its above resolvent as in

$$\widehat{P}\Psi_{\nu,res}^{(+)} = \left(E^{(+)} - \widehat{P}\widehat{H}\widehat{P}\right)^{-1}\widehat{P}\widehat{H}\widehat{Q}\sum_{k}\frac{|\zeta_k(E)\rangle\langle\zeta_k^{*}(E)|}{E - f_k(E)}\widehat{Q}\widehat{H}\widehat{P}\Psi_{\nu,dir}^{(+)}$$

This shows that the scattering resonance term has energy poles at complex roots $E_s - \Gamma_s/2$ of the equation $E = f_k(E)$. The time-dependent partition treatment we have introduced allows us to Fourier transform from E to time t to find the temporal behavior of the resonance state, which for an isolated resonance s gives[12]

$$\widehat{P}\Psi_{\nu,res}(t) = \widehat{P}X(t)\exp\left[-(iE_s + \Gamma_s/2)t/\hbar\right]$$

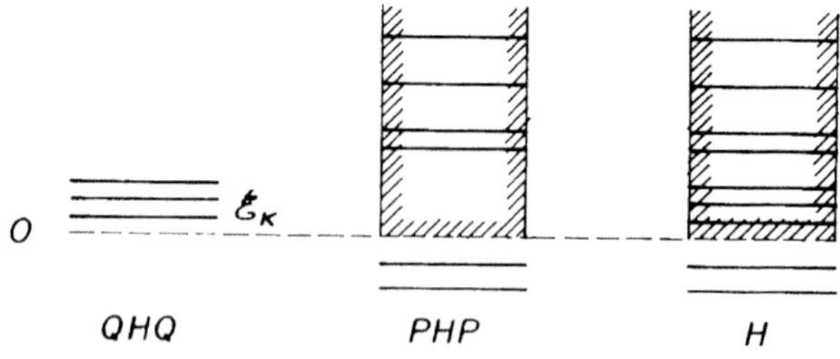

Fig. 1 Pictorial representation of the energy spectra of Hamiltonians *PHP*, *QHQ*, and *H*. *Reproduced with permission from Micha, D.A. Effective Hamiltonian Methods for Molecular Collisions.* Adv. Quantum Chem. ***1974,*** 8, *231. Copyright 1974, Wiley.*

leading to a probability of exponential decay with the expected lifetime $\tau_s = \hbar/\Gamma_s$.

The previous treatment relates to scattering phenomena and shows how resonance states in the Q subspace contribute to the asymptotic form of a total scattering wavefunction through the P–Q couplings. A related treatment for photodissociation following excitation by light can also be based on projection operators, but with decay boundary conditions instead of scattering boundary conditions. A general procedure for a molecular system in a state α coupled to a radiation field with n_L photons and using molecule-field states $\nu = (\alpha, n_L)$ has been developed introducing projectors over open and closed dissociation channels defined by the energy conditions $E \geq E_\nu^{(0)}$, the asymptotic energy of state ν, and $E < E_\nu^{(0)}$, respectively.[36]

As done for effective molecular potentials, specific choices of projectors can be made for resonance dynamics in distortion states or in adiabatic states.

(a) Distortion long-lived states

When the interaction of molecules A and M is not strong, it is expected that the long-lived states will be well described by a combination of a restricted set of internal states u_n, $n_1 \leq n \leq n_2$, functions of internal electronic and nuclear coordinates $\boldsymbol{R}_{int}$ and we can choose $\widehat{Q} = \sum_n |u_n\rangle\langle u_n|$. Introducing bound-state eigenfunctions $\phi_{k,\,dis}^{(n)}\left(\vec{R}\right)$ of the distortion Hamiltonians we displayed before, the resonance eigenstates of $\widehat{\mathcal{H}}_Q(E)$ are linear combinations of basis functions $\phi_{k,\,dis}^{(n)}\left(\vec{R}\right)u_n(\boldsymbol{R}_{int})$. The way these resonance states decay into the P subspace is governed by the P–Q couplings in $\widehat{P}\Psi_{\nu,\,res}^{(+)}$.

(b) Adiabatic long-lived states

For strongly interacting molecules A and M forming states which change with their mutual distance R, it is possible to describe resonance states with adiabatic states $\left|z_\rho\left(\vec{R}\right)\right\rangle$, functions of all electronic and nuclear variables over which one must integrate, with $\rho_1 \leq \rho \leq \rho_2$ and asymptotic energies $\mathcal{E}_\rho^{(0)}$ above the total energy E and therefore energetically closed. The projector is then

$$\widehat{Q}\left(\vec{R}\right) = \sum_\rho \widehat{Q}_\rho\left(\vec{R}\right),\ \widehat{Q}_\rho\left(\vec{R}\right) = \left|z_\rho\left(\vec{R}\right)\right\rangle\left\langle z_\rho\left(\vec{R}\right)\right|$$

and resonance states $\zeta_k\left(\vec{R}, \boldsymbol{R}_{int}\right)$ can be constructed as linear combinations of bound states $\phi_{k,\,ad}^{(\varrho)}\left(\vec{R}\right)z_\rho\left(\boldsymbol{R}_{int}; \vec{R}\right)$ where the bound-state wavefunction

$\phi_{k,ad}^{(\varrho)}\left(\vec{R}\right)$ is obtained from the adiabatic potential $\mathcal{E}_\varrho(R)$. As before for adiabatic phenomena, P–Q couplings giving the decay of resonance states into open channels involve nonadiabatic couplings

$$C_{\alpha\rho}\left(\vec{R},\vec{\nabla}_{\vec{R}}\right).$$

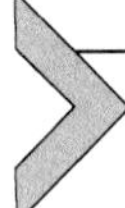

3. DISSIPATIVE MANY-ATOM DYNAMICS FROM PARTITIONING METHODS

The dynamics of a many-atom system under thermodynamical constraints presents special challenges to computational modeling. Electrons and nuclei must be included in calculations of properties such as cohesive energy, transport of thermal energy and electrical charges, light absorbance, and photoconductivity. Projection operator methods can help organize and simplify model calculations when properties of interest involve only a portion of the total system, which can be then mathematically broken into a primary (or p-) region of interest, to be treated starting from the atomic composition of the relevant region and its immediate environment, interacting with a remaining secondary (or s-) region treated as a medium or bath, within an statistical approximation in terms of its important degrees of freedom. The total Hamiltonian is then decomposed as $\widehat{H}=\widehat{H}_p+\widehat{H}_s+\widehat{H}_{ps}$.

The treatment must be done combining quantal and statistical mechanics, and can be generally developed introducing the statistical density operator of von Neumann,[19] called here $\widehat{\Gamma}(t)$, and its equation of motion of Liouville–von Neumann (L-vN) equation. The partitioning of this density operator leads to the introduction of a RDOp for each region. The formalism has been extensively reviewed in the literature of nonequilibrium statistical mechanics for open systems,[20–22] and has been applied to molecular spectroscopy in solutions,[23,24] to charge and energy transfer dynamics,[25] and to condensed phase dynamics.[26] In several papers and a book,[37–39] Löwdin presented a superoperator algebra suitable for the mathematical treatment of the L-vN equation and outlined its partition, for the special case of a time-independent Hamiltonian of an isolated system. In what follows a treatment is developed suitable for an open many-atom system, with both electrons and nuclei included, with thermodynamical constraints and undergoing a dissipative dynamics.

The partition treatment can be presented in a compact way introducing Liouvillian superoperators (shown in script fonts) induced by commutators such as $\widehat{\mathcal{H}}\widehat{A} = \left[\widehat{H}, \widehat{A}\right]$. The L-vN equation for the whole system is then

$$i\hbar\partial\widehat{\Gamma}/\partial t = \widehat{\mathcal{H}}\widehat{\Gamma}(t)$$

to be solved with initial condition $\widehat{\Gamma}(t_0) = \widehat{\Gamma}^{(0)}(\Lambda)$, with Λ a set of initial thermodynamical parameters.

3.1 Contracted Density Operator from a Partitioned Functional Space

Implementation of the above formulation must be done with specific projection operators. One frequent choice is to partition the functional space of the whole system using a finite basis set for the projector $\widehat{P}$ and another for $\widehat{Q} = \widehat{I} - \widehat{P}$. This is frequently presented in reviews of the subject.[22] The equation can be split into two coupled equations for diagonal and nondiagonal parts of the DOp, constructed from projectors $\widehat{P}$ and $\widehat{Q} = \widehat{I} - \widehat{P}$, with $\widehat{\Gamma} = \widehat{\Gamma}_D + \widehat{\Gamma}_N$, $\widehat{\Gamma}_D = \widehat{P}\widehat{\Gamma}\widehat{P} + \widehat{Q}\widehat{\Gamma}\widehat{Q}$, $\widehat{\Gamma}_N = \widehat{P}\widehat{\Gamma}\widehat{Q} + \widehat{Q}\widehat{\Gamma}\widehat{P}$. Applying these definitions to operators in the L-vN equation, with $\widehat{H} = \widehat{H}_D + \widehat{H}_N$, it becomes the set

$$\begin{aligned} i\hbar\,\partial\widehat{\Gamma}_D/\partial t &= \widehat{\mathcal{H}}_D\widehat{\Gamma}_D(t) + \widehat{\mathcal{H}}_N\widehat{\Gamma}_N(t) \\ i\hbar\,\partial\widehat{\Gamma}_N/\partial t &= \widehat{\mathcal{H}}_D\widehat{\Gamma}_N(t) + \widehat{\mathcal{H}}_N\widehat{\Gamma}_D(t) \end{aligned}$$

which can be formally solved for $\widehat{\Gamma}_N$ with the initial condition $\widehat{\Gamma}(t_0) = \widehat{\Gamma}^{(0)}$ introducing the time-evolution superoperator $\widehat{\mathcal{U}}_D(t,t') = \exp_T\left[(-i/\hbar)\int_{t'}^{t} ds\widehat{\mathcal{H}}_D(s)\right]$ to write

$$\widehat{\Gamma}_N(t) = \widehat{\Gamma}_N(t_0) + (i\hbar)^{-1}\int_{t_0}^{t} dt'\widehat{\mathcal{U}}_D(t,t')\widehat{\mathcal{H}}_N\widehat{\Gamma}_D(t')$$

and replacing it in the first line to obtain the integrodifferential equation

$$\begin{aligned} \partial\widehat{\Gamma}_D/\partial t = {} & (i\hbar)^{-1}\widehat{\mathcal{H}}_D\widehat{\Gamma}_D(t) + (i\hbar)^{-1}\widehat{\mathcal{H}}_N\widehat{\Gamma}_N(t_0) \\ & + (i\hbar)^{-2}\int_{t_0}^{t} dt'\widehat{\mathcal{H}}_N\widehat{\mathcal{U}}_D(t,t')\widehat{\mathcal{H}}_N\widehat{\Gamma}_D(t') \end{aligned}$$

Here the second term to the right gives a fluctuation rate around the initial statistical density, and the third term is a delayed dissipative rate containing a memory kernel. The equation for the component $\widehat{\Gamma}_P = \widehat{\Gamma}_D \widehat{P}$ follows by multiplying both sides of the equation times the $\widehat{P}$ operator to the right. The equation terms parallel the ones in the treatment of wavefunctions, but now the DOp equation accounts also for an initial statistical distribution in the system subject to thermodynamical constraints. This is an alternative form of the Nakajima–Zwanzig equation.[22] It can be simplified averaging both sides over the initial distribution of parameters Λ, in which case the fluctuation term can usually be set to zero, and the memory kernel can be approximated. The formulation applies equally to a time-dependent Hamiltonian.

3.2 Reduced Density Operator from a Partitioned Physical Space

A second choice useful in atomistic modeling is to partition the physical space of the whole system into p- and s-regions with given atomic compositions containing electrons and nuclei in each one. It is described here in more detail to clarify how related projectors can be defined for a many-electron system.

Let $\boldsymbol{X}$ stand for the set of all electronic position and spin variables, with $\boldsymbol{X}_p$ and $\boldsymbol{X}_s$ including only p- or s-electronic variables. Similarly, $\boldsymbol{Y}_p$ and $\boldsymbol{Y}_s$ stand for the ion-core position coordinates in the two regions, and all variables are collected in $\boldsymbol{R} = (\boldsymbol{X}, \boldsymbol{Y}) = (\boldsymbol{R}_p, \boldsymbol{R}_s)$. The density operator in the coordinate representation is a function $\left\langle \boldsymbol{R} | \widehat{\Gamma}(t) | \boldsymbol{R}' \right\rangle$, which satisfies electron exchange permutational symmetry for all electrons in the two sets of electronic variables. A RDOp for the p-region is obtained from a trace over s-variables as $\Gamma^{(p)}(\boldsymbol{R}_p, \boldsymbol{R}'_p, t) = \int d\boldsymbol{R}_s \left\langle \boldsymbol{R}_p, \boldsymbol{R}_s | \widehat{\Gamma}(t) | \boldsymbol{R}'_p \boldsymbol{R}_s \right\rangle$, with a similar expression for $\Gamma^{(s)}(\boldsymbol{R}_s, \boldsymbol{R}_{s'}, t)$ of the s-region.

The choice of a projection operator for the many-atom system must be done keeping in mind that in principle electrons in p- and s-regions are all permuting among themselves and must be related in a quantum mechanical sense. The permutational exchange of electrons between the two regions may however be assumed to have negligible effect if the two regions are large and their electron densities remain constant on average over a statistical distribution for the constrained s-system.

To extract a RDOp for the p-region it is convenient to formally introduce a set of s-wavefunctions $\Psi_J^{(s)}(\boldsymbol{X}_s, \boldsymbol{Y}_s)$ with a statistical distribution $w_J(\Lambda)$ of states J where Λ stands for a set of initial values of collective physical variables such as electronic density or net charge and ion-core positions, and a s-region statistical density operator $\widehat{\Gamma}^{(s)}(\Lambda) = \sum_J w_J(\Lambda)|\Psi_J^{(s)}\rangle\langle\Psi_J^{(s)}|$ normalized so that its trace over s-variables gives $tr_s\left[\widehat{\Gamma}^{(s)}(\Lambda)\right] = 1$. This could be used in the definition of a projection superoperator $\widehat{\mathcal{P}}(\Lambda)$ (with superoperators show in script fonts here and in what follows) which acts on an operator $\widehat{A}$ of all variables to give another operator, as in $\widehat{\mathcal{P}}(\Lambda)\widehat{A} = \widehat{A}^{(p)}(\Lambda)\widehat{\Gamma}^{(s)}(\Lambda)$ with $\widehat{A}^{(p)}(\Lambda) = tr_s\left[\widehat{A}\right]$, taking the trace over s-variables. This satisfies the superprojector idempotency condition $\widehat{\mathcal{P}}(\Lambda)^2 = \widehat{\mathcal{P}}(\Lambda)$. The result is a projector in product form containing only p-variables in the first factor and only s-variables in the second one. This however does not satisfy permutation symmetry over all electronic variables, and could lead to inaccurate values for physical properties. A more reliable choice involves an additional average over the parameter range $\Delta\Lambda$, shown with a bar on top, which can be expected to correctly describe physical properties which are needed only on the average. Let now $\overline{\widehat{\Gamma}^{(s)}} = \int_{\Delta\Lambda} d\Lambda \widehat{\Gamma}^{(s)}(\Lambda)/\Delta\Lambda$ and introduce the average superprojector $\widehat{\mathcal{P}}$ in

$$\widehat{\mathcal{P}}\widehat{A} = \overline{\widehat{A}^{(p)}}\,\overline{\widehat{\Gamma}^{(s)}},\quad \overline{\widehat{A}^{(p)}} = \int_{\Delta\Lambda} d\Lambda\,\mathrm{tr}_s[\widehat{A}]/\Delta\Lambda$$

which factors the projected operator into p- and s-components as in a mean field treatment. Idempotency is again satisfied, and a complementary projector is defined by $\widehat{\mathcal{Q}} = \widehat{\mathcal{I}} - \widehat{\mathcal{P}}$. The total density operator can be decomposed as a product of p- and s-factors with $\widehat{\Gamma}(t) = \widehat{\mathcal{P}}\widehat{\Gamma}(t) + \widehat{\mathcal{Q}}\widehat{\Gamma}(t) = \widehat{\Gamma}_P(t) + \widehat{\Gamma}_Q(t)$, where $\widehat{\Gamma}_P(t) = \overline{\widehat{\Gamma}^{(p)}}(t)\overline{\widehat{\Gamma}^{(s)}}(t)$, and we identify $\overline{\widehat{\Gamma}^{(p)}}(t) = \widehat{\rho}(t)$ as the RDOp of the p-region to be found. We also have that $\widehat{\rho}(t) = tr_s\left[\widehat{\mathcal{P}}\widehat{\Gamma}(t)\right]$ by construction with the normalized s-RDOp. The remaining DOp $\widehat{\Gamma}_Q(t)$ gives a correction to the factorized mean field form of the leading DOp.

The next step is to partition the L-vN equation of motion and to solve for the p-region RDOp. The partition formalism for the DOp and L-vN equations are done by analogy to the derivation above, with the result

$$\begin{aligned}\partial\widehat{\Gamma}_P/\partial t &= (i\hbar)^{-1}\widehat{\mathcal{P}}\widehat{\mathcal{H}}\widehat{\mathcal{P}}\widehat{\Gamma}_P(t) + (i\hbar)^{-1}\widehat{\mathcal{P}}\widehat{\mathcal{H}}\widehat{\mathcal{Q}}\widehat{\Gamma}_Q(t_0) \\ &\quad + (i\hbar)^{-2}\int_{t_0}^{t} dt'\widehat{\mathcal{P}}\widehat{\mathcal{H}}\widehat{\mathcal{Q}}\widehat{\mathcal{U}}_Q\left(t,t'\right)\widehat{\mathcal{Q}}\widehat{\mathcal{H}}\widehat{\mathcal{P}}\Big)\widehat{\Gamma}_P(t')\end{aligned}$$

where $\widehat{\mathcal{U}}_Q(t,t') = \exp_T\left[(-i/\hbar)\int_{t'}^{t} dt''\widehat{\mathcal{Q}}\widehat{\mathcal{H}}\widehat{\mathcal{Q}}\right]$, which is valid also for a time-dependent Hamiltonian. This equation again displays a fluctuation rate in the second term and a delayed dissipation rate in the third term. They can now be physically interpreted as involving p- and s-region phenomena where electrons and nuclei undergo their dynamics while the two regions interact. Taking the trace of both sides over s-variable and averaging over initial conditions gives the equation for the RDOp $\widehat{\rho}(t)$. The resulting equation is

$$\frac{\partial\widehat{\rho}}{\partial t} = -\frac{i}{\hbar}\left[\widehat{F}_p,\widehat{\rho}(t)\right] - \frac{1}{\hbar^2}\int_{t_0}^{t} dt'\widehat{\mathcal{M}}_p(t,t')\widehat{\rho}(t')$$

where $\widehat{F}_p = \widehat{H}_p + tr_s\left(\widehat{H}_{ps}\overline{\widehat{\Gamma}^{(s)}}\right)$ is an effective Hamiltonian in the p-variables, and it has been assumed that the fluctuation term averages to zero. The second term contain a delayed dissipation rate with a p-region memory kernel superoperator

$$\widehat{\mathcal{M}}_p(t,t') = tr_s\left[\widehat{\mathcal{P}}\widehat{\mathcal{H}}\widehat{\mathcal{Q}}\widehat{\mathcal{U}}_Q(t,t')\widehat{\mathcal{Q}}\widehat{\mathcal{H}}\widehat{\mathcal{P}}\overline{\widehat{\Gamma}^{(s)}}\right]$$

which yet contains the p–s interaction to all orders in the time-evolution superoperator $\widehat{\mathcal{U}}_Q(t,t')$. However, in many cases it can be approximated by the unperturbed form $\widehat{\mathcal{U}}_Q^{(0)}(t,t') = \exp_T\left[(-i/\hbar)\int_{t'}^{t} dt''\widehat{\mathcal{Q}}\left(\widehat{\mathcal{H}}_p + \widehat{\mathcal{H}}_s\right)\widehat{\mathcal{Q}}\right]$ which does not contain the p–s interaction and is easier to calculate. The resulting memory kernel to second order in the p–s couplings can be obtained from time-correlation functions of physical operators in the s-region.[25] The integrodifferential equation for the RDOp can then be solved numerically for memory kernels corresponding to instantaneous, long lasting, and intermediate dissipation in the p-region. This has recently been done within a model containing electronic excitation and vibrational displacements in the s-region.[40] After the initial averaging, dissipative rates from fast electronic motions and slow atomic motions are usually separated into fast and slow terms in $\widehat{\mathcal{M}}_p(t,t') = \widehat{\mathcal{M}}_p^{(el)}(t,t') + \widehat{\mathcal{M}}_p^{(at)}(t,t')$.

Fast (instantaneous) electronic dissipation is due to electronic fluctuations (excitons) in the medium and can be described by the Lindblad rate expression constructed from transition rates between vibronic states in the p-region induced by interactions with the s-region. Slow (delayed) dissipation is due to atomic lattice vibrations. It can be obtained for bilinear displacement couplings of p- and s-regions in terms of time-correlation functions of s-region displacements.

For measurement times long compared with TCF relaxation times, it is given by the Redfield approximation for dissipative rates, which leads to differential equations containing the RDOp only at time t.[41,42] The dissipative rate becomes time independent for phenomena observed after long times compared with relaxation times of medium excitons and phonons, when the whole physical system has settled into a steady state in the absence of external forces, and is given by a superoperator $\widehat{\mathcal{R}}$. It can be decomposed into two terms corresponding to vibronic excitations from the ground state to excited electronic states plus vibrational transitions in the ground electronic state, giving the decomposition $\widehat{\mathcal{R}} = \widehat{\mathcal{R}}^{(el)} + \widehat{\mathcal{R}}^{(at)}$.

The RDop is yet a many-electron object with all the position and spin variables of N electrons in the p-region. To proceed with model calculations, it has been convenient to further introduce one-electron RDOps. Letting $\boldsymbol{X_p} = (x, X^{N-1})$ the one-electron density operator for a given atomic conformation $\boldsymbol{Y}_p$ follow from the contraction $\rho^{(1)}(x; x') = tr_{N-1}\left[\rho\left(x, X^{N-1}; x', X^{N-1}\right)\right]$. It has been expanded in a basis set of Kohn–Sham orbitals generated within density functional theory, and incorporated in a treatment which adds vibrational states of the lattice and photon field states, both of which have been treated as a medium or s-region, to generate dissipative rates affecting electronic dynamics. Combined with ab initio treatments of electronic structure and for the long-lasting dissipation form of the memory kernel, the equation of motion for the RDOp has been solved in several recent studies to model and calculate the optical and photoconductivity properties of semiconductor slabs.[43–45,33,46–48]

4. CONCLUSIONS

This chapter contains an overview of partitioning of functional spaces for the treatment of molecular phenomena, and of partitioning of statistical density operators for many-atom systems. Distortion, adiabatic, and fast motion states have been introduced to construct projection operators and equations of motion for projected wavefunctions, and to describe molecular effective potentials and long-lived states. The projectors provide guidelines on how to model physical properties, how to systematically improve models, and how to estimate the errors introduced by using finite basis sets. The treatment of partitioning for many-atom systems leads to contracted density operators and to RDOp and their equations of motion, adding some new results to what has been previously published. The equations display the form of fluctuation and dissipation rates which are essential in the description of the properties of materials and have been shown in forms suitable for computational work.

The formalisms in this chapter have involved only time-independent projection operators, but they can be extended to include time-dependent projectors, which would then add another rate term to the equations of motion.[25,29] Alternatively, partitioning can be done in a symmetric way for P and Q spaces, or for p- and s-regions, so that the leading term in projected wavefunctions or density operators are products of two time-dependent factors, each with its own equation of motion.[30] This is needed for example when photoexcitation occurs in a medium instead of a primary region of interest.[49]

Other aspects of great interest but not described here relate to the quantum theory of measurement, quantum decoherence, and time irreversibility. This has been treated in the context of a physical system decomposed into observed object and measuring instrument. The partitioning approach is a natural way to deal with the subject, and related treatments can be found in the literature.[20,42,50–52]

ACKNOWLEDGMENTS

Work partly supported by the Chemistry Division of the National Science Foundation of the USA, grants NSF 1011967 and NSF 1445825, and by the Dreyfus Foundation.

REFERENCES

1. Löwdin, P.-O. Studies in Perturbation Theory Part I. An Elementary Iteration-Variation Procedure for Solving the Schroedinger Equation by Partitioning Techniques. *J. Mol. Spectrosc.* **1963**, *10*, 12.
2. Löwdin, P.-O. Studies in Perturbation Theory Part IV. Solutions of Eigenvalue Problem by Projection Operator Formalism. *J. Math. Phys.* **1962**, *3*, 969.
3. Löwdin, P.-O. Studies in Perturbation Theory Part VI. Contraction of Secular Equations. *J. Mol. Spectrosc.* **1964**, *14*, 112.
4. Löwdin, P.-O. Studies in Perturbation Theory Part VII. Localized Perturbation. *J. Mol. Spectrosc.* **1964**, *14*, 119.
5. Feshbach, H. Unified Theory of Nuclear Reactions. *Ann. Phys.* **1958**, *5*, 357.
6. Goldberger, M. L.; Watson, K. M. *Collision Theory*; J. Wiley: New York, 1964.
7. Messiah, A. *Quantum Mechanics*, Vol. II. North-Holland: Amsterdam, 1962.
8. Micha, D. A. Compound State Resonances in Atom-Diatomic Molecule Collisions. *Phys. Rev.* **1967**, *162*, 88.

9. Micha, D. A.; Brandas, E. Variational Methods in the Wave-Operator Formalism. A Unified Treatment of Bound and Quasi-Bound Electronic and Molecular States. *J. Chem. Phys.* **1971**, *55*, 4792.
10. Micha, D. A. Long-Lived States in Atom-Molecule Collisions, Accounts. *Chem. Res.* **1973**, *6*, 138.
11. Micha, D. A. Effective Hamiltonian Methods for Molecular Collisions. *Adv. Quantum Chem.* **1974**, *8*, 231.
12. Redmon, M. J.; Micha, D. A. A Computational Method for Multi-Channel Scattering Calculations. Applications to Rotational Excitation and Long-Lived States of He-N_2. *Chem. Phys. Lett.* **1974**, *28*, 341.
13. Micha, D. A. Optical Models in Molecular Collision Theory. In *Modern Theoretical Chemistry*; Miller, W. H., Ed.; Dynamics of Molecular Collisions, Vol. IA; Plenum Publishing Co: New York, 1976; pp 81–129.
14. Kuruoglu, Z. C.; Micha, D. A. Calculation of Resonances in the H+H_2 Reaction Using the Faddeev-AGS Method. *Int. J. Quantum Chem. Symp.* **1989**, *23*, 103.
15. Micha, D. A. A Coupled-Channels Approach to Molecular Photodissociation Using Decay Boundary Conditions. *J. Phys. Chem.* **1991**, *95*, 8082.
16. Levine, R. D. Adiabatic Approximation for Nonreactive, Subexcitation Molecular Collisions. *J. Chem. Phys.* **1968**, *49*, 51.
17. Miller, W. H. Coupled Equations and the Minimum Principle for Collisions of an Atom and a Diatomic Molecule. *J. Chem. Phys.* **1969**, *50*, 2758.
18. George, T. F.; Ross, J. Quasistatistical Complexes in Chemical Reactions. *J. Chem. Phys.* **1972**, *56*, 5786.
19. Von Neumann, J. *Mathematical Foundation of Quantum Mechanics*; Princeton University Press: New Jersey, 1955.
20. Prigogine, I. *Non-Equilibrium Statistical Mechanics*; Wiley: New York, 1962.
21. Kubo, R.; Toda, M.; Hashitsume, N. *Statistical Physics II*, 2nd ed.; Springer-Verlag: Berlin, 1985.
22. Zwanzig, R. *Non-Equilibrium Statistical Mechanics*; Oxford University Press: Oxford (Engl), 2001.
23. Cohen-Tannoudji, C.; Dupont-Roc, J.; Grynberg, G. *Atom–Photon Interactions*; Wiley: New York, 1992.
24. Mukamel, S. *Principles of Non-Linear Optical Spectroscopy*; Oxford University Press: New York, 1995.
25. May, V.; Kuhn, O. *Charge and Energy Transfer Dynamics in Molecular Systems*, 2nd ed.; Wiley-VCH: Weinheim (Germany), 2004.
26. Nitzan, A. *Chemical Dynamics in Condensed Phases*; Oxford University Press: Oxford (Engl), 2006.
27. Micha, D. A.; Burghardt, I., Eds.; *Quantum Dynamics of Complex Molecular Systems*; Springer-Verlag: Berlin, 2007.
28. Burghardt, I.; May, V.; Micha, D. A.; Bittner, E. R., Eds.; *Energy Transfer Dynamics in Biomaterial Systems*; Springer-Verlag: Heidelberg, 2009.
29. Micha, D. A. Density Matrix Treatment of Electronic Rearrangement. *Adv. Quantum Chem.* **1999**, *35*, 317.
30. Micha, D. A. Density Matrix Theory and Computational Aspects of Quantum Dynamics in an Active Medium. *Int. J. Quantum Chem.* **2000**, *80*, 394.
31. Leathers, A.; Micha, D. A. Density Matrix Treatment of the Non-Markovian Dissipative Dynamics of Adsorbates on Metal Surfaces. *J. Phys. Chem. A* **2006**, *110*, 749.
32. Micha, D. A.; Leathers, A.; Thorndyke, B. Density Matrix Treatment of Electronically Excited Molecular Systems: Applications to Gaseous and Adsorbate Dynamics. In: *Quantum Dynamics of Complex Molecular Systems*; Micha, D. A., Burghardt, I., Eds.; Springer-Verlag: Heidelberg, 2007; pp 165–194.

33. Micha, D. A. Density Matrix Treatment of Non-Adiabatic Photoinduced Electron Transfer at a Semiconductor Surface. *J. Chem. Phys.* **2012**, *137*, 22A521.
34. Mott, N. F.; Massey, H. S. W. *Theory of Atomic Collisions*. Oxford University Press: London, 1965.
35. Micha, D. A. Optical Models in Molecular Collision Theory. In: *Dynamics of Molecular Collisions Part A*; Miller, W. H. Ed.; Plenum Press: New York, 1976. Chapter 3.
36. Micha, D. A. A Coupled-Channel Approach to Molecular Photodissociation Using Decay Boundary Conditions. *J. Phys. Chem.* **1991**, *95*, 8082.
37. Löwdin, P. O. Quantum Theory as a Trace Algebra. *Intern. J. Quantum Chem.* **1977**, *12*(Suppl. 1), 197.
38. Löwdin, P. O. On Operators, Superoperators, Hamiltonians, and Liouvillians. *Int. J. Quantum Chem.* **1982**, *16*, 485. Quantum Chem. Symp.
39. Löwdin, P. O. *Linear Algebra for Quantum Theory*; Wiley: New York, 1998.
40. Leathers, A. S.; Micha, D. A. Density Matrix for non-Markovian Dissipative Dynamics: A Numerical Method. *Chem. Phys. Lett.* **2005**, *415*, 46.
41. Redfield, A. G. The Theory of Relaxation Processes. *Adv. Magn. Reson.* **1965**, *1*, 1.
42. Lindblad, G. *Non-Equilibrium Entropy and Irreversibility*; Reidel: Dordrecht (Holland), 1983.
43. Kilin, D. S.; Micha, D. A. Surface Photovoltage at Nanostructures on Si Surfaces: Ab Initio Results. *J. Phys. Chem. C* **2009**, *113*, 3530.
44. Leathers, S.; Micha, D. A.; Kilin, D. S. Direct and Indirect Electron Transfer at a Semiconductor Surface With an Adsorbate: Theory and Application to $Ag_3Si(111)$:H. *J. Chem. Phys.* **2010**, *132*, 114702-1.
45. Kilin, D. S.; Micha, D. A. Relaxation of Photoexcited Electrons at a Nanostructured Si(111) Surface. *J. Phys. Chem. Lett.* **2010**, *1*, 1073.
46. Hembree, R. H.; Micha, D. A. Photoinduced Electron Transfer at a Si(111) Nanostructured Surface: Effect of Varying Light Wavelength, Temperature, and Structural Parameters. *J. Chem. Phys.* **2013**, *138*, 184708.
47. Vazhappilly, T.; Micha, D. A. Computational Modeling of the Dielectric Function of a Silicon Slab With Varying Thickness. *J. Phys. Chem. C* **2014**, *118*, 4429.
48. Vazhappilly, T.; Hembree, R. H.; Micha, D. A. Photoconductivities from Band States and a Dissipative Electron Dynamics: Si(111) Without and With Adsorbed Ag Clusters. *J. Chem. Phys.* **2016**, *144*, 024107-1.
49. Yi, Z.-G.; Micha, D. A.; Sund, J. Density Matrix Theory and Calculations of Nonlinear Yields of CO Photodesorbed from Cu(001) by Light Pulses. *J. Chem. Phys.* **1999**, *110*, 10562.
50. Zurek, W. H. Decoherence and the Transition from Quantum to Classical. *Phys. Today* **1991**, *44*, 36. Revisited in *Los Alamos Science* **2002,** *27*, 2.
51. Obcemea, Ch.; Brandas, E. Analysis of Prigogine's Theory of Subdynamics. *Annu. Rev. Plant Physiol. Plant Mol. Biol.* **1983**, *151*, 383.
52. Brandas, E. Partitioning Technique for Open Systems. *Mol. Phys.* **2010**, *108*, 3259.

Vibrational Quantum Squeezing Induced by Inelastic Collisions[☆]

Manuel Berrondo[*,1], **Jose Récamier**[†]
[*]Brigham Young University, Provo, UT, United States
[†]Instituto de Ciencias Físicas, Universidad Nacional Autónoma de México, Cuernavaca, Morelos, Mexico
[1]Corresponding author: e-mail address: berrondo@byu.edu

Contents

Abstract

The energy transfer between vibrational and translational degrees of freedom during an inelastic collision between two molecules induces quantum squeezing in the vibrational molecular coordinate under very diverse circumstances. In this chapter, we present the relevant calculation for the very simple case of an atom–diatomic collinear transition, the Landau–Teller model. Both the uncertainty of the vibrational coordinate and the Husimi function show clear evidence of quantum squeezing. Our model treats the relative translation of the colliding species as a classical variable. The vibrational motion of the diatomic molecule is treated quantum mechanically in terms of the evolution operator and coherent states. The corresponding classical and quantum equations of motion are coupled. We first consider the squeezing of a coherent vibrational state where we include the dynamic evolution of the Husimi function. The second case corresponds to an initial thermal distribution of vibrational states where we plot the position uncertainty for the squeezed vibrational state.

[☆]Dedicated to the bright memory of Prof. Per Olov Löwdin.

Advances in Quantum Chemistry, Volume 74
ISSN 0065-3276
http://dx.doi.org/10.1016/bs.aiq.2016.05.002

1. INTRODUCTION

The Landau–Teller model[1] is a very simple, though nontrivial, model of vibrational–translational transfer in molecular collisions. It has traditionally been used to calculate state-to-state transition probabilities in various approximations.[2–5] A single collinear dimension is used for the relative motion of an atom and a diatomic molecule. For our purposes the translational degree of freedom can be treated classically to a good approximation. The relative molecular distance is thus taken as a classical translational coordinate, while vibrational motion of the diatomic molecule is quantized. Hence we adopt a quasi-classical formalism to describe the response of a diatomic molecule, treated as a quantum harmonic oscillator, to an external atomic interaction with an implicit time dependence. We then consider explicitly the coupling between Hamilton equations of motion for translation and the quantum equations of motion involving the diatomic molecular vibrations.[5,6] Instead of solving the translational equation of motion independent of the vibrational motion[3,7–9] in our case we solve the coupled set of equations simultaneously. The atom–diatomic repulsive interaction modifies the harmonic character of the vibrational motion resulting in an effective time-dependent frequency and an additional perturbation responsible for the squeezing of the coherent vibrational states.

In this chapter we will consider coherent states[10] instead of the number states appearing in the usual state-to-state transition probabilities. In the case of quadrature squeezing we can take $\hat{y}$ and $\hat{p}_y$ as the quadrature operators. Assuming $\hbar = 1$ quantum squeezing will appear whenever $(\Delta\hat{O})^2 < 1/2$ with either $\hat{O} = \hat{y}$ or $\hat{O} = \hat{p}_y$. For a coherent state $|\alpha\rangle = \hat{D}(\alpha)|0\rangle$ with $\hat{D}(\alpha) = \exp(\alpha\hat{a}^\dagger - \alpha^*\hat{a})$ the displacement operator, both quadratures have the minimal dispersion allowed by Heisenberg's principle, that is $(\Delta\hat{y})^2 = (\Delta\hat{p}_y)^2 = 1/2$. The coherent state represents the quantum state closest to classical motion, given its minimum uncertainty product.

For a squeezed state the fluctuations in one quadrature are smaller than those of a coherent state: they are squeezed. Of course, the fluctuations in the other quadrature must be enhanced in order that the uncertainty relation is not violated.[11,12] Much interest has been devoted to the production of squeezed quantum states, ie, states where the noise in one observable is less than that of the vacuum state. The majority of these studies have focused on the production of squeezed radiation states for the usual coherent states

$|\alpha\rangle$[13–15] and more recently for *deformed* Morse-like nonlinear coherent states $|\alpha, f\rangle$.[16] A lesser amount of work has been devoted to the squeezing of material systems.[17]

Many molecular properties stem directly from the vibrational wave function; thus the control of the position and width of the wave packet has played a fundamental role in quantum control of molecules with laser pulses. A squeezed wave packet of vibrational states can be obtained by means of the electronic excitation of a molecule by ultrashort coherent light pulses whose length must be small in comparison with the vibrational period.[18] With the development of the technology for the production of short laser pulses, it has been possible to use wave packets for investigating intraatomic and intramolecular processes in real time.[17,19] In Ref. 20 the authors present a method to design laser fields for generation of spatially squeezed molecular wave packets. Their approach is based on optimal control theory and possible applications to laser femtosecond chemistry are discussed.

In Ref. 21 the authors consider the squeezing of minimal width vibrational wave packets of diatomic molecules by means of laser schemes that couple the ground and excited electronic configurations of the molecule. This chapter presents an alternative way of producing vibrational squeezed states: during the molecular collision the interaction of the incoming atom with the diatomic molecule acts as the driving force (instead of the interaction with the laser field). As a result the squeezing involves a fixed (usually the ground) electronic state of the diatomic molecule.

Mathematically, a squeezed state is generated by the action of a squeeze operator $\hat{S}(\zeta)$ defined as

$$\hat{S}(\zeta) = \exp\left[\frac{1}{2}(\zeta^{*}\hat{a}^{2} - \zeta\hat{a}^{\dagger 2})\right],$$

where the number ζ is the complex squeeze parameter. The squeeze operator $\hat{S}(\zeta)$ can be understood as a two-photon generalization of the displacement operator $\hat{D}(\alpha)$.

The Husimi Q distribution is the Fourier transform of the anti-normal-order characteristic function and can be simply written as

$$Q_{\alpha}(z,t) = \frac{1}{\pi}\langle z|\varrho_{\alpha}(t)|z\rangle,$$

where ϱ_{α} is the density operator and $|z\rangle$ is a coherent state. The Q function is the average of the projector onto the vacuum state in the field displaced in phase space by $-z$. The Q distribution is positive, bounded by $1/\pi$, and normalized. For a coherent state, the Husimi distribution and the Wigner quasi-distribution function are Gaussian functions. However, the Wigner quasi-distribution function is more suitable for the analysis of nonclassical states since it attains negative values at those points in phase space where nonclassicality is present: indeed, the negativity of the Wigner function is a signature of nonclassicality.[11,12] In order to analyze the quantum squeezing it is enough to consider the Husimi distribution function which is easier to evaluate.

In order to show the vibrational squeezing produced by the collision, we follow the time evolution of a coherent initial state $|\alpha\rangle$ according to the Landau–Teller model representing the atom–diatomic dynamics. After the collision takes place, the vibrational state changes character and does not remain coherent. For the harmonic case considered in this chapter, the Heisenberg uncertainty product retains its minimum value, but manifested as a squeezed state. In the explicit examples we provide, both the direct plot of the uncertainty $\Delta\hat{y}$ and the Husimi function $Q_\alpha(z, t)$ show clearly and explicitly the appearance of squeezing.

Coherent initial states are the right choice for the case of individual atom–molecule collisions. If the system we are considering involves multiple collisions, as in a gas, it is more appropriate to define the initial condition through a density matrix to include incoherent superposition of states. In this case the initial distribution does not have a minimal dispersion product as for coherent states. The results we obtain for a thermal initial distribution show, however, that squeezing of the corresponding quadrature with respect to its initial value does occur. This is amply shown in the plot of dispersion vs time.

The chapter is organized as follows. In Section 2 we present the algebraic form of the Landau–Teller model for the collision of an atom with a diatomic molecule, the evolution operator for this model, as well as the resulting coupled differential equations for the corresponding time-dependent group parameters. Section 3 presents the results for four different sets of initial states and physical parameters. The results show explicitly the postcollisional appearance of quantum squeezing in the vibrational modes starting from a purely coherent state. Section 4 shows results for squeezed vibrational states taking a thermal distribution as the initial density matrix. Finally, Section 5 contains a short discussion of the method employed and the results obtained.

2. ALGEBRAIC FORM OF THE LANDAU–TELLER MODEL

We consider a collinear collision of an atom A with a diatomic molecule B–C.[1,2,5] The coordinates for the positions of atoms A, B, and C is illustrated in Fig. 1A. The Hamiltonian for this system has three degrees of freedom in the laboratory frame and is given by

$$H_{lab} = \frac{p_A^2}{2m_A} + \frac{p_B^2}{2m_B} + \frac{p_C^2}{2m_C} + V_{BC}(x_C - x_B) + V_{AB}(x_B - x_A), \quad (1)$$

where *ps* are the momenta and *ms* the masses for the three atoms. We assume the vibrational motion of B–C to be harmonic, the interaction between A and B to be purely repulsive, and the interaction between A and C to be negligible. Separating the center of mass motion and using reduced mass coordinates, the problem boils down to two dimensions as seen in Fig. 1B.

The resulting Jacobi coordinates[1,2] are y', the interatomic B–C distance, and x', the distance between the atom A and the center of mass of B and C, thus eliminating the global center of mass. A further transformation to dimensionless variables x and y allows us to construct a potential energy surface for the dynamics of the collision. The potential appropriate for the harmonic motion of the diatomic system is given in terms of the natural oscillation frequency ω_0 and the reduced mass μ_{BC} as

$$V_{BC} = \frac{1}{2}\hbar\omega_0 y^2 \quad (2)$$

and

$$y = \sqrt{\frac{\omega_0 \mu_{BC}}{\hbar}}(y' - y_0). \quad (3)$$

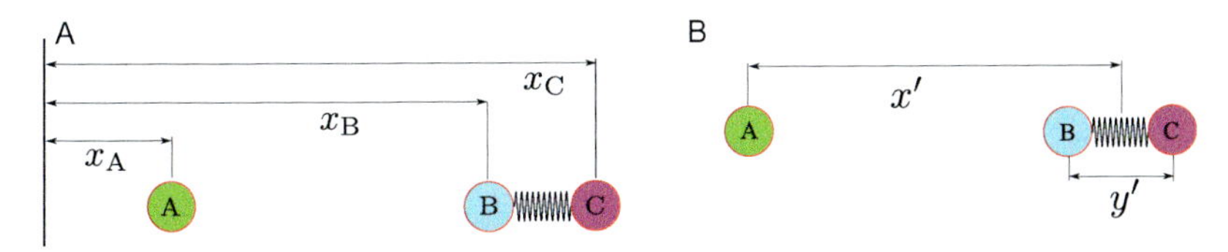

Fig. 1 The Landau–Teller model in laboratory coordinates (A) and in Jacobi relative coordinates (B).

For the collisional part of the potential, we choose a repulsive exponential of dimensionless strength V_0 and inverse range γ:

$$V_{AB} = \hbar\omega_0 V_0 e^{-\gamma(x-y)}. \tag{4}$$

These two functions form a potential energy surface for the collinear collision. The system takes on a new interpretation: it is now reduced to a particle of dimensionless reduced mass

$$m = \frac{m_A m_C}{m_B(m_A + m_B + m_C)} \tag{5}$$

at position x and a particle of unit mass at y.[1,2]

We now choose x to be a classical translation coordinate, while the oscillator coordinate $\hat{y}$ is quantized.[3–5,7,22] The full Hamiltonian is divided out by $\hbar\omega_0$ resulting in the final dimensionless Hamiltonian:

$$H = \frac{1}{2m}p_x^2 + \frac{1}{2}\hat{p}_y^2 + \frac{1}{2}\hat{y}^2 + V_0 e^{-\gamma(x-\hat{y})}. \tag{6}$$

The translation is described by the classical equation of motion

$$m\frac{d^2x}{dt^2} = \gamma V_0 e^{-\gamma x} e^{\gamma\langle\hat{y}\rangle}, \tag{7}$$

where we have taken the expectation value of the vibrational coordinate. It couples to the quantum equation through this time-dependent expectation value $\langle\hat{y}\rangle$.

Regarding the quantum motion, we can expand the exponential in Eq. (6) to second order in $\hat{y}$, thus introducing a new quadratic term. The linear term in $\hat{y}$ acts as an effective driving term. We hence take advantage of the quasi-classical variable separation and focus on the quantum Hamiltonian

$$H_q = \frac{1}{2}\hat{p}_y^2 + \frac{1}{2}(1 + \gamma^2 V(x))\hat{y}^2 + \gamma V(x)\hat{y} + V(x), \tag{8}$$

where $V(x) = V_0 e^{-\gamma x}$.

Defining the usual ladder operators

$$\begin{aligned} a &= \frac{1}{\sqrt{2}}(\hat{y} + i\hat{p}_y) \\ a^\dagger &= \frac{1}{\sqrt{2}}(\hat{y} - i\hat{p}_y) \end{aligned} \tag{9}$$

the final form of the quantum Hamiltonian is

$$H_q = \left(N + \frac{1}{2}\right)\omega(x) + \frac{1}{4}\gamma^2 V(x)(a^2 + a^{\dagger 2}) + \frac{1}{\sqrt{2}}\gamma V(x)(a + a^{\dagger}) + V(x), \tag{10}$$

with N the number operator and $\omega(x) = 1 + \gamma^2 V_0 e^{-\gamma x}/2$. In this approximation the variable x is the classical trajectory, and it is a time-dependent function whose explicit form is obtained by solving Newton's equations. The Hamiltonian H_q is then a time-dependent Hamiltonian corresponding to a parametric oscillator with frequency $\omega(x(t))$ and linear and quadratic time-dependent forcing terms. The coefficient for the quadratic terms is of second order in the parameter γ in contrast with the coefficient for the linear terms which is linear in that parameter. The quadratic terms generate squeezing.

In order to find the evolution operator[23] for this time-dependent Hamiltonian, we introduce the Lie algebra generated by $\{a, a^2, a^{\dagger}, a^{\dagger 2}, N, I\}$ including the number operator N and the identity I. The primary commutation relations are the usual $\left[a, a^{\dagger}\right] = I$, $[a, N] = a$, and its Hermitian conjugate. The Hamiltonian H_q in Eq. (10) is an element of the Lie algebra by construction.

The Lie algebra can be exponentiated to form a Lie group[24] with the time evolution operator $U(t)$ being an element of this group. The resulting Lie group parameters are functions of time through the $x(t)$ dependence of the classical translation coordinate.[25] The quantum dynamics can then be rewritten matching the coefficients of the Lie algebra basis in the Schrödinger equation:

$$i\left(\frac{d}{dt}U(t)\right)U(t)^{-1} = H_q. \tag{11}$$

In this equation the dimensionless time requires a scaling of $t = t'\omega_0$ to stay consistent with the scaled energy units. The time dependence for $U(t)$ is transferred completely into six Lie group parameters $\{g_i(t)\}$ corresponding to the six basis elements of the Lie algebra. Once we have determined the evolution operator, we can get the corresponding Husimi function[11,12] after solving the six ordinary differential equations. We express the evolution operator as a product of exponentials, each corresponding to one of the generators of the Lie algebra, chosen in normal order[26]:

$$U(t) = e^{g_1(t)a^\dagger} e^{g_2(t)a^{\dagger 2}} e^{g_3(t)a} e^{g_4(t)a^2} e^{g_5(t)N} e^{g_6(t)I}. \tag{12}$$

This Wei–Norman[27,28] form for the time evolution operator simplifies its application to coherent states. The left-hand side of Eq. (11) is evaluated applying the Baker–Hausdorff formula.[10,25] The resulting coupled equations of motion for each Lie parameter are:

$$i\dot{g}_1 = \gamma^2 V g_1 g_2 + \sqrt{2}\gamma V g_2 + \omega g_1 + \frac{\gamma V}{\sqrt{2}}, \tag{13}$$

$$i\dot{g}_2 = \gamma^2 V g_2^2 + 2\omega g_2 + \frac{\gamma^2 V}{4}, \tag{14}$$

$$i\dot{g}_3 = \frac{1}{2}\gamma^2 V g_1 - \gamma^2 V g_2 g_3 - \omega g_3 + \frac{\gamma V}{\sqrt{2}}, \tag{15}$$

$$i\dot{g}_4 = -2\gamma^2 V g_2 g_4 - 2\omega g_4 + \frac{\gamma^2 V}{4}, \tag{16}$$

$$i\dot{g}_5 = \omega + \gamma^2 V g_2, \tag{17}$$

$$i\dot{g}_6 = V + \frac{1}{2} + \frac{1}{4}\gamma^2 V g_1^2 + \frac{1}{2}\gamma^2 V g_2 + \frac{\gamma V}{\sqrt{2}} g_1 + \frac{\gamma^2 V}{4}, \tag{18}$$

where

$$\omega(x(t)) = \frac{1}{2}\left(1 + \gamma^2 V(x(t))\right) \tag{19}$$

and

$$V(x(t)) = V_0 e^{-\gamma x(t)}. \tag{20}$$

As a result, the translational equation of motion Eq. (7) becomes quasi-classically coupled to Eqs. (13)–(18) resulting in a system of eight first-order differential equations which may be solved numerically. The initial condition $U(t = t_0) = I$, the identity, implies a vanishing initial value for the six functions $g_i(t = t_0) = 0$, while the initial position and velocity for the classical variable are chosen as to have an asymptotically incoming atom A.

3. SQUEEZING COHERENT STATES

In order to detect the vibrational squeezing after the collision takes place, we calculate the uncertainty in the vibrational coordinate $\Delta\hat{y}(t)$ defined through

$$(\Delta\hat{y})^2 = \langle\hat{y}^2\rangle - \langle\hat{y}\rangle^2 \tag{21}$$

as the collision progresses in time. The expectation value $\langle\hat{y}(t)\rangle$ represents an average oscillator trajectory in the Ehrenfest sense. The expectation value is taken with respect to an evolving initial coherent state of the oscillator. In other words, the initial quantum state is assumed to be a coherent state labeled by its complex eigenvalue α,

$$|\alpha(t)\rangle = U(t)|\alpha\rangle \tag{22}$$

and

$$\langle\hat{y}(t)\rangle = \langle\alpha(t)|\hat{y}|\alpha(t)\rangle. \tag{23}$$

The relative translational motion has been assumed to be classical, so its trajectory is well defined.

In Fig. 2A we show a parametric plot of the vibrational motion in the x–y plane assuming an initial ground state $\alpha = 0$. The triatomic mass ratio is 1:1:1, and the initial position of the atom A with respect to the center of mass of B–C is $x_0 = 60$, with velocity $v_0 = -12$ at initial time $t_0 = -5$. For the repulsive A–B potential we have chosen a unit strength $V_0 = 1$ and an inverse range $\gamma = 0.4$. Fig. 2B shows a plot of the time evolution of the uncertainty of the vibrational coordinate, Eq. (21).

From Fig. 2A we see that the collision takes place essentially when $x < 0$. The outgoing vibrational motion does no longer correspond to the ground state. In order to decide whether it remains coherent, we look at Fig. 2B where we clearly appreciate the squeezing that takes place after the collision.

A better visualization of the actual squeezing is achieved by following the time evolution of the Husimi function $Q(z, t)$. Q is a distribution function dependent on the complex variable z. It is defined as[11,12]

$$Q_\alpha(z,t) = \frac{1}{\pi}\langle z|\varrho_\alpha(t)|z\rangle, \tag{24}$$

where $\varrho_\alpha(t) = |\alpha(t)\rangle\langle\alpha(t)|$ is the time-evolved density matrix corresponding to the initial coherent state $|\alpha\rangle$. For our present case the Q function can be calculated explicitly as $Q = |A|^2/\pi$, where

$$A_\alpha(z,t) = \exp\left[-\frac{1}{2}\left(|\alpha|^2 + |z|^2\right) + B^2 g_4(t) + z^*(B + g_1(t)) + Bg_3(t) + z^{*2} g_2(t) + g_6(t)\right]$$

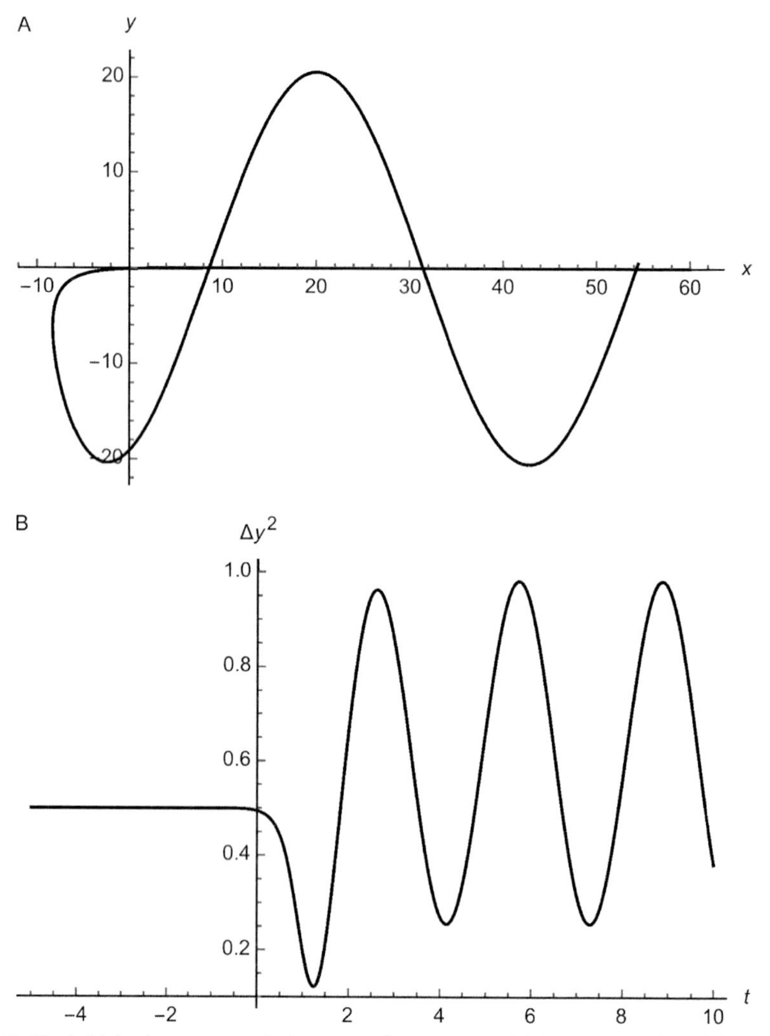

Fig. 2 The initial coherent state is the ground state $\alpha = 0$. (A) Parametric plot of the vibrational motion with $m = 1/3$, $x_0 = 60$, $v_0 = -12$, and $V_0 = 1$, $\gamma = 0.4$. (B) Plot of $(\Delta\hat{y})^2$ vs t. Before the collision the value of 1/2 corresponds to a coherent state. After collision with the atom the vibrational coordinates wanders below this value, clearly showing the squeezing.

and

$$B = \alpha e^{g_5(t)}.$$

In Fig. 3 we present five snapshots of the animation of the Husimi function as a function of the complex phase space z for the same set of parameters chosen in Fig. 2. Each contour plot represents the function at a different time.

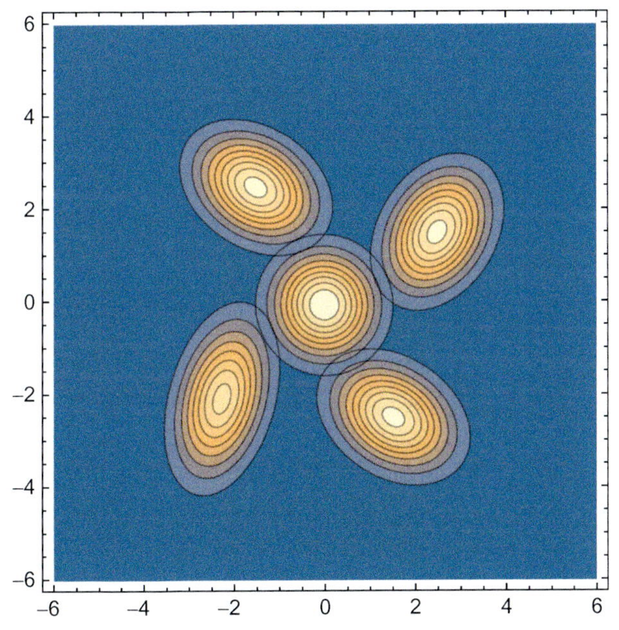

Fig. 3 Superposition of snapshots of the time evolution of the Husimi function corresponding to times $t = 0$ (*center*), $t = \pi/2$ (*lower left*), $t = \pi$ (*upper left*), $t = 3\pi/2$ (*upper right*), and $t = 2\pi$ (*lower right*). The collisional parameters are the same as in Fig. 2.

The initial state of the system is the ground state, corresponding to a coherent state with $\alpha = 0$. Its Husimi function is a Gaussian centered at the origin. As the time evolves, the state is displaced (due to the linear terms in the interaction Hamiltonian) and squeezed. From Fig. 2 we see that the collision takes place near $t = 0$ and the system attains a final state shortly afterwards as indicated by the plot of the dispersion as a function of time corresponding to oscillations with constant amplitude. In Fig. 3 we see that the amount of squeezing at $t = \pi/2$ remains practically constant; that is, the ratio between the axes of the ellipses is almost the same at the other values of t shown in the figure. The usual rotation in phase space of the distribution is also clearly seen.

In Fig. 4 we show a superposition of five snapshots of the animation of the Husimi function as a function of the complex phase space z for the same set of parameters chosen in Fig. 2. In this case the initial state of the system is a coherent state with $\alpha = 2$ corresponding to a state with average occupation number $\langle N \rangle = 4$. At the initial time the Husimi function is a displaced Gaussian. As the time evolves, the distribution is further displaced and squeezed presenting a similar behavior as that found in Fig. 3.

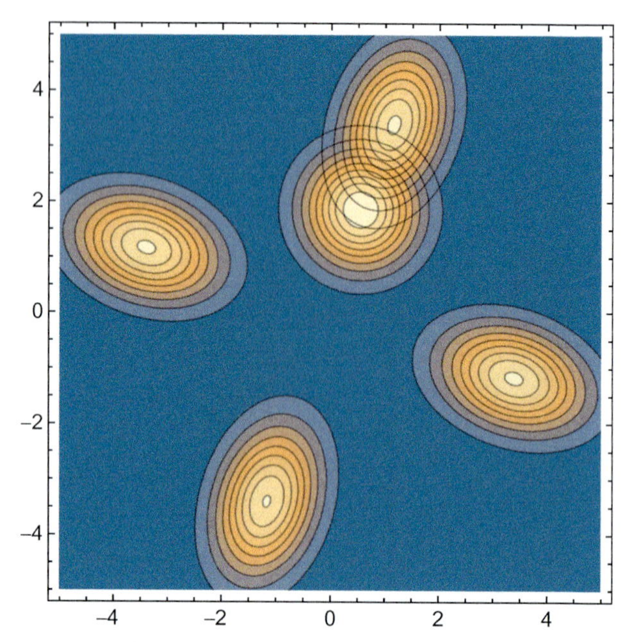

Fig. 4 Superposition of snapshots of the time evolution of the Husimi function corresponding to times $t = 0$ (*circle*), $\pi/2$ (*bottom*), π (*upper left*), $3\pi/2$ (*upper right*), and 2π (*right*) for an initial coherent state with $\alpha = 2$. The collisional parameters are the same as in Fig. 2.

In Fig. 5 we show the Husimi function for $\gamma = 0.75$ and the rest of the parameters the same as those used in Fig. 2. Here the distribution at time $t = 0$ is a Gaussian centered at the origin. We now let the time evolve and consider an instant of time where the collision is still taking place, $t = \pi/5$. We see that the distribution has been displaced (the distribution shown in the lower part of the plot) and the amount of squeezing is much larger than that shown at $t = \pi$ (the distribution at the upper part of the plot) where the collision has already taken place and the amount of squeezing remains fixed.

In Fig. 6A we show a parametric plot of the vibrational motion in the $x - y$ plane assuming an initial state $\alpha = \sqrt{2}(1 + i)$. The triatomic mass ratio is 1:1:1, the initial position of the atom A with respect to the center of mass of B–C is $x_0 = 60$, with velocity $v_0 = -12$ at initial time $t_0 = -5$. For the repulsive A–B potential we have chosen a unit strength $V_0 = 1$ and an inverse range $\gamma = 0.4$. Fig. 6B shows a plot of the time evolution of the uncertainty of the vibrational coordinate, Eq. (21).

From Fig. 6A we see that the collision takes place essentially when $x < 0$. The outgoing vibrational motion does correspond to a state with a larger

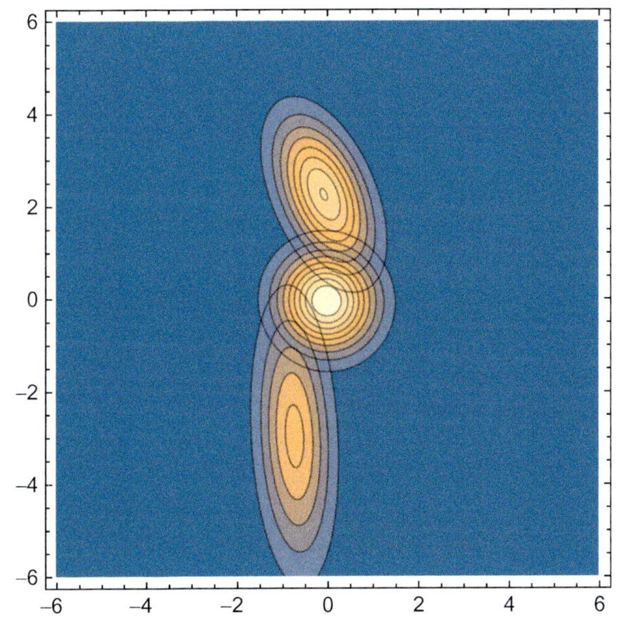

Fig. 5 Superposition of snapshots of the time evolution of the Husimi function corresponding to times $t = 0$ (*center*), $\pi/5$ (*down*), and π (*up*), for an initial coherent state with $\alpha = 0$. The parameters are: $m = 1/3$, $x_0 = 10$, $v_0 = -12$, $V_0 = 1$, $\gamma = 0.75$.

oscillation amplitude. In order to decide whether it remains coherent, we look at Fig. 6B where we clearly appreciate the presence of squeezing after the collision. Notice also that the dispersion attains larger values than those shown in Fig. 2B. where we dealt with the ground state as an initial coherent state.

Finally in Fig. 7 we show two 3D plots of the Husimi function at times $t = 0$ (A) and $t = \pi/2$ (B). At the initial time the distribution is a Gaussian function centered at the point $x = -\sqrt{2}$, $y = \sqrt{2}$. At $t = \pi/2$ the distribution has been displaced so that the center is now localized roughly at $x = -\sqrt{2}, y = -\sqrt{2}$ and it is evident that it has been squeezed.

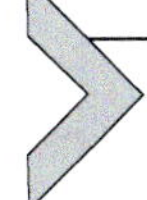

4. SQUEEZING FROM AN INITIAL THERMAL DISTRIBUTION

In the case of an initial thermal distribution at temperature T the uncertainty in the vibrational coordinate $\Delta\hat{y}(t)$ is still defined as in Eq. (21) with

$$\langle \hat{A}(t) \rangle = \mathrm{Tr}(\varrho(t)\hat{A}) \tag{25}$$

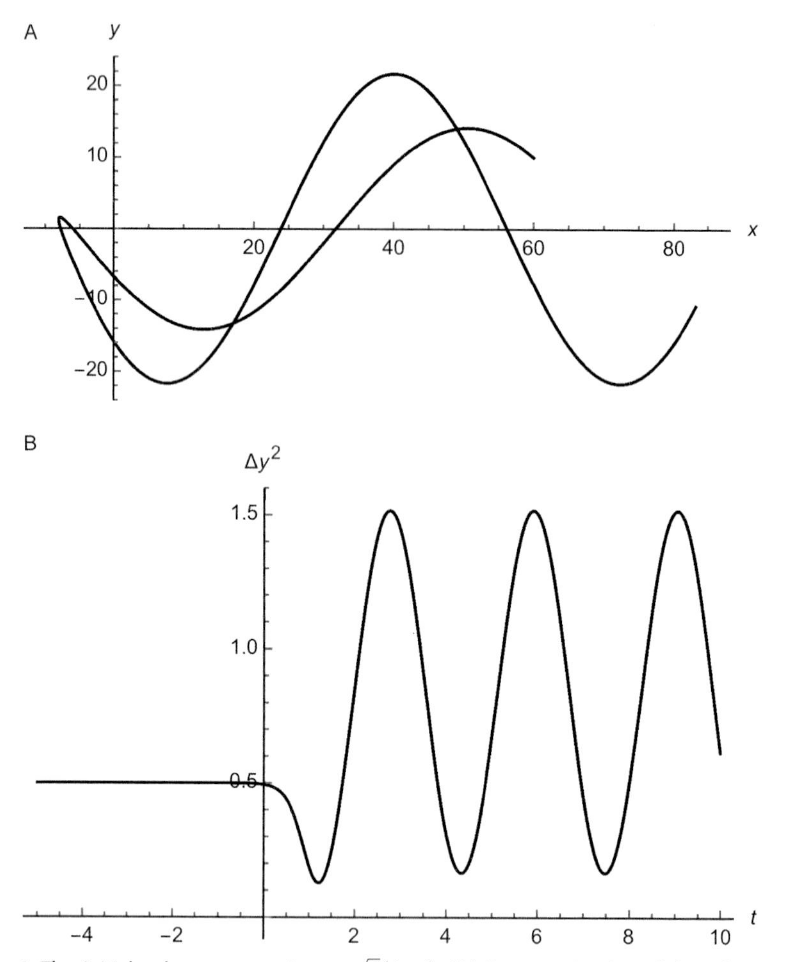

Fig. 6 The initial coherent state is $\alpha = \sqrt{2}(1+i)$. (A) Parametric plot of the vibrational motion with $m = 1/3$, $x_0 = 60$, $v_0 = -12$, and $V_0 = 1$, $\gamma = 0.4$. (B) Plot of $(\Delta\hat{y})^2$ vs t. Before the collision the value of 1/2 corresponds to a coherent state. After collision with the atom the vibrational coordinates wanders below this value, clearly showing the squeezing.

as a function of time. The expectation value $\langle\hat{y}(t)\rangle$ represents an average oscillator trajectory with respect to the thermal distribution ϱ. The initial normalized distribution at temperature T is

$$\varrho(0) = (1 - e^{-\beta})e^{-\beta H_0} \tag{26}$$

with the usual notation for inverse temperature, $\beta = 1/kT$. Its evolution in time is defined by the unitary operator $U(t)$ as

$$\varrho(t) = U(t)\varrho(0)U^{\dagger}(t). \tag{27}$$

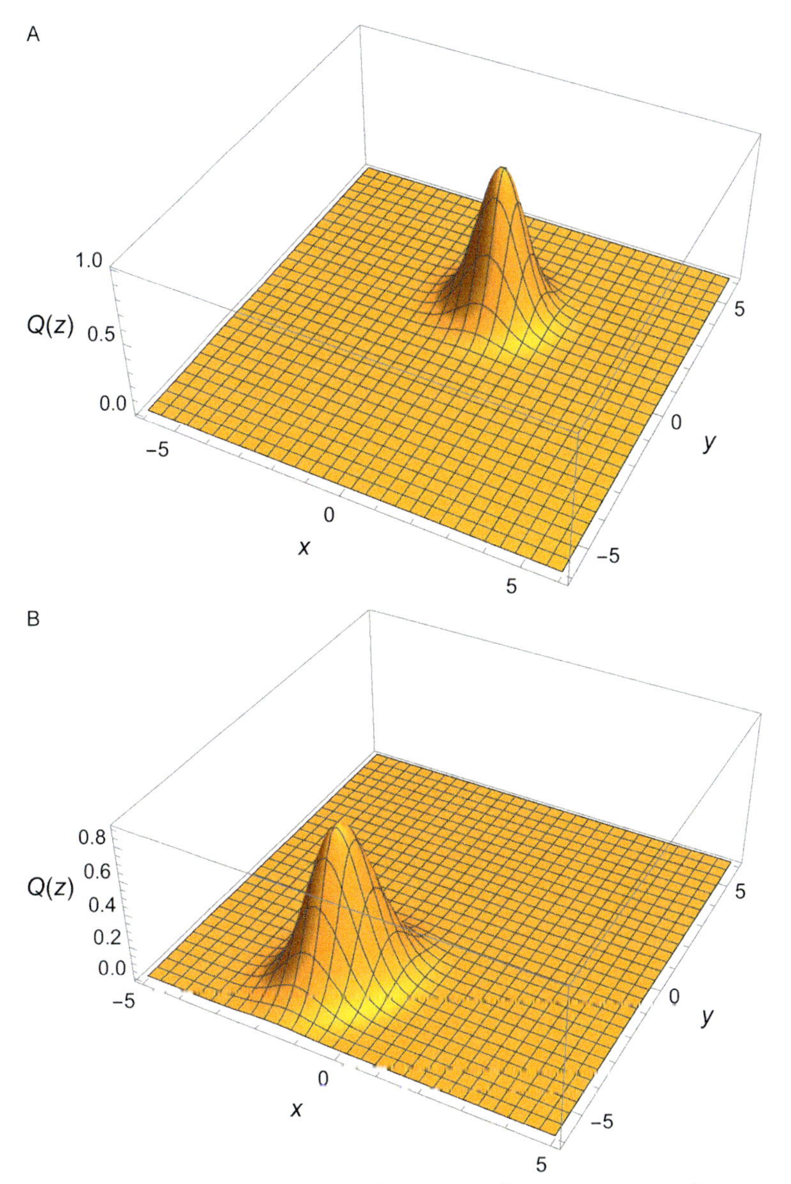

Fig. 7 Snapshots of the time evolution of the Husimi function corresponding to times $t = 0$ (A) and $t = \pi/2$ (B), for an initial coherent state with $\alpha = \sqrt{2}(1 + i)$. The parameters are: $m = 1/3$, $x_0 = 60$, $v_0 = -12$, $V_0 = 1$, $\gamma = 0.4$.

In the Heisenberg picture we can rewrite Eq. (25) as

$$\langle \hat{A}(t)\rangle = (1-e^{-\beta})\mathrm{Tr}(e^{-\beta H_0}U^\dagger(t)\hat{A}U(t)) \tag{28}$$

and evaluate the trace with respect to the vibrational number states corresponding to a unit frequency $\omega_0 = 1$.

For the particular case of the dispersion $(\Delta\hat{y})^2$ we get:

$$\langle(\Delta\hat{y})^2\rangle = |G_1 + G_2^*|^2\left(\frac{1}{e^\beta - 1} + \frac{1}{2}\right), \tag{29}$$

where the coefficients G_1, G_2 are given by

$$G_1 = e^{g_5}(1 - 4g_2g_4), \quad G_2 = 2g_2e^{-g_5}.$$

Fig. 8 shows a plot of the time evolution of the uncertainty of the vibrational coordinate given by Eq (29). At the initial time, the functions $g_i(t)$ appearing in the time evolution operator are such that $g_i(t_0) = 0$ so that $U(t_0) = 1$. The uncertainty in the vibrational coordinate at the initial time is then $\langle(\Delta\hat{y})^2\rangle = \left(\frac{1}{e^\beta - 1} + \frac{1}{2}\right)$, dependent only on the value of the temperature. For $kT \ll 1$ it approaches the coherent state value, that is,

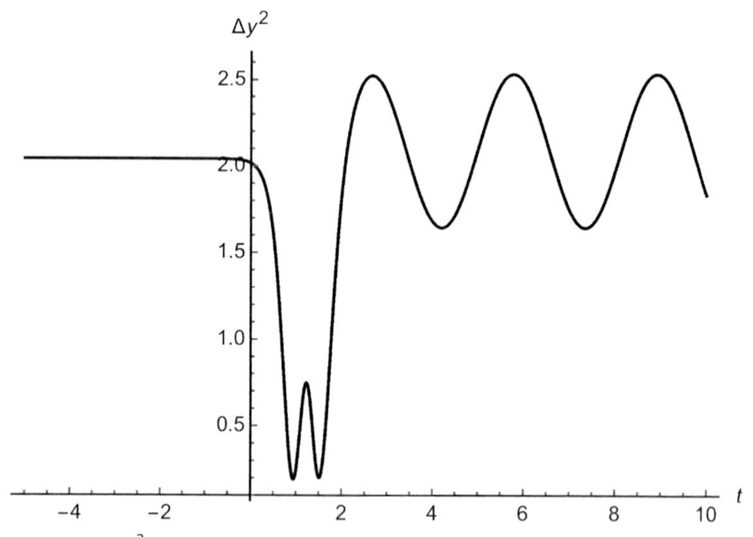

Fig. 8 Plot of $(\Delta\hat{y})^2$ vs t. The initial distribution corresponds to a temperature of $kT = 2$. After collision with the atom the uncertainty of the vibrational coordinate attains values smaller than its initial value clearly showing the presence of squeezing.

$\langle(\Delta\hat{y})^2\rangle \simeq 1/2$; for $kT \gg 1$ it is close to $kT + 1/2$ so that it grows in proportion with the temperature. The parameters used for Fig. 8 are: $kT = 2$, the triatomic mass ratio is 1:1:1, the initial position of the atom A with respect to the center of mass of B–C is $x_0 = 60$, with velocity $v_0 = -12$ at initial time $t_0 = -5$, and inverse range $\gamma = 0.4$. We immediately notice that the initial value for $\langle(\Delta\hat{y})^2\rangle$ is close to 2. This corresponds to the fact that we are no longer dealing with an initial coherent state as was the case in Section 3. Here, the initial value of the uncertainty in the vibrational coordinate is a function of the temperature only. However, as the collision takes place, the functions $g_i(t)$ evolve and we can observe squeezing below the initial dispersion. Just after the collision takes place $\langle(\Delta\hat{y})^2\rangle < 1/2$.

5. DISCUSSION

In this work we have applied Lie algebraic methods to construct the time evolution operator for a system composed of an atom and a diatomic molecule modeled by a harmonic oscillator, in a head-on collision. This is the simplest nontrivial system where the effects of vibrational quantum squeezing can be observed. We make use of a semiclassical approximation solving the classical equations of motion for the atom–molecule center of mass coordinate obtaining a well defined trajectory. This amounts to a time-dependent effective frequency for the oscillator, and as a consequence, we have to deal with a time-dependent quantum Hamiltonian. This one is composed of a parametric oscillator with linear and bilinear forcing terms, its structure being similar to that of a dynamical Casimir effect plus linear forcing terms.[29] The time evolution operator can be written exactly as a product of exponentials and the evolution of the vibrational coordinate and its dispersion is monitored as the collision takes place. In order to be specific we first consider the case of an initial ground state for the molecule, ie, a coherent state with amplitude $\alpha = 0$. The dispersion before the collision is constant corresponding to a coherent state, taking the smallest value consistent with Heisenberg's uncertainty principle. After the collision the molecule is excited and the dispersion oscillates taking values smaller than the corresponding one to the initial coherent state. In other words, the vibrational motion gets squeezed. Squeezed states are mainly used in order to increase the signal to noise ratio. As is evident from the Husimi plots presented in Section 3, the squeezing is accompanied by a rotation of the squeezed states in phase space.

In order to profit from the squeezing in the vibrational position variable, we have to demodulate this rotational motion.

The results presented in Section 3 refer to the squeezing of individual coherent vibrational states. In Section 4 we have also considered the case of a thermal distribution at temperature T. Here, the uncertainty of the vibrational coordinate is a function of temperature and evolves in time due to the temporal evolution of the functions $g_i(t)$ appearing in the time evolution operator. At the initial time, the uncertainty in the vibrational coordinate is a function of the temperature only. In the limit $T \to 0$ it takes the value corresponding to a coherent state, namely the ground state. But for the more general case of finite temperature, it will not correspond to a minimum uncertainty state. Following the evolution of the system as the collision takes place, we have shown that for the case of a thermal distribution there is indeed squeezing present. For the example illustrated in Section 4, squeezing similar to that of a coherent state $\langle(\Delta\hat{y})^2\rangle \simeq 1/2$ is evident in a small region of time just after the collision has taken place.

Inclusion of anharmonicities in the vibrational motion can be important in the case of higher energy initial coherent states or higher temperatures. In this event, a Morse oscillator would be more appropriate than the simple harmonic motion assumed in this chapter. Beyond squeezing, additional quantum effects would appear in this anharmonic case.[30]

ACKNOWLEDGMENTS

We acknowledge the partial support from DGAPA-UNAM project IN108413.

REFERENCES

1. Child, M. S. *Molecular Collision Theory*; Dover: New York, 1996.
2. Secrest, D.; Johnson, B. R. Exact Quantum Mechanical Calculation of a Collinear Collision of a Particle with a Harmonic Oscillator. *J. Chem. Phys.* **1966**, *45*, 4556.
3. Rapp, D.; Kassal, T. Exact Quantum-Mechanical Calculation of Vibrational Energy Transfer to an Oscillator by Collision with a Particle. *J. Chem. Phys.* **1968**, *48*, 5287.
4. Rapp, D. Complete Classical Theory of Vibrational Energy Exchange. *J. Chem. Phys.* **1960**, *32*, 735.
5. Benjamin, I. Semiclassical Algebraic Description of Inelastic Collisions. *J. Chem. Phys.* **1986**, *85*, 5611.
6. Wendler, T.; Récamier, J.; Berrondo, M. Collinear Inelastic Collisions of an Atom and a Diatomic Molecule Using Operator Methods. *Rev. Mex. Fis.* **2013**, *59*, 296.
7. Gazdy, B.; Micha, D. A. The Linearly Driven Parametric Oscillator—Its Collisional Time-Correlation Function. *J. Chem. Phys.* **1985**, *82*, 4937.
8. Récamier, J. An Algebraic Method for the Study of Collisions with an Anharmonic Oscillator. *Int. J. Quantum Chem.* **1990**, *24*, 655.
9. Récamier, J.; Gazdy, B.; Micha, D. A. Energy Transfer in Collisions Between Two Vibrating Molecules. *Chem. Phys. Lett.* **1985**, *119*, 383.

10. Öhrn, Y. Time Dependent Treatment of Molecular Processes. *Adv. Quantum Chem.* **2015**, *70*, 69–109.
11. Haroche, S.; Raimond, J. M. *Exploring the Quantum*; Oxford University Press: Oxford, 2006.
12. Gerry, C.; Knight, P. *Introductory Quantum Optics*; Cambridge University Press: New York, 2005.
13. Román-Ancheyta, R.; Berrondo, M.; Récamier, J. Parametric Oscillator in a Kerr Medium: Evolution of Coherent States. *JOSA-B* **2015**, *32*, 1651–1655.
14. Agarwal, G. S.; Arun Kumar, S. Exact Quantum Statistical Dynamics of an Oscillator with Time-Dependent Frequency and Generation of Nonclassical States. *Phys. Rev. Lett.* **1991**, *67*, 3665.
15. Lo, C. F. Generating Displaced and Squeezed Number States by a General Driven Time-Dependent Oscillator. *Phys. Rev. A* **1991**, *43*, 404.
16. Morse-Like Squeezed Coherent States and Some of Their Properties. *J. Phys. A Math. Theor.* **2013**, *46*, 375303.
17. Alber, G.; Zoller, P. Laser Excitation of Electronic Wave-Packets in Rydberg Atoms. *Phys. Rep. Rev. Sect. Phys. Lett.* **1991**, *199*, 232–280.
18. Vinogradov, A. V.; Janszky, J. Excitation of Squeezed Vibrational Wave Packets Associated with Franck-Condon Transitions in Molecules. *Sov. Phys. JETP* **1991**, *72*, 211.
19. Zewail, A. Laser Femtochemistry. *Science* **1988**, *242*, 1645–1653.
20. Averbuck, I.; Shapiro, M. Optimal Squeezing of Molecular Wave-Packets. *Phys. Rev. A* **1993**, *47*, 5086.
21. Chang, B. Y.; Lee, S.; Sola, I. R.; Santamaría, J. Coherent Control of Photochemical and Photobiological Processes. *J. Photochem. Photobiol. A Chem.* **2006**, *180*, 241–247.
22. Récamier, J.; Berrondo, M. Vibration-Translation Energy Transfer in a Collision Between an Atom and a Morse Oscillator. *Mol. Phys.* **1991**, *73*, 831.
23. Löwdin, P. O. Quantum Theory of Time-Dependent Phenomena Treated by the Evolution Operator Technique. *Adv. Quantum Chem.* **1967**, *3*, 323.
24. Yang, B.; Han, K.; Ding, S. Application of Dynamical Lie Algebraic Method to Atom-Diatomic Molecule Scattering. *J. Math. Chem.* **2000**, *28*, 247.
25. Berrondo, M.; Récamier, J. Dipole Induced Transitions in an Anharmonic Oscillator: A Dynamical Mean Field Model. *Chem. Phys. Lett.* **2010**, *503*, 180.
26. Alhassid, Y.; Levine, R. D. Connection Between the Maximal Entropy and the Scattering Theoretic Analyses of Collision Processes. *Phys. Rev. A* **1978**, *18*, 89.
27. Wei, J.; Norman, E. Lie Algebraic Solution of Linear Differential Equations. *J. Math. Phys.* **1963**, *4*, 575.
28. Wei, J.; Norman, E. On Global Representations of the Solutions of Linear Differential Equations as a Product of Exponentials. *Proc. Am. Math. Soc.* **1964**, *15*, 327.
29. Dodonov, A. V.; Dodonov, V. V. Dynamical Casimir Effect in Two-Atom Cavity QED. *Phys. Rev. A* **2012**, *85*, 055805.
30. de los Santos-Sánchez, O.; Récamier, J. Nonlinear Coherent States for Nonlinear Systems. *J. Phys. A* **2011**, *44*, 145307.

CHAPTER EIGHT

Resonances in the Continuum, Field-Induced Nonstationary States, and the State- and Property-Specific Treatment of the Many-Electron Problem

Cleanthes A. Nicolaides[1]
Theoretical and Physical Chemistry Institute, National Hellenic Research Foundation, Athens, Greece
[1]Corresponding author: e-mail address: caan@eie.gr

Contents

Preface

I start by thanking Erkki Brändas and Jack Sabin for inviting me to contribute to this special volume commemorating the 100th birthday of Per-Olov Löwdin. Their initiative adds to previous ones involving conferences and books that have been dedicated to him, all expressing the respect and admiration that Löwdin inspired throughout his scientific career in his associates and professional colleagues.

Although there are many people who are better qualified to comment on Löwdin's personality and achievements, I take this opportunity to state briefly my impressions of him.

Advances in Quantum Chemistry, Volume 74
ISSN 0065-3276
http://dx.doi.org/10.1016/bs.aiq.2016.03.001

I met Löwdin only a few times during the 1970s and 1980s, in conferences and in Sanibel symposia, starting with the conference on "The Future of Quantum Chemistry" that was held in Dalseter, Norway, Sept. 1–5, 1976, organized by J.-L. Calais and O. Goscinski to celebrate his 60th birthday. Those encounters (which included a couple of cocktail parties and soccer games where he played goalie) led to a nice rapport, even though I was much younger. They were sufficient to leave me with the best of impressions about his openness, about his interest in assisting young scientists from all over the world, and about his scientific inquisitiveness and aim for mathematical clarity and justification.

Löwdin's scientific and organizational achievements were instrumental in advancing the cause of quantum chemistry in Sweden as well as internationally, especially during the 1950s and 1960s. For example, the Uppsala summer schools and the Sanibel winter symposia became institutions. He is remembered with respect and affection not only for his research papers but also for his exceptional activity which accelerated the recognition of quantum chemistry as a distinct scientific discipline with a diverse community of theoretical scientists. As a member of this community, I feel lucky for the opportunity given to me by the Editors to contribute the paper which follows.

Abstract

The paper summarizes elements of theories and computational methods that we have constructed and applied over the years for the nonperturbative solution of *many-electron problems* (MEPs), in the absence or presence of *strong* external fields, concerning *resonance/nonstationary* states with a variety of electronic structures.

Using brief arguments and comments, I explain how these MEPs are solvable in terms of practical time-independent or time-dependent methods, which are based on single- or multistate Hermitian or non-Hermitian formulations. The latter result from the *complex eigenvalue Schrödinger equation* (CESE) theory. The CESE has been derived, for field free as well as for field-induced resonances, by starting from Fano's 1961 discrete-continuum standing-wave superposition, and by imposing outgoing-wave boundary conditions on the resulting solution. Regularization is effected via the use of complex coordinates for the orbitals of the outgoing electron(s) in each channel. The Hamiltonian coordinates remain real.

The computational framework emphasizes the use of appropriate *forms* of the trial wavefunctions and the choice of function spaces according to the *state- and property-specific* methodology, using either nonrelativistic or relativistic Hamiltonians. In most cases, the bound part of excited wavefunctions is obtained via state-specific "*HF or MCHF plus selected parts of electron correlation*" schemes. This approach was first introduced to the theory of multiply excited and inner-hole autoionizing states in 1972, and its feasibility was demonstrated even in cases of multiply excited negative-ion scattering resonances.

For problems of states interacting with strong and/or ultrashort pulses, the *many-electron time-dependent Schrödinger equation* is solved via the *state-specific expansion approach*.

Applications have produced a plethora of numerical data that either compare favorably with measurements or constitute testable predictions of properties of *N-electron* field-free and field-induced *nonstationary* states.

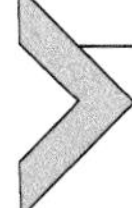

1. QUANTUM CHEMISTRY AND MANY-ELECTRON PROBLEMS IN THE HIGH-LYING PORTIONS OF THE "EXCITATION AXIS"

For every complex problem there is a solution that is clear, simple and wrong.
H. L. Mencken (1880–1956)

The emergence of Quantum Chemistry (QC) as a distinct discipline has been associated with and driven by the requirement of tackling efficiently and quantitatively the *many-electron problem* (MEP) and its multifarious manifestations in a multitude of electronic structures, of properties, and of cases of dynamics of physical/chemical relevance.

Following Wigner's use of the term in his 1934 theory of electrons in metals,[1] the MEP also came to be known among quantum chemists as the problem of "*electron correlation*," using as zero-order reference the single-configurational restricted Hartree–Fock (HF) solution and focusing on the accurate calculation of the total energy of the ground state. This definition seems to have been introduced into the literature of QC in 1952 by Taylor and Parr,[2] in connection with their discussion on the convergence of the method of "*superposition of configurations*" in the Helium ground state.

Even though the pursuit of the accurate determination of the total energy of the ground state continues to characterize nearly all publications on many-electron methods and algorithms, it should be kept in mind that when it comes to properties and dynamics involving highly excited N-electron states, this is not necessarily rewarding or even feasible.

In Fig. 1 I have sketched two major directions/categories for modern QC which I consider to be distinct in terms of the fundamental theories that govern them and the scientific information which they seek to produce. While the MEP is omnipresent, the physics, the nature of experimental possibilities, and the corresponding fundamentals of many electron quantum mechanics in these two "directions" are different.

The horizontal arrow of Fig. 1 represents the achievements and the continuing extensive activity by practitioners all over the world, which have been pushing research mainly toward larger and larger sizes of atomic and molecular systems in their ground state, with occasional excursions to low-lying excited discrete states. It expresses what has become by far the dominant component of QC in terms of numbers of publications, authors, and financial support, namely "computational quantum chemistry" (CQC).

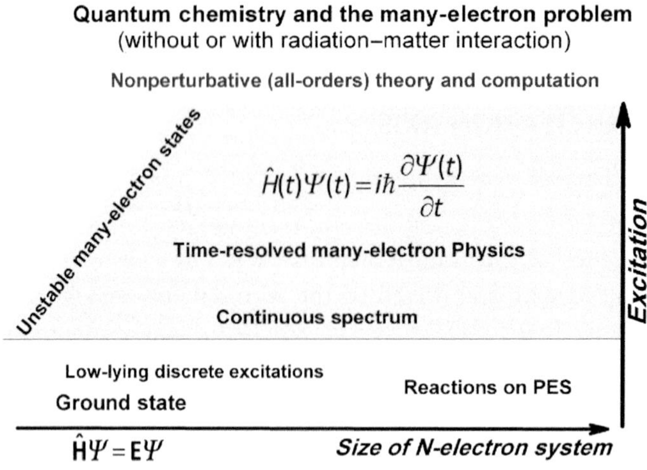

Fig. 1 Sketch depicting the two "directions" of Quantum Chemistry referred to in the introduction. The chapter discusses topics in the high-lying portions of the "excitation axis." The corresponding *many-electron time-independent or time-dependent problems* involve field-free or field-induced *nonstationary* electronic states relaxing into the continuous spectrum. *Reprinted from Nicolaides, C. A. Quantum Chemistry and Its "Ages" Int. J. Quantum Chem.* ***2014****, 114, 963; Nicolaides, C. A. Quantum Chemistry on the Time Axis: Electron Correlations and Rearrangements on Femtosecond and Attosecond Scales. Mol. Phys.* ***2016****, 114, 453. http://dx.doi.org/10.1080/00268976.2015.1080870, with permission from J. Wiley and Sons, and from Taylor and Francis, publishers.*

Here, the fundamental equation is Schrödinger's *real eigenvalue* equation for stationary discrete states:

$$\mathbf{H}\Psi_n = E_n\Psi_n, \quad \langle\Psi_n|\Psi_m\rangle = \delta_{nm} \tag{1}$$

When it comes to applications (countless systems), the utility of CQC extends to materials science and even to biology. Using the relevant approximate solutions $(\tilde{\Psi}_n, \tilde{E}_n)$ and additional theory, electronic/vibrational/rotational spectra (positions and intensities), models of electron distributions, of bonding, and of reactivity, other properties, etc., are computed (computable). For some classes of cases, density functional theory is, in principle, also applicable.

As a result of the dominance of CQC, the general perception has been that CQC is essentially the same as QC, with the front moving along the direction where methods and algorithms are improved in order to produce reliable solutions of Eq. (1) for electronic ground/low-lying discrete states of molecules with large numbers of electrons. Indeed, using as reference the evolution of methods and of types of applications in CQC, distinct "ages of QC" have been suggested. However, this identification of QC with CQC can be challenged, on the basis that modern QC is also concerned with novel MEPs in the high-lying portions of the "excitation axis" of Fig. 1, *near* (below and above) and *far* from the threshold of the continuous spectrum. Arguments supporting this description of QC and MEPs were propounded and documented in Ref. 3, see also Ref. 4.

The present paper has to do with theories and methods that enable the solution of broad categories of MEPs in the aforementioned realm of the "excitation axis" of Fig. 1. The topics are stated in the next section. These refer to *resonance* states of finite energy width as well as to discrete *doubly excited states* (DESs) and *Rydberg* series very close to threshold. Furthermore, they involve properties and phenomena induced by *strong* and/or *ultrashort* (few-cycles) electromagnetic fields. In all cases, Hermitian or non-Hermitian nonperturbative (variational) formalisms have been constructed and implemented.

The many-electron quantum mechanics for the correct formal and computational treatment of the plethora of situations that these topics contain is much more sophisticated and demanding than that which applies to ground or low-lying discrete states. Now, given the level of complexity of the general formalisms which have to account for the continuous spectrum of scattering states in conjunction with a multitude of open-shell electronic structures, the difficulties and challenges of the MEP not only persist but are accentuated considerably. For example, consider the *nonstationary* states that are created by the interaction, $\boldsymbol{V}(t)$, of atomic/molecular ground or excited states with *strong/ultrashort* electromagnetic pulses which drive the system into the continuous spectrum. In order to understand quantitatively such phenomena, one must pursue the *nonperturbative* solution of the *many-electron time-dependent Schrödinger equation* (METDSE):

$$\boldsymbol{H}(t)\Psi(t) = i\hbar\frac{\partial\Psi(t)}{\partial t}, \qquad \boldsymbol{H}(t) = \mathbf{H}_{A(M)} + V(t) \tag{2}$$

Now, the MEP becomes time dependent, and issues of the interplay between dynamics and electron correlations are resolved on the time axis.[4]

I close this introduction by noting that, because of the nature and scope of the paper, the presentation is necessarily condensed. Furthermore, the reference list is inevitably dominated by publications of the author, without or with coauthors. This is necessary in order to support the brief arguments and to guide the reader accordingly. The work by other researchers, which is cited, is directly connected to the statements and comments.

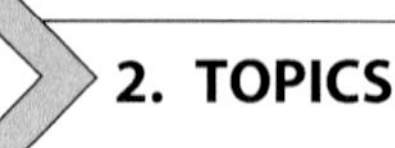

2. TOPICS

The paper comments briefly on the topics I, II, and III listed later in the chronological order in which they were introduced in the literature and on the formalisms and computational methods which we have proposed and applied toward the solution of the corresponding MEPs for states, properties, and processes. Extensive recent discussions, including characteristic results of calculations that we have published on prototypical systems and problems, are presented in Refs. 3–7.

(I) Theory and calculation of the properties of N-electron *field-free resonance states* ("autoionizing," or "Auger," or "negative-ion compound" states, etc.), of the "*Feshbach*," or the "*shape*" type. In principle, these states can be (and have been) created and measured in excitation/collision processes of various types. In addition, this topic includes the *multistate* theory and calculation of spectra just below and above threshold, which, however, cannot be discussed here, due to the constraints on the length of the article (see Section 5.1).

(II) Theory and calculation of the *nonperturbative* response (energy shifts and ionization rates) of ground and excited discrete states, as well as of resonance states, to *strong static or periodic* electric or magnetic fields.

(III) Theory and calculation of the *time-resolved* physics of N-electron nonstationary wave packets, determined based on the *nonperturbative* solution of the METDSE, for time-independent Hamiltonians (as is the case of autoionization or of decay due to perturbation by a static external field), and, principally, for time-dependent Hamiltonians describing the perturbation of N-electron ground or excited states by *strong and/or ultrashort* electromagnetic pulses.

3. STATE- AND PROPERTY-SPECIFIC QUANTUM CHEMISTRY

The goal in the design and execution of our treatments of topics I, II, and III has been, on the one hand, the formulation of general as well as computationally tractable theoretical frameworks and, on the other hand, the consistent and reliable tackling of issues of electronic structure, of electron correlation, and of the continuous spectrum. The choices and optimization of basis sets (analytic as well as numerical orbitals), and the calculation of the wavefunctions and of the matrix elements that are dictated by theory, are done according to the "*state- and property-specific*" (SPS) QC (Ref. 7 also 3–6).

The basic argument of the SPS theory has been that, in order to render the solution of a variety of N-electron time-independent or time-dependent problems economic or even feasible, of primary importance is the choice of *forms* of trial wavefunctions and the optimization of the function spaces that are relevant to the state(s) and to the property of interest. This means that, in many cases, it is not necessary to pursue a very accurate number for the total energy. Instead, the degree of accuracy which is expected from the approximate solution of the pertinent Schrödinger equation(s) with respect to the total energy varies depending on the state(s) and on the problem.

For example, as proposed and demonstrated computationally in the 1970s,[7] the calculation of radiative and radiationless transition probabilities can very often be accomplished reliably by analyzing and understanding directly the off-diagonal matrix elements, the state-specific orbitals and mixing coefficients, and the possible N-electron overlaps, rather than by first aiming at a very accurate calculation of the total energies and wavefunctions.

Critical to the success of this endeavor, especially when it comes to excited states, is the definition and accurate calculation of a zero-order "*Fermi-sea*" wavefunction, normally obtained via the solution of the state-specific multiconfiguration Hartree–Fock (MCHF) equations.[7] The concept of the *Fermi-sea* self-consistent orbitals and corresponding symmetry-adapted configurations was introduced in the 1970s, with prototypical applications to simple excited states of atoms.[7]

In other words, the focus in the treatment of the various MEPs is *not* merely on the construction of formalism by manipulating various operator–matrix expressions, expecting that these can, somehow, be computed accurately under the *assumption* that there exists a convenient (nearly), complete set (Hilbert space) on which they are defined. Instead, having established a formally sound framework, the emphasis is first on how the total wavefunction of each state can be discerned in terms of the significance of its main components, and of corresponding basis sets, regarding the problem and the property under investigation. This means that one focuses mainly on mixing coefficients of symmetry-adapted configurations and on the numerical accuracy of appropriate zero-order and correlation orbitals.

Beyond the HF or the MCHF zero-order approximation, the contributions of symmetry-adapted components describing electron correlation are computed via diagonalization of Hamiltonian matrices that are normally small. In practice, this leads to reliable *state-specific* wavefunctions that are generally compact, numerically accurate, and transparent. Consequently, they are suitable for applications, and, often, for transferring information from one system to another.

A case in point is the calculation of wavefunctions and properties of resonance states. Starting in 1972,[8] a series of publications have demonstrated the critical importance of properly optimized (according to theory) function spaces, especially when computer facilities are very limited.[5–7]

It is worth pointing out that from the choice of the *Fermi-sea* wavefunction depends on the further separation of the remaining electron correlations into "*nondynamical*" and "*dynamical*." These concepts, which are often discussed and used in the recent literature of molecular electronic structure calculations, were introduced and first applied to low-lying discrete states of small atoms by Sinanoğlu and coworkers in the 1960s, who used the more restricted concept of the "*Hartree–Fock sea*" of spin-orbitals, which had been based on the filling of single hydrogenic or "minimal basis" shells.[9] The discussion in Ref. 7 on these concepts includes as an example the analysis of a rather difficult bonding situation, namely that which characterizes the weakly bound Be_2 molecule.

The essentials of the SPS theory are discussed in Refs. 5–7. Many of its applications have been concerned with atomic properties and phenomena, with emphasis on topics I, II, and III. Aspects of its methods have also been demonstrated in problems with diatomics and polyatomics.[5–7] For the latter, there are open and challenging issues for theory as well as for computational methods.

4. BACKGROUNDS

The following paragraphs recall the state of affairs with theory and calculations at the time of the introduction of the analyses, formalisms, and SPS methods for the quantitative treatment of MEPs in the topics I, II, and III. Any other approaches which may have been published in subsequent and recent years are not in the scope of this review, regardless of the degree of their overlap with the main ideas and results of the many-electron approaches discussed here.

For topic I, the reference period is up to about the end of 1971, which is when the *decaying-state, state-specific theory*, and results were completed, and published in Ref. 8. For topic II, the reference period is up to 1988, which is when we began publishing our first papers on the stationary "*many-electron, many-photon theory*."[10–15] For topic III, the reference period is up to 1994, which is when our first papers on the "*state-specific expansion approach*" (SSEA) to the solution of the METDSE were published.[16–18]

4.1 Topic I: Field-Free Resonance States and the MEP

As one can verify from a perusal of the QC literature up to 1971, the achievements of quantum chemists during the decades of the 1950s and 1960s on the MEP were not in this area. Instead, the focus was on the development and analysis of methods that solve Eq. (1) for $(\tilde{\Psi}_n, \tilde{E}_n)$ of ground and low-lying discrete states, based on the scheme "*HF plus electron correlation energy*." For example, see Ref. 19, whose 13 review articles embody the bulk of proposals, trends, and progress, up until the end of the 1960s, toward the understanding of the MEP associated with the solution of Eq. (1). While a PhD candidate at Yale (1969–71), I familiarized myself with the substance and the details of most of those articles. The experience, which I acquired from the study and implementation of Roothaan's analytic HF method[20] and of Sinanoğlu's theory of electron correlation in ground and low-lying discrete states,[9] was instrumental in my taking the first steps of the *state-specific* treatment of inner-hole and of multiply excited states (even of negative ions), in the decaying-state framework of Ref. 8 (see Section 5).

At the same time (1950s and, mostly, 1960s), the theme of "resonances" in atomic and molecular physics was emerging both on the experimental and on the theoretical front (eg, see the reviews 21,22). The bulk of formalisms and methods followed the path of scattering theory to the extent that they could handle few problems of low-lying resonance states in two- and three-electron systems.[21–26] In addition, bound-state-type methods, generally known as the "*diagonalization/stabilization method*" (DSM), had also been introduced and studied.[27–31] The common characteristic of those approaches was that, regardless of the level of rigor of the formalism, the implementation was done by using a single basis set, either for just the bound wavefunctions or for both the bound and the scattering components (as is the case of the DSM).

Out of the many possible examples of work from that period, of interest to the reader may be the CI calculations and conclusions of Hylleraas[27] on a couple of low-lying DESs of H^-. This 1950 paper is probably the origin of publications using some version of DSM. Hylleraas[27] dealt with the problem by using only discrete basis functions and searching for the "correct" roots of the diagonalized energy matrix. The uncertainties, the concerns, and the approximations which can be found in that paper, by the man who 20 years earlier had produced the remarkably accurate calculations on the ground state of helium and its isoelectronic ions, are testimony to the subtleties and difficulties (in addition to lack of computer power) with which the few people in the 1950s–60s who had tackled the problem of actual calculation of resonance states were faced.

In summary, as explained and documented in Ref. 8 and also recently in Ref. 5, the theories and/or methods of calculation of wavefunctions, energies, and widths (total and partial) of resonance/autoionizing states, which had been published by 1971, were characterized either by formal incompleteness or by serious practical difficulties concerning their applicability to N-electron structures (with electrons excited either from the valence or from the inner shells), while incorporating the corresponding electron correlations.

Given this background, the approach, which was introduced in Ref. 8 and since then improved and expanded,[5] showed how it is possible to incorporate advanced polyelectronic methods for arbitrary electronic structures and for electron correlations in resonance (autoionizing) states of N-electron atoms and molecules ($N > 2$), into theory that draws from the general theory of scattering and accounts for the presence of the continuous spectrum.

4.2 Topics II and III: Ground or Excited States Perturbed by *Strong* Static or Periodic or Pulsed Fields, and the Nonperturbative Solution of the MEP

In topics II and III, V of Eq. (2) may or may not be time dependent. In principle, the theory for the quantitative treatment of the perturbation of ground or excited states by *strong* fields must be either nonperturbative with respect to the total Hamiltonian, $\mathbf{H}_{A(M)} + V$, or capable of summing accurately the time-independent or time-dependent series of perturbation theory to all orders. Hence, it necessarily engages, directly or indirectly, a multitude of states from both the discrete and the continuous spectrum, whose degree of mixing and the concomitant effects on observables depend on the spectrum of $\mathbf{H}_{A(M)}$ and on the parameters of the external field. The main result of this field-induced mixing is that the system is driven into the continuous spectrum; ie, it releases one or more electrons.

For static and periodic fields, the METDSE is reducible to stationary forms. The energies of the state(s) of interest not only are shifted but are also broadened because of the mixing with components of the continuous spectrum. Thus, discrete states obeying Eq. (1) turn into states with characteristics of *resonances*, where the energy shifts and widths are field dependent. At the same time, new features in the spectrum may appear, as a function of, say, the intensity, for fixed frequency.

For ultrashort and/or strong interactions caused by pulsed fields of few periods, the physics is necessarily time dependent, and the changes in the old levels (and the appearance of new ones) occur at each moment during the interaction. Now, an originally stationary state becomes a field-dependent nonstationary wave packet, $\Psi(t)$, obeying Eq. (2). Projection of $\Psi(t)$ onto stationary states of the discrete and/or the continuous spectrum after the end of the pulsed interaction gives the occupation probabilities of these states.

The phenomena of electron emission corresponding to the above situations are known as *tunneling* or *(multi)photoionization*. Associated with them are shifts of the energies of the field-free states (for weak fields, these yield static and dynamic (hyper)polarizabilities) and time-independent rates or time-dependent probabilities of transition into the continuous spectrum.

Because of the severe complications that result from the addition of strong V to $\mathbf{H}_{A(M)}$, until the late 1980s no practical theory and computed results existed showing how to perform nonperturbative (or, all orders in perturbation theory) calculations for field-induced resonances and/or nonstationary involving arbitrary closed- or open-shell many-electron states, including the effects of electron correlation.

In the context of the foregoing comments, two theoretical advances from the 1970s and 1980s must be recalled. These concern the nonperturbative theory and calculation of properties in both topics. Their introduction was done at a time when it had become clear that the experimental progress with intense laser at various wavelengths would require the development and application of nonperturbative theories and methods.

4.2.1 Topic II: Complex-Energy Variational Methods for Static and Periodic Strong Fields

Following the 1971 mathematical results of Aguilar, Balslev, and Combes, as clarified by Simon, regarding the spectrum of the "dilatationally analytic" Coulomb Hamiltonian upon "rotation" of its coordinates which are allowed to be complex, $r \rightarrow re^{i\theta}$,[32] Doolen[33] proposed the variational *complex coordinate rotation* (CCR) procedure, whose capacity he demonstrated by computing "a resonant eigenvalue in the $e^{-}H$ singlet s wave" using a basis set of Hylleraas functions.

In principle, this "rotation" reveals the full spectrum of discrete, scattering, and resonance states. Reinhardt and coworkers[34–37] found via computation on hydrogen that the CCR method also works for the electric dipole operator of strong dc and ac fields, which is not dilatationally analytic. As regards "hydrogen in ac fields," Chu and Reinhardt[36] invoked the work of Shirley[38] on the reduction of problems with periodic time-dependent Hamiltonians to stationary problems in the "semiclassical" approximation ("Floquet methods") and demonstrated the physical relevance of complex eigenvalues of the electric dipole "Floquet Hamiltonian," $\boldsymbol{H}_F(\theta)$, by computing widths of field-induced resonances representing multiphoton ionization rates.

The approaches and findings by Reinhardt and coworkers during the 1970s constituted a significant step forward in the theory of atom-strong field interactions, since they were based on a variational method which takes into account the continuous spectrum. However, when it comes to the spectra of nearly all of atoms and molecules, the use of the CCR method is unrealistic, since the recipe demands a large number of diagonalizations of $\boldsymbol{H}(\theta)$ on a large basis set and the search for the properties of all roots as a function of θ and of other variations in the basis.

The practical limitations of the CCR method regarding the solution of MEPs for arbitrary electronic structures and for the calculation of partial widths, for complex Hamiltonians without or with an external field, were pointed out in Refs. 5,39,40. One of the main reasons is that the use of a common basis set in terms of which the Hamiltonian matrix is constructed cannot represent economically and accurately the function spaces which characterize on resonance the "*localized*" and the "*asymptotic*" parts of the correct solution.[5] In other words, just as in the ordinary case of N-electron ground states the brute-force diagonalization of $\mathbf{H}$ on a big basis set runs quickly into trouble, the requirements associated with the CCR method could not (cannot) render it suitable for the treatment of MEPs for arbitrary electronic structures. In fact, when the N-electron Hamiltonian includes an external ac field, the practical difficulty of the CCR task increases beyond reasonable levels, since, in addition, huge "Floquet blocks" would have to be constructed and tested, via repeated diagonalizations, seeking convergence of many roots.

The aforementioned limiting difficulties regarding the solution of the MEP cease to exist in the polyelectronic theory which was proposed in the 1970s as a natural extension of the work that had started in Ref. 8. This is the *state-specific "complex eigenvalue Schrödinger equation"* (CESE) *approach*,[5] features of which are explained in Section 5.

The basis of the CESE method is the theory-guided emphasis on recognizing the two-part *form* of the resonance eigenfunction and on pursuing the separate optimization of the corresponding function spaces. Now, the coordinates of the Hamiltonian are real, and so are the coordinates of the correlated wavefunction for the *localized* part, Ψ_0. Only the *asymptotic* part of the trial wavefunction in the appropriate channel is made complex in order to regularize the solution on resonance. Its extension to problems with external static or ac fields, named the *many-electron, many-photon theory*, uses similar arguments and methods with a time-independent Hamiltonian that includes the interaction. (The matrix elements are, of course, different.)[5,10–15] Computational comparisons of the CCR and the CESE methods which demonstrate the efficiency of the CESE method even in simple cases, where the CCR method is applicable, were published in Refs. 41,42.

4.2.2 Topic III: "Grid" Methods

As the science and technology of the production and spectroscopic use of electromagnetic pulses that are intense and/or ultrashort is advancing, the horizon of the scientific area of light–matter interactions is widened significantly, especially with respect to prospects of obtaining new information on multielectron systems. The comprehensive quantitative understanding, via theory, of corresponding effects (mainly multiphoton and pump-probe processes) presupposes the capacity of obtaining and using the $\Psi(t)$ of Eq. (2), describing the nonstationary state of interest. In the 1980s, Kulander[43,44] implemented the "grid method" for integrating numerically the METDSE, via two approximate schemes which invoke the independent-electron model. The first was the time-dependent HF approximation, which was applied to He.[13] The second was the "*single active electron*" approximation, where the time-dependent wavefunction of only one electron moving in an average local potential is computed. It was applied to the calculation of the time-dependent multiphoton ionization of Xenon.[44] The calculations and analysis of Kulander provided for the first time useful information on specific multiphoton processes in multielectron atoms. Yet, both of these schemes were limited to closed-shell, single determinantal states (eg, noble gas atoms). They are "one-shot" approaches which ignore the MEP even for closed-shell states, since they do not take into account the electron correlation and interchannel-coupling components of electronic structures and of transition amplitudes.

In 1994, the Athens team published a polyelectronic theory (the SSEA) with numerical results, which shows how the METDSE can be solved systematically for arbitrary electronic structures.[6,16,17] The SSEA is not a "grid" method (see Section 5). The initial applications involved two proof-of-principle cases (apart from hydrogen). The first was to the correlated wavefunction of the Li^- $1s^2 2s^2$ ground state, with two ionization coupled channels ($1s^2 2s\, ^2S$, $1s^2 2p\, ^2P^o$).[16] The second was to He $1s2s\, ^1S$, which is the prototypical open-shell, metastable excited state having a lower state of the same symmetry.[17] Problems of field-induced time-dependent dynamics in diatomic molecules were also solved at that time.[6]

Furthermore, by implementing the SSEA, it also became possible for the first time to compute from first principles the *nonexponential decay* of real, multiparticle unstable states, where the matrix elements involve only the time-independent $\mathbf{H}_{A(M)}$ and appropriately computed N-electron wavefunctions.[18]

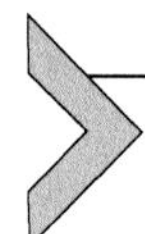

5. OVERVIEW AND ELEMENTS OF THE SPS THEORY ON TOPICS I, II, AND III

The contents of the previous sections have much to do with the "why" our SPS work on topics I, II, and III was undertaken, in view of the desideratum to treat the corresponding MEPs quantitatively within rigorous theoretical frameworks. In this last section, I will outline only a few elements of "what" has been done and "how" it was (is) done. The extent of the presentation is very limited, focusing mostly on topic I, because of space restrictions. Apart from the original publications, much of the important information can be found in the recent reviews.[3–7]

5.1 Topic I

Hermitian as well as non-Hermitian (CESE theory) formulations have been constructed and applied for the prediction of energies, widths (partial and total), and oscillator strengths of resonances due to singly or multiply excited states and to inner-hole (Auger) states, of neutrals, and of positive and negative ions, mainly of atoms, but also for characteristic cases of negative-ion resonances and of *diabatic* states in diatomics such as He_2^- and He_2^+.[5]

In addition, certain of the features and results of the SPS approach can also be useful for treating problems in nuclear and in solid state physics.

Indeed, such applications have already taken place. For example, in Ref. 45 we reported energies and analysis of the wavefunctions for inner-hole states of clusters simulating the real metal. In Ref. 46, the implementation of the CESE theory produced tunneling rates in multibarrier solid state structures relevant to properties of devices. Recently[47] the results from the time-dependent decaying-state model of Ref. 39 proved relevant to an analysis of nuclear decays, by answering the challenge[48] to the long-honored method of radioactive carbon dating which has been based on the assumption of exponential decay of the ${}^{14}_{6}C$ isotope.

The problems for which quantitative solutions have been produced include not only cases of isolated states but also complex spectra due to *multistate* interactions in multichannel continua, even in cases where the binding is weak and the widths are very small. I give two examples:

(1) The CESE work and the analysis of results reported in Refs. 49–51 achieved the first definitive resolution and interpretation of the spectrum of the complex eigenvalues representing the low- as well as the high-lying resonance states of H^- (up to the $n=5$ hydrogen threshold), for which the zero-order model had been proposed about 40 years earlier (dipole resonances).[52] It is worth stressing that this is the first real system where essentially the complete resonance spectrum has been resolved. Among other things, this achievement required the appropriate use of (very) diffuse functions that went out to 8000 a.u., a fact which, in the context of the CESE theory, allowed the calculation of widths down to 10^{-9} a.u.

(2) In a series of papers,[53–57] and section 5.2 of Ref. 6, a CI, K-matrix formalism which unifies, in a computationally tractable manner, the treatment of *multistate* interactions in the discrete and the continuous spectra, including multiply excited states, was proposed and applied, with emphasis on the possibility of calculating wavefunctions and spectroscopic data just below and just above the fragmentation thresholds.

For example, Ref. 54 reported the cross-sections for the simultaneous photoionization and photoexcitation of He, around and at the He^+ $n=2$ threshold, as well as the positions and autoionization widths of three channels of ${}^1P^o$ Rydberg series of resonances.

Another example can be found in Refs. 55,56, where, using the Breit–Pauli Hamiltonian and MCHF-based CI wavefunctions, we made the first quantitative predictions of the wavefunction character, the quantum defects, the oscillator strengths, and the fine structure of the $A\ell$ $KL3s^2nd\,{}^2D$ Rydberg series perturbed by the $KL3s3pnp(\varepsilon p)$ ^2D channel and, especially, the valence "$KL3s3p^2$" ^2D correlated state in the presence of the $KL3s^2\varepsilon d$ continuum.

Regarding the earlier multielectron, multistate problem, it is worth noting the following: The valence configuration $KL3s3p^2$ was found to be spread over the lower portion of the discrete spectrum and not to be located just above the $A\ell^+$ threshold, as concluded by Taylor et al.[58] based on the results of standard "full CI" calculations. The significance of this disagreement goes beyond the fact that two methods gave different results. Indeed, it has to do with the fundamentals of quantum mechanical calculations regarding strongly perturbed spectra, and with the degree of validity of standard methods of CQC with common basis sets, such as the one applied in Ref. 58, which ignore the particularities and details of the Rydberg series and of the continuous spectrum. Our predictions were later confirmed experimentally in Ref. 59.

The CI-K-matrix SPS theory of Refs. 53–57 was developed by drawing from the quantum defect theories of Seaton,[60] Fano,[61] and Greene et al.[62] However, it differs from them in essential ways, especially as regards the calculation of bound and scattering wavefunctions and K-matrices for N-electron systems. One of them is that it is formulated and implemented without having to refer to irregular Coulomb functions, as do the formalisms in Refs. 60–62. Such functions are not used in the CI-type calculations which are necessary when it comes to the solution of problems of spectra involving the mixing of N-electron basis wavefunctions.

5.1.1 Decaying-State Theory as Framework for the Description of N-Electron Resonance States

Paper[8] is the origin of this research program, having as immediate sequels.[39,40,63] It is characterized by two main proposals, which are outlined in the following paragraphs. The first is the general theoretical framework for the formal and practical analysis, understanding, and calculation of resonance (autoionizing) states. The second is the incorporation into this framework of electronic structure methods based on the scheme of "*state-specific, open-shell, multideterminantal Hartree–Fock plus electron correlation*," for the description of the localized part, Ψ_0, of such states, even though the Ritz minimum-energy principle is not applicable.

Following a critical discussion on the concepts and on the then existing methods,[8] it was concluded that the time-dependent concept and formalism of *decaying states* can provide the framework which justifies a unified treatment for arbitrary electronic structures (with electron correlation) whose energies are in the continuous spectrum. The formalism produces the result of the complex-energy simple pole to which a resonance state corresponds, $z_0 = E_0 + \Delta - \frac{i}{2}\Gamma$, in terms of well-defined matrix elements for the energy of the initially $(t=0)$ localized wave packet, Ψ_0, $E_0 = \langle \Psi_0 | \mathbf{H} | \Psi_0 \rangle$, the energy shift, Δ, and the width, Γ, in analogy with the resonance scattering theory of Feshbach.[64,65] Using this formalism, physically relevant results have been obtained on the energy—as well as on the time— axis, for a model in which the "*self-energy*" of the autoionizing state, $A(z)$, is approximated by $A(z) \approx A(E_0)$. For example, the analysis of the time dependence of the isolated decaying states presented in Ref. 39 has led to conclusions regarding the possible observation of nonexponential decay for unstable states very close to threshold, as well as the foundations of the theory and the possible manifestation of the connection of nonexponential decay to the issue of time asymmetry.[5,47,66]

The formal description of the decaying state in Refs. 8,39 was adapted from that of the theory of radiation damping of Heitler[67] and of Schönberg,[68] and from the presentation of Goldberger and Watson.[69] In spite of the fact that the formal approach to the decay due to atom (molecule)–field interaction and to autoionizing states is nearly the same (eg, the choices of the density of states differ), there is a critical difference, conceptually and practically, which adds difficulty to the case of autoionization. This is the following: whereas in the problem of atom–field interaction the initial and final states are defined precisely as the eigenfunctions of the unperturbed, field-free Hamiltonian (ie, these are stationary states of Eq. 1), in the case of N-electron resonances no such separation is offered, since, upon preparation of the initial state, the interaction causing the decay is intrinsic to the system. In other words, the initially prepared (Ψ_0, E_0) is an *N-electron wave packet* and not a stationary state of the exact operator $\mathbf{H}$, and the question is how to construct it and compute it as efficiently and accurately as possible.

Finally, it is worth pointing out that in the context of the decaying-state theory, the quantum motion is obtained to all orders of time-dependent perturbation theory, since the survival amplitude for decay is defined by the Fourier transform, over the energy spectrum, of $\langle \Psi_0 | 1/(z - \mathbf{H}) | \Psi_0 \rangle$. Provided that Ψ_0 is computable, the theory is formally consistent and avoids the "mathematical" questions of convergence of the time-dependent perturbation series for autoionization which became the object of study in Refs. 37,70, around the time the work of Ref. 8 was published. In fact, this is done without resorting to the unphysical and computationally naïve mathematical assumption of a zero-order Hamiltonian which is hydrogenic, as was done in Refs. 37,70.

5.1.2 Essentials of the SPS Approach

The introductory discussion of Ref. 8 first pointed to some of the peculiarities and difficulties that characterize resonance states of N-electron atoms and molecules. For example, I quote: "*...In addition, the useful property of one-to-one correspondence between reality and configurational assignment may in some cases be lost as one moves up in energy and the density of states increases. Also, from the mathematical point of view, they do not have the useful property of square-integrability, having outgoing radiation boundary conditions. Thus, they do not form a complete orthonormal set and variational or perturbation theories dealing with such states must essentially be non-Hermitian in character*" (Ref. 8, p. 2079).

In accordance with the last statement, the theory of Ref. 8 starts by *asserting* a CESE which the many-electron atom is expected to obey in a resonance state:

$$\mathbf{H}\Psi = W\Psi, \quad W = E_r - i\Gamma/2 \text{ (equation 1 of Ref. 8)} \tag{3}$$

$$\Psi_{r\to\infty} \sim e^{ik_m r} \tag{4}$$

k_m is the complex momentum for decay channel m (equation 2 of Ref. 8).

The appearance of a complex eigenvalue in a Schrödinger equation when $\mathbf{H}$ is formally Hermitian need not cause concern as regards the principles of quantum mechanics. It is explained by the fact that the asymptotic boundary condition on the solution of Eq. (3) is not that of square-integrability (Hilbert space), which is the quantum mechanical condition for the function space on which the real eigenvalue Eq. (1) for discrete states is valid. Instead, assuming one outgoing electron in a particular channel, the eigenfunction satisfies the Sommerfeld outgoing radiation boundary condition[71] with a complex energy (complex momentum), which Siegert published in the context of scattering theory.[72] Siegert's result emerged as a complex pole of a model of an s-wave scattered by a short-range potential (nuclear physics), which, however, has limited relation to dynamics involving N-electron systems with Coulomb interactions.

The association of resonance states with a CESE does not gainsay the fact that, following Fano's Hermitian formalism,[73] they can also be treated in terms of real functions and real energies. Later, I will explain how the non-Hermitian Eq. (3) can be obtained rigorously, together with a two-part form of the complex eigenfunction, by using as the starting point Fano's theory, which does not depend on models (such as that of Siegert) and is not restricted by the type of the potential or by the types of excitation and electronic structures.

The next initial step in the theory of Ref. 8 was the projection of Ψ onto Ψ_0 defined earlier. This means that what is left out is the *asymptotic* component, X_{as}, of the resonance eigenfunction, Ψ_r, which carries the information of the complex eigenvalue, according to Eqs. (3) and (4). Therefore, the argument can be made that, whether in wavefunction or in operator–matrix representation, the *form* of the N-electron Ψ_r that best describes the physics *on resonance* is

$$\Psi_r = \alpha\Psi_0 + X_{as} \tag{5}$$

The square-integrable Ψ_0 results from "*dynamic localization*" and contains most of the information pertaining to the character of the system, including those coming from the continuum, before the residual interaction causes the mixing with other states. Its accurate knowledge is critical for the reliability of the overall calculation of the energy and, especially, of the total and partial widths and of other matrix elements. Depending on the problem, the coefficient α and the "asymptotic" part, X_{as}, are functions of the energy, real or complex. The symbol X_{as} represents both the terms of the wavefunction expansion and their coefficients. The additional mixings with other states can be included in terms of multistate formalisms.[49–51,53–57]

The form (5) is fundamental. The SPS formalisms are rendered computationally tractable by focusing on the theory-guided appropriate choices of function spaces for Ψ_0 and X_{as}. These are different and are optimized separately, while keeping them as *state-specific* as possible.

The label "*state-specific*" for the approach introduced in Ref. 8 refers mainly to the calculation of the correlated (Ψ_0, E_0). By bypassing the insurmountable obstacle which characterizes any method that depends on the brute-force diagonalization of the Hamiltonian (real or complex) on a single basis, it introduced and demonstrated the feasibility for obtaining the wavefunctions of such states (any open-shell electronic structure) in terms of the scheme "*HF plus localized and asymptotic electron correlation*" and, soon afterward, in terms of the scheme "*MCHF plus remaining localized and asymptotic electron correlation.*"[5,40]

In most cases, the decay comes from pair correlations, and these have both localized and asymptotic ("*hole-filling*") components. For "*shape*" resonances, the open channel is represented by a single-electron correlation function.[5]

In this context, crucial step is the direct solution (no root searching) of the *state-specific* HF or MCHF equations for the electronic configuration(s) of interest, expected to represent, *in zero-order*, Ψ_0. If properly converged, the square-integrable solution does so into a local minimum, satisfying the virial theorem inside the continuous spectrum, and, if necessary, is obtained subject to appropriate orbital orthogonality constraints.[5,8,74,75]

Thus, by breaking down the MEP in terms of a zero-order HF or MCHF wavefunction and remaining correlation, it has been argued and demonstrated that it is possible to gauge the calculation according to fundamental elements of polyelectronic theory. For example, in the abstract of Ref. 8 one reads: *This method requires projection onto known one-electron zeroth-order functions and it thus overcomes the difficulties of the well-known P, Q methods which require projection onto exact wave functions.*

That the HF and MCHF equations can be solved reliably for a variety of highly excited open-shell electronic structures even for weakly bound systems, such as He^- "$2s^2 2p$" $^2P^o$ and " $2s2p^2$" ^2D, was first demonstrated in Ref. 8, at a time when this was far from obvious and computational possibilities of trial and error were very limited. I stress that such HF or MCHF calculations require proper care and numerical accuracy in order to ascertain *localization*, on which the remaining localized correlation is built variationally in order to produce the final Ψ_0 of Eq. (5).[5,75]

Of course, for the much simpler cases of relatively isolated configurations with strong attractive potentials, eg, $1s2s^2 2p^n$ ($n = 1, 2, \ldots$) Auger states, construction of the corresponding HF equations and convergence of their solution are straightforward, either nonrelativistically or relativistically. This fact has been utilized in a series of many-electron calculations of one-electron energies, of Auger energies and widths, and of radiative decay (fluorescence yield), where electron correlation and relativistic effects are accounted for Refs. 45,76–79. Characteristic example, with comparisons with other theoretical methods and with experiment, is the study of the wavefunction, electron correlation, energy position, and width of the Be^+ $1s2s^2$ ^2S Auger state.[76]

In more recent years, the calculation of inner-hole single- or multiconfigurational HF or Dirac–Fock atomic wavefunctions and energies is used routinely. Nevertheless, of significance are specific correlation effects beyond such approximations.[79]

I point out that, in computing an optimal Ψ_0, contributions from electronic configurations representing portions of the continuum can also be included without destroying its localization. In the context of the SPS theory, this is often done at the zero-order MCHF level, by including into the equations "*open-channel like*" configurations.[5,75] It was first done approximately in Ref. 8 via diagonalization of small matrices where nearly degenerate configurations representing *nondynamical* correlations were included.

X_{as} contains bound $(N-1)$-electron core wavefunctions coupled to open-channel, term-dependent orbitals representing the outgoing electron in each channel. It may or may not be optimized in the presence of Ψ_0, depending on the problem and on the methodology that is implemented. The coefficients of Ψ_0 and X_{as} in the resonance state are determined by their mixing in a final step of the calculation. Depending on the problem, either Ψ_0 alone or both Ψ_0 and X_{as} are used for the calculation of matrix elements and corresponding properties. For total and partial widths, the state-specific nature of the overall calculation allows, among other things, quantitative analyses that reveal the dominant effects of strongly mixing zero-order configurations, of overlaps due to the nonorthonormality of orbitals, and of quantum destructive and constructive interference and cancelation effects, without and with interchannel coupling.

Calculations of energies and of partial and total widths of a variety of low- and of high-lying resonances have been carried out on the real-energy axis and in the complex-energy plane, for nonrelativistic and relativistic autoionization of atoms[5,40–51,75–77,80–88] as well as for predissociation of diatomics.[89]

I note that, for some problems, it suffices to compute and use only (Ψ_0, E_0), if it is deemed that it carries to a good approximation the information which is needed for the calculation of the quantity of interest. For example, this is the case of the total energy when the energy shift due to the interaction with the part of the continuum represented by X_{as} is expected to be relatively unimportant for the problem of interest. This understanding has allowed systematic work on various properties such as:

- Analysis of wavefunctions and regularities of energy spectra of doubly, triply, or even quadruply excited states.[90]
- Analysis and computation of one-electron binding and Auger energies.[45,78,79]
- Calculation of potential energy curves of "Feshbach" or "shape" resonances of negative ions of diatomics,[91], etc.

5.1.3 The Complex Eigenvalue Schrödinger Equation

I now return to the CESE, Eq. (3). Its adoption leads to an issue which is important for the understanding and the calculation of resonance states and whose resolution was presented in 1981.[40] I explain:

According to quantum mechanics, the Hermitian **H** has a complete set of stationary states, consisting of the union of the discrete states of Eq. (1) and the scattering states of the continuous spectrum. The latter satisfy the real-energy Schrödinger equation, but not as solutions of an eigenvalue equation. Instead, they have scattering boundary conditions obeying Dirac normalization:

$$\mathbf{H}\Psi(E) = E\Psi(E), \quad \left\langle \Psi(E)|\Psi(E')\right\rangle = \delta\left(E - E'\right) \tag{6}$$

In his classic 1961 paper, Fano[73] presented a Hermitian formalism of CI in the continuum, where the energy-dependent phenomenology of resonances in terms of matrix elements emerges rigorously on the real-energy axis. The questions of the MEP and ab initio calculation were avoided in Ref. 73, since the basis N-electron wavefunctions and their energies were assumed known. The derivations in Ref. 73 observed the conditions of Eq. (6). The complex eigenvalues for resonances that had occasionally been discussed in the scattering-theory literature of nuclear physics[92] were not even mentioned by Fano.

So, a question which has to do with the formal "nature" of resonance states as well as with criteria that must be considered when attempting their calculation in N-electron systems is the following: Given the rigor and physical relevance of Fano's formalism which is done on the real-energy axis, and given that the complete spectrum consists of states that satisfy Eqs. (1) and (6), where are the complex eigenvalues of **H** and the corresponding resonance wavefunctions "hidden," and how can they be revealed and calculated? (I recall that many-body systems cannot be subjected to the "easy," even analytic, theoretical descriptions that physicists often apply in terms of simple model potentials).

The above question intrigued me for the first time when, based on the existing theory of poles of the S-matrix,[69] on the result of Siegert,[72] and on the previous work of Herzenberg et al.,[26] I wrote equation (1) in Ref. 8 (Eq. 3), as an *assertion*. At the time, I was unable to find a formally rigorous answer. This was achieved in collaboration with Komninos in 1980,[40] as outlined later. The same approach was later used for the analysis of the resonance states that are induced by an external field.[13]

Fano[73] used the stationary superposition for prediagonalized basis functions with real energies,

$$\Psi(r,E) = a(E)\Psi_0(r) + \int_0 b_E\left(E'\right)\varphi\left(r,E'\right)dE' \tag{7}$$

and formally solved for the energy-dependent coefficients, $a(E)$ and $b_E(E')$.

As pointed out in Ref. 8, of key importance in the theory of resonances is the asymptotic boundary condition in the description of the relevant wavefunctions. Accordingly, in Refs. 13,40 we started by recognizing that Eq. (7) is a *standing wave* and showed how, for large *r*, it is reduced to an equivalent solution in terms of the sum of two adjoint complex *traveling waves*, in the form,

$$\Psi(r,E) \xrightarrow[r\to\infty]{} -\sqrt{\frac{\pi}{2k}}Va(E)\left[\left(1-\frac{\lambda(E)}{i\pi}\right)e^{iN} + \left(1+\frac{\lambda(E)}{i\pi}\right)e^{-iN}\right] \tag{8}$$

where V is the matrix element $\langle\Psi_0|\mathbf{H}|\varphi(E)\rangle$, and the phase N corresponds to the different result for each potential. The value of $\lambda(E)$ is obtained by imposing the outgoing-wave boundary condition that a decaying state must have. Using Eq. (8), this means that on resonance, the coefficient of the incoming wave must be zero, ie, $\lambda(E) = -i\pi$. Therefore, using Fano's expression for $\lambda(E) = (E - E_0 - \Delta(E))/|V(E)|^2$, both the complex eigenvalue on resonance and the asymptotic part of the eigenfunction emerge naturally, expressed in terms of computable quantities:

$$E_r = E_0 + \Delta(E_r) - i\Gamma(E_r)/2, \quad \Psi_r^{as} \sim \sqrt{\frac{2\pi}{k_r}}Va(E_r)e^{iN} \tag{9}$$

So, it is the sum of Ψ_r^{as} with Ψ_0 that produces the two-part form of the complex eigenfunction on resonance, Eq. (5).[5,13,40] This form provides a guiding recipe for calculations, regardless of formalism and computational details. Apart from this, the asymptotic expression of Ψ_r^{as} has been used for the derivation of a condition on resonance which has been applied for the optimization of "small" trial wavefunctions.[82]

The derivation of the many-electron CESE outlined earlier is rigorous and justifies Eq. (3) which was stated as an assertion in Ref. 8. The remaining problem has to do with the correct and consistent computation of matrix elements using resonance wavefunctions, since they are not square-integrable. A practical method is the use of complex coordinates, introduced in 1961 for short-range potentials (nuclear physics) by Dykhne and Chaplik.[93] A related discussion can be found in Ref. 5.

According to the foregoing discussion, and to results regarding the invariance of $\langle \Psi_0|\mathbf{H}|\Psi_0 \rangle$ under coordinate rotations in $\mathbf{H}$ and Ψ_0,[39] the SPS form of trial wavefunctions for the nonperturbative calculation of resonance complex eigenvalues (CESE method) is

$$\tilde{\Psi} = a_0 \Psi_0 + \sum_n a_n \tilde{u}_n \text{ (equation 7.7 of Ref. 39)} \tag{10}$$

where $\tilde{u}_n$ are N-electron configurations for asymptotic correlation, containing $(N-1)$-electron terms with real coordinates coupled to complex functions for the outgoing electron in each open channel.[5] Unlike the case of the theory and computational implementation of the CCR method, in the CESE method the coordinates of $\mathbf{H}$ are real.

The choice of the function spaces representing Ψ_0 and $\tilde{u}_n$ depends on the problem. In our treatment of atomic resonances, Ψ_0 normally consists of both MCHF numerical and correlation analytic orbitals. When the MCHF solution is impossible, as in the case of the H^- resonances,[49–51] a systematic choice of diffuse functions replaces the numerical HF ones, reaching the asymptotic region.

5.1.4 Exterior Complex Scaling and Analytic Continuation Involving the Asymptotic Region

The proposal discussed in Ref. 63, concerning the systematic calculation of matrix elements of $\mathbf{H}$ and $\mathbf{H}^2$ involving resonance wavefunctions, later named the method of "*exterior complex scaling*," is in the same spirit, namely, that of recognizing in practice the different roles played by Ψ_0 and by the asymptotic component of the resonance eigenfunction.

Specifically, the ECS regularization procedure proposed in Ref. 63 was written as:

$$\int_{\text{all space}} \psi^2 dr = \int_0^R \psi^2 dr + \int_C \psi^2 ds \text{ (equation 3 of Ref. 63)} \tag{11}$$

where R is a point on the real axis at the edge of the inner region and $R < Res < \infty$.

I quote from Ref. 63 "*Construction (3) shows that, as the resonance width tends to zero, the contour integral is brought to the real axis and the integral* $\int_{\text{all space}}$ *is finite on the real line because it is taken over bound states. This suggests that* *the square-integrable function Ψ_0 and Gamow's [resonance] function can be thought of as being related* via *analytic continuation.*"

For early numerical applications to various potentials, including that of the "volcanic" ground state of He_2^{2+}, see Ref. 94.

5.2 Topic II

The solution of MEPs for this topic has been achieved via the construction and implementation of the *many-electron, many-photon theory*, which is a non-perturbative, non-Hermitian, SPS formulation in the spirit of the CESE theory for field-free resonances. Lack of space does not permit elaboration. The reader is referred to section 11 of Ref. 5 and Refs. 10–15,95. Hamiltonians with electric (static and periodic) and magnetic static fields have been used in prototypical applications. For example:

(1) Energy shifts and single-electron ionization transition rates for absorption of one or more photons have been calculated, for one-, two-, and three-color fields.[5,10–15]

(2) Quantitative studies were carried out on DESs perturbed by strong fields, whereby many continua are mixed, and positions and widths change due to the perturbation. For example, see Refs. 14 for DESs of H^- and He in ac and dc electric fields, and 95 for DESs of H^- in a static magnetic field.

(3) Cross-sections for the photoejection of two-electrons from the closed-shell $1s^2\ {}^1S$ state of He and H^-, as well as from the open-shell metastable state $1s2s2p\ {}^4P^o$ of He^-, were determined for energies very close to threshold.[96,97]

5.2.1 Field-Induced Resonances from Analytic Continuation of Large-Order Perturbation Series

A formally possible alternative to nonperturbative variational methods for the calculation of complex energies of resonances is the construction and implementation of methods of *large-order perturbation theory* (LOPT) in conjunction with techniques of summation. The subject of LOPT is rather esoteric. Even though such approaches have features of mathematical elegance, the reality of the MEP has restricted their application to one-electron systems. Reviews of LOPT methods as applied to resonances created by the application of electric fields to hydrogen were given by Silverstone[98] and Silverman and Nicolaides[99] in the book.[100]

In Section 5.1.4, I quoted a statement from Ref. 63 regarding the possibility of direct connection between the square-integrable Ψ_0 and the resonance eigenfunction, Ψ_r, via the change of boundary conditions and analytic continuation.

Within a different framework, Reinhardt[101] produced an enlightening analysis and justification regarding analytic continuation for the calculation of the ground state resonance of hydrogen in a strong electric field in terms of Padé summation of the divergent LOPT Rayleigh–Schrödinger series.

In the approach of Silverman and Nicolaides,[99,102,103] complex energies were calculated to high accuracy not only for the ground state but also for hundreds of degenerate excited levels of hydrogen. The analytic continuation is accomplished by shifting the origin of the real eigenvalue series into the complex plane, where the relevant divergent series are summed by a twofold application of Padé approximants. Because this formalism is based on expansions over state-specific function spaces, the formalism is applicable to N-electron problems.

5.3 Topic III

The solution of the METDSE is achievable in terms of the SSEA.[4,6,16–18] The SSEA is conceptually simple, as well as general. For each problem of interest, we first make an approximate analysis of the apparent requirements set by the pulse characteristics and the spectrum of the stationary states. Then, the SSEA solution, $|\Psi(t)\rangle_{\mathrm{SSEA}}$, is constructed and computed in the form (I omit the index for each possible channel):

$$|\Psi(t)\rangle_{\mathrm{SSEA}} = \sum_m a_m(t)|m\rangle + \int_0 b_\varepsilon(t)|\varepsilon\rangle d\varepsilon \tag{12}$$

The expansion (12) holds for atoms as well as for molecules. $|m\rangle$ represents the relevant discrete *state-specific* wavefunctions and $|\varepsilon\rangle$ are energy-normalized state-specific scattering states. Resonance states are represented by their localized component. The mixing complex coefficients are time dependent. The Hamiltonian is either $\mathbf{H}_{A(M)}$ for autoionization, with Ψ_0 being the initial ($t=0$) wave packet, or the full $\boldsymbol{H}(t)$ of Eq. (2) for atom (molecule)–pulse interaction. With these Hamiltonians, and with the state-specific wavefunctions constituting $|\Psi(t)\rangle_{\mathrm{SSEA}}$, numerically accurate bound–bound, bound–free, and free–free matrix elements are calculated.

Substitution of (12) into the METDSE produces a system of coupled equations containing energies, matrix elements, and the unknown coefficients. In the process of its solution, one can monitor and evaluate with transparency and economy the dependence of the evolution of $\Psi(t)$ of

Eq. (2) on each $|m\rangle$ and $|\varepsilon\rangle$. Since these stationary wavefunctions are state specific, the solution of the METDSE according to the SSEA takes into account:

- The zero-order features of electronic structures of initial, intermediate, and final states. These can be calculated at the HF or MCHF levels.
- The dominant electron correlations, for those states where this is necessary.
- The presence of perturbed or unperturbed Rydberg levels and multiply excited states.
- The contribution of the continuous spectrum of energy-normalized, channel-dependent scattering states, without or with resonance states.
- The interchannel coupling.

See Refs. 4,6 and their references.

A rather spectacular spectroscopic result in this direction has been the experimental and theoretical discovery that, at the fundamental level of electronic structure and dynamics of atoms and molecules, there is a *relative time delay* in the emission of different electrons upon absorption of an energetic photon.[104] The theory and many-electron approach used in Ref. 104 for the treatment of the dynamics of the photoejection of the $2s$ and $2p$ electrons of Neon by an attosecond pulse are explained again in Ref. 4.

6. EPILOGUE

Topics I, II, and III of Section 2, with which this paper has dealt, belong to the high-lying portion of the "excitation axis" of Fig. 1 and contain broad categories of *resonance/nonstationary* states and of electronic dynamics involving the continuous spectrum near and far from threshold, for Hamiltonians without and with external strong fields.

After a brief review of the situation that existed with theory and calculations on N-electron systems when our approaches were introduced (Section 4), I outlined how the requirements for the solution of a variety of MEP have been satisfied in terms of efficient SPS formalisms and computational methodologies. These have been developed and implemented within generally applicable single- and multistate Hermitian (real energies) and non-Hermitian (complex energies), time-dependent and time-independent frameworks, which place emphasis on the proper *forms* of the trial wavefunctions, as in Eqs. (5), (10), and (12), and on the choice and use of corresponding function spaces.

The methods of calculation of correlated wavefunctions beyond the zero-order HF or MCHF levels are *variational* and constitute the standard SPS approach. In addition, in the case of time-independent formulations of problems of field-induced resonances, it has been demonstrated that the implementation of *large-order perturbation theory* is also possible, using a formalism which incorporates analytic continuation and efficient summation of divergent series, and is in principle applicable to many-electron ground and excited states (see Section 5.2.1).

The core ideas of the SPS theory are stated in Section 3. Reviews of the SPS methodology and applications to the topics discussed here can be found in Refs. 3–7.

The corresponding results for wavefunctions, properties, and phenomena are numerous, covering a broad spectrum of prototypical cases. Where experimental data are available, the comparison between them and those obtained from SPS calculations shows very good agreement.

ACKNOWLEDGMENTS

I am thankful to Yannis Komninos and Theodoros Mercouris for our decades-old collaboration on the topics discussed in this chapter.

REFERENCES

1. Wigner, E. On the Interaction of Electrons in Metals. *Phys. Rev.* **1934**, *46*, 1002.
2. Taylor, G. R.; Parr, R. G. Superposition of Configurations: The Helium Atom. *Proc. Natl. Acad. Sci. U. S. A.* **1952**, *38*, 154.
3. Nicolaides, C. A. Quantum Chemistry and Its "Ages" *Int. J. Quantum Chem.* **2014**, *114*, 963.
4. Nicolaides, C. A. Quantum Chemistry on the Time Axis: Electron Correlations and Rearrangements on Femtosecond and Attosecond Scales. *Mol. Phys.* **2016**, *114*, 453. http://dx.doi.org/10.1080/00268976.2015.1080870.
5. Nicolaides, C. A. Theory and State-Specific Methods for the Analysis and Computation of Field-Free and Field-Induced Unstable States in Atoms and Molecules. *Adv. Quantum Chem.* **2010**, *60*, 163.
6. Mercouris, Th.; Komninos, Y.; Nicolaides, C. A. The State-Specific Expansion Approach and Its Application to the Quantitative Analysis of Time-Dependent Dynamics in Atoms and Small Molecules. *Adv. Quantum Chem.* **2010**, *60*, 333.
7. Nicolaides, C. A. State- and Property-Specific Quantum Chemistry. *Adv. Quantum Chem.* **2011**, *62*, 35; Nicolaides, C. A.; Beck, D. R. Approach to the Calculation of the Important Many-Body Effects of Photoabsorption Oscillator Strengths. *Chem. Phys. Lett.* **1975**, *36*, 79.
8. Nicolaides, C. A. Theoretical Approach to the Calculation of Energies and Widths of Resonant (Autoionizing) States in Many-Electron Atoms. *Phys. Rev. A* **1972**, *6*, 2078.
9. Sinanoğlu, O. Electron Correlation in Atoms and Molecules. *Adv. Chem. Phys.* **1969**, *14*, 237.
10. Mercouris, Th.; Nicolaides, C. A. Polyelectronic Theory of Atoms in Strong Laser Fields. CO_2-Laser Seven-Photon Ionization of H^-. *J. Phys. B* **1988**, *21*, L285.

11. Nicolaides, C. A.; Mercouris, Th. Multiphoton Ionization of Negative Ions in the Presence of a dc-field. Application to Li^-. *Chem. Phys. Lett.* **1989**, *159*, 45.
12. Nicolaides, C. A.; Mercouris, Th.; Aspromallis, G. Many-Electron, Many-Photon Theory of Non-Linear Polarizabilities. *J. Opt. Soc. Am. B* **1990**, 7, 494.
13. Nicolaides, C. A.; Themelis, S. I. Theory of the Resonances of the LoSurdo-Stark Effect. *Phys. Rev. A* **1992**, *45*, 349.
14. Mercouris, Th.; Nicolaides, C. A. Computation of the Widths of Doubly Excited States Coupled by External AC or DC Fields. *J. Phys. B* **1991**, *24*, L557; Nicolaides, C. A.; Themelis, S. I. Doubly Excited Autoionizing States in a DC Field. Widths, Polarizabilities and Hyperpolarizabilities of the He $2s^2\ {}^1S$ and 2s2p ${}^3P^o$ states. *J. Phys. B* **1993**, *26*, L387.
15. Mercouris, Th.; Themelis, S. I.; Nicolaides, C. A. Nonperturbative Theory and Computation of the Nonlinear Response of He to dc- and ac-Fields. *Phys. Rev. A* **2000**, *61*, 013407.
16. Mercouris, Th.; Komninos, Y.; Dionissopoulou, S.; Nicolaides, C. A. Computation of Strong-Field Multiphoton Processes in Polyelectronic Atoms. State-Specific Method and Application to H and Li^-. *Phys. Rev. A* **1994**, *50*, 4109.
17. Nicolaides, C. A.; Dionissopoulou, S.; Mercouris, Th. Time-Dependent Multiphoton Ionization from the He 1s2s ${}^1S$ Metastable State. *J. Phys. B* **1996**, *29*, 231.
18. Nicolaides, C. A.; Mercouris, Th. On the Violation of the Exponential Decay Law in Atomics Physics: Ab Initio Calculation of the Time-Dependence of the He^- $1s2p^2\ {}^4P$ Nonstationary State. *J. Phys. B* **1996**, *29*, 1151; Mercouris, Th.; Nicolaides, C. A. Time Dependence and Properties of Nonstationary States in the Continuous Spectrum of Atoms. *J. Phys. B* **1997**, *30*, 81.
19. Lefebvre, R., Moser, C., Eds. Correlation Effects in Atoms and Molecules. In *Adv. Chem. Phys.* **1969**, *14. Articles written by Brueckner, K. A.; Čizek, J.; Grimaldi, F.; Jucys, A. P.; Judd, B. R.; Kelly, H. P.; Löwdin, P.-O.; Nesbet, R. K.; Sandars, P. G. H.; Sinanoğlu, O.; Tolmachev, V. V.*
20. Roothaan, C. C. F.; Bagus, P. S. Atomic Self-Consistent Field Calculations by the Expansion Methods. In Alder, B., Fernbach, S., Rotenberg, M., Eds.; Methods in Computational Physics, Vol. 2; Academic Press: New York, **1963**. p 47; Roos, B.; Salez, C.; Veillard, A.; Clementi, E. A General Program for Calculation of Atomic SCF Orbitals by the Expansion Method. *IBM Technical Report RJ 518*; 1968.
21. Burke, P. G. Resonances in Electron Scattering and Photon Absorption. *Adv. Phys.* **1965**, *14*, 521.
22. Smith, K. Resonant Scattering of Electrons by Atomic Systems. *Rep. Prog. Phys.* **1966**, *29*, 373.
23. O'Malley, T. F.; Geltman, S. Compound-Atom States for Two Electron Systems. *Phys. Rev.* **1965**, *137*, A1344.
24. Berk, A.; Bhatia, A. K.; Junker, B. R.; Temkin, A. Projection-Operator Calculations of the Lowest e^-–He Resonance. *Phys. Rev. A* **1986**, *34*, 4591.
25. Altick, P. L.; Moore, E. N. Configuration Interaction in the Helium Continuum. *Phys. Rev.* **1996**, *147*, 59.
26. Herzenberg, A.; Kwok, K. L.; Mandl, F. Resonance Scattering Theory. *Proc. Phys. Soc.* **1964**, *84*, 477; Bardsley, J. N.; Herzenberg, A.; Mandl, F. Electron Resonances of the H_2^- Ion. *Proc. Phys. Soc.* **1966**, *89*, 305.
27. Hylleraas, E. A New Stable State of the Negative Hydrogen Ion. *Astrophys. J.* **1950**, *111*, 209.
28. Holøien, E. The $(2s^2)^1S$ Solution of the Non-Relativistic Schrödinger Equation for Helium and the Negative Hydrogen Ion. *Proc. Phys. Soc. A* **1958**, *71*, 357.
29. Holøien, E.; Midtdal, J. New Investigations of the ${}^1S$ Autoionizing States of He and H^-. *J. Chem. Phys.* **1966**, *45*, 2209; Perkins, J. F. Variational-Bound Method for Autoionizing States. *Phys. Rev.* **1969**, *178*, 89.

30. Taylor, H. S.; Nazaroff, G. V.; Golebiewski, A. Qualitative Aspects of Resonances in Electron-Atom and Electron-Molecule Scattering, Excitation and Reactions. *J. Chem. Phys.* **1966**, *45*, 2872.
31. Eliezer, I.; Pan, Y. K. Calculations on Some Excited States of He^-. *Theor. Chim. Acta (Berlin)* **1970**, *16*, 63; Fels, M. F.; Hazi, A. U. Calculation of Energies and Widths of Compound-State Resonances in Elastic Scattering: Stabilization Method. *Phys. Rev. A* **1971**, *4*, 662.
32. Simon, B. Resonances in N-Body Quantum Systems with Dilatation Analytic Potentials and the Foundations of Time-Dependent Perturbation Theory. *Ann. Math.* **1973**, *97*, 247.
33. Doolen, G. D. A Procedure for Calculating Resonance Eigenvalues. *J. Phys. B* **1975**, *8*, 525; Doolen, G. D. Complex Scaling: An Analytic Model and Some New Results for e^++H Resonances. *Int. J. Quantum Chem.* **1978**, *14*, 523.
34. Reinhardt, W. P. Method of Complex Coordinates: Application to the Stark Effect in Hydrogen. *Int. J. Quantum Chem. Symp.* **1976**, *10*, 359.
35. Cerjan, C.; Hedges, R.; Holt, C.; Reinhardt, W. P.; Scheibner, K.; Wendoloski, J. J. Complex Coordinates and the Stark Effect. *Int. J. Quantum Chem.* **1978**, *14*, 393.
36. Chu, S.-I.; Reinhardt, W. P. Intense Field Multiphoton Ionization via Complex Dressed States: Application to the H Atom. *Phys. Rev. Lett.* **1977**, *39*, 1195.
37. Maquet, A.; Chu, S.-I.; Reinhardt, W. P. Stark Ionization in dc and ac Fields: An L^2 Complex-Coordinate Approach. *Phys. Rev. A* **1983**, *27*, 2946.
38. Shirley, J. H. Solution of the Schrödinger Equation with a Hamiltonian Periodic in Time. *Phys. Rev.* **1965**, *138*, B979.
39. Nicolaides, C. A.; Beck, D. R. Complex Scaling, Time-Dependence and the Calculation of Resonances in Many-Electron Systems. *Int. J. Quantum Chem.* **1978**, *14*, 457.
40. Nicolaides, C. A.; Komninos, Y.; Mercouris, Th. Theory and Calculation of Resonances Using Complex Coordinates. *Int. J. Quantum Chem. Symp.* **1981**, *15*, 355.
41. Bylicki, M.; Nicolaides, C. A. Computation of Resonances by Two Methods Involving the Use of Complex Coordinates. *Phys. Rev. A* **1993**, *48*, 3589.
42. Bednarz, E.; Bylicki, M. Fast Convergent Approach for Computing Atomic Resonances. *Int. J. Quantum Chem.* **2002**, *90*, 1021.
43. Kulander, K. C. Time-Dependent Hartree-Fock Theory of Multiphoton Ionization: Helium. *Phys. Rev. A* **1987**, *36*, 2726.
44. Kulander, K. C. Time-Dependent Theory of Multiphoton Ionization of Xenon. *Phys. Rev.* **1988**, *38*, 778.
45. Nicolaides, C. A.; Zdetsis, A.; Andriotis, A. State-Specific Many-Electron Theory of Core Levels in Metals: The 1s Binding Energy of Be Metal. *Solid State Commun.* **1984**, *50*, 857.
46. Bylicki, M.; Jaskólski, W.; Oszwałdowski, R. Resonant Tunneling Lifetimes in Multi-Barrier Structures—A Complex Coordinate Approach. *J. Phys. Condens. Matter* **1996**, *8*, 6393.
47. Nicolaides, C. A. Comment on "Is Radioactive Decay Really Exponential?" by Aston P. J. *Eur. Phys. Lett.* **2013**, *101*, 42001.
48. Aston, P. J. Is Radioactive Decay Really Exponential? *Eur. Phys. Lett.* **2012**, *97*, 52001.
49. Bylicki, M.; Nicolaides, C. A. Theoretical Resolution of the H^- Resonance Spectrum up to the n = 4 Threshold, I: States of $^1P^o$, $^1D^o$, and $^1F^o$ Symmetries. *Phys. Rev. A* **2000**, *61*, 052508.
50. Bylicki, M.; Nicolaides, C. A. Theoretical Resolution of the H^- Resonance Spectrum up to the n = 4 Threshold, II: States of ^1S and ^1D Symmetries. *Phys. Rev. A* **2000**, *61*, 052509.
51. Bylicki, M.; Nicolaides, C. A. Theoretical Resolution of the H^- Resonance Spectrum up to the n = 5 Threshold: States of $^3P^o$ Symmetry. *Phys. Rev. A* **2000**, *65*, 012504.

52. Gailitis, M.; Damburg, R. The Influence of Close Coupling on the Threshold Behaviour of Cross Sections of Electron-Hydrogen Scattering. *Proc. Phys. Soc. Lond.* **1963**, *82*, 192.
53. Komninos, Y.; Nicolaides, C. A. Multi-Channel Reaction Theory and Configuration-Interaction in the Discrete and in the Continuous Spectrum. Inclusion of Closed Channels and Derivation of Quantum Defect Theory. *Zeits. Phys. D* **1987**, *4*, 301.
54. Komninos, Y.; Nicolaides, C. A. Many-Electron Approach to Atomic Photoionization: Rydberg Series of Resonances and Partial Photoionization Cross Sections in Helium, Around the n = 2 Threshold. *Phys. Rev. A* **1986**, *34*, 1995.
55. Komninos, Y.; Aspromallis, G.; Nicolaides, C. A. Theory and Computation of Perturbed Spectra. Application to the Al ^{2}D Relativistic J = 5/2, 3/2 Spectrum. *J. Phys. B* **1995**, *28*, 2049.
56. Komninos, Y.; Nicolaides, C. A. Quantum Defect Theory for Coulomb and Other Potentials in the Framework of Configuration Interaction and Implementation to the Calculation of ^{2}D and ^{2}F^o Perturbed Spectra of Al. *J. Phys. B* **2004**, *37*, 1817.
57. Komninos, Y.; Nicolaides, C. A. Effects of Configuration Interaction on Photoabsorption Spectra in the Continuum. *Phys. Rev. A* **2004**, *70*, 042507.
58. Taylor, P. R.; Bauschlicher, C. N.; Langhoff, S. R. The ^{2}D Rydberg Series of AlI. *J. Phys. B* **1988**, *21*, L333.
59. Dyubko, S. F.; Efremov, V. A.; Gerasimov, V. G.; MacAdam, K. B. Microwave Spectroscopy of Al Rydberg States: ^{2}F^o Terms. *J. Phys. B* **2003**, *36*, 3797; Dyubko, S. F.; Efremov, V. A.; Gerasimov, V. G.; MacAdam, K. B.; Dyubko, S. F.; Efremov, V. A.; Gerasimov, V. G.; MacAdam, K. B. Microwave Spectroscopy of Al Atoms in Rydberg States: D and G Terms. *J. Phys. B* **2003**, *36*, 4827.
60. Seaton, M. J. Quantum Defect Theory. *Rep. Prog. Phys.* **1983**, *46*, 167.
61. Fano, U. Connection Between Configuration-Mixing and Quantum Defect Treatments. *Phys. Rev. A* **1978**, *17*, 93.
62. Greene, C. H.; Fano, U.; Strinatti, G. General Form of the Quantum Defect Theory. *Phys. Rev. A* **1979**, *19*, 1485; Greene, C. H.; Rau, A. R. P.; Fano, U. General Form of the Quantum Defect Theory. II. *Phys. Rev. A* **1982**, *26*, 2441.
63. Nicolaides, C. A.; Beck, D. R. The Variational Calculation of Energies and Widths of Resonances. *Phys. Letts.* **1978**, *65A*, 11.
64. Feshbach, H. A Unified Theory of Nuclear Reactions, I. *Ann. Phys. (N.Y.)* **1958**, *5*, 357.
65. Feshbach, H. A Unified Theory of Nuclear Reactions, II. *Ann. Phys. (N.Y.)* **1958**, *19*, 357.
66. Nicolaides, C. A. Physical Constraints on Nonstationary States and Non-Exponential Decay. *Phys. Rev. A* **2002**, *66*, 022118.
67. Heitler, W. *The Quantum Theory of Radiation*, 3rd ed.; Clarendon Press: Oxford, **1954**.
68. Schönberg, M. On the General Theory of Damping in Quantum Mechanics. *Nuovo Cim.* **1951**, *8*, 817.
69. Goldberger, M. L.; Watson, K. M. *Collision Theory*; John Wiley & Sons: New York, **1964**.
70. Simon, B. Convergence of Time-Dependent Perturbation Theory for Autoionizing States of Atoms. *Phys. Lett.* **1971**, *36A*, 23.
71. Sommerfeld, A. *Partial Differential Equations in Physics*; Academic Press: New York, **1964**, Section 28.
72. Siegert, A. J. F. On the Derivation of the Dispersion Formula for Nuclear Reactions. *Phys. Rev.* **1939**, *56*, 750.
73. Fano, U. Effects of Configuration Interaction on Intensities and Phase Shifts. *Phys. Rev.* **1961**, *124*, 1866.

74. Nicolaides, C. A. Hole-Projection, Saddle Points and Localization in the Theory of Autoionizing States. *Phys. Rev. A* **1992**, *46*, 690.
75. Nicolaides, C. A.; Piangos, N. A. State-Specific Approach and Computation of Resonance States. Identification and Properties of the Lowest $^2P^o$ and ^2D Triply Excited States of He^-. *Phys. Rev. A* **2001**, *64*, 052505.
76. Nicolaides, C. A.; Komninos, Y.; Beck, D. R. The K-Shell Binding Energy of Be and Its Fluorescence Yield. *Phys. Rev. A* **1983**, *27*, 3044.
77. Nicolaides, C. A.; Mercouris, Th. Partial Widths and Interchannel Coupling in Autoionizing States in Terms of Complex Eigenvalues and Complex Coordinates. *Phys. Rev. A* **1985**, *32*, 3247.
78. Nicolaides, C. A.; Beck, D. R. On the Theory of KLL Auger Energies. *Chem. Phys. Lett.* **1974**, *27*, 269.
79. Beck, D. R.; Nicolaides, C. A. Theory of Auger Energies in Free Atoms: Application to the Alkaline Earths. *Phys. Rev. A* **1986**, *33*, 3885.
80. Piangos, N. A.; Komninos, Y.; Nicolaides, C. A. He^- ^2D Weakly Bound Triply Excited Resonances: Interpretation of Previously Unexplained Structures in the Experimental Spectrum. *Phys. Rev. A* **2002**, *66*, 032721.
81. Komninos, Y.; Aspromallis, G.; Nicolaides, C. A. Resonance Scattering Theory: Application to the Broad He^- 1s2s2p $^2P^o$ Resonance. *Phys. Rev. A* **1983**, *27*, 1865.
82. Mercouris, Th.; Nicolaides, C. A. Localized and Asymptotic Electron Correlation in Autoionizing States in Terms of Complex Coordinates. *J. Phys. B* **1984**, *17*, 4127.
83. Komninos, Y.; Makri, N.; Nicolaides, C. A. Electronic Structure and the Mechanism of Autoionization for Doubly Excited States. *Z. Phys. D* **1986**, *2*, 105.
84. Chrysos, M.; Aspromallis, G.; Komninos, Y.; Nicolaides, C. A. Partial Widths of the He^- ^2S Two-Electron Ionization Ladder Resonances. *Phys. Rev. A* **1992**, *46*, 5789.
85. Sinanis, Ch.; Aspromallis, G.; Nicolaides, C. A. Electron Correlation in the Auger Spectra of the Ne^+ $K2s2p^5$ $(^{3,1}P^o)3p$ ^2S Satellites. *J. Phys. B* **1995**, *28*, L423.
86. Themelis, S. I.; Nicolaides, C. A. Energies, Widths and l-Dependence of the H^- ^3P and He^- ^4P TEIL States. *J. Phys. B* **1995**, *28*, L379.
87. Aspromallis, G.; Sinanis, Ch.; Nicolaides, C. A. The Lifetimes of the Fine Structure Levels of the Be^- $1s^22s2p^2$ ^4P Metastable State. *J. Phys. B* **1996**, *29*, L1.
88. Sinanis, Ch.; Komninos, Y.; Nicolaides, C. A. Computation of the Position and the Width of the B^- $1s^22s^22p^2$ ^1D Shape Resonance. *Phys. Rev. A* **1998**, *57*, R3158.
89. Petsalakis, I. D.; Mercouris, Th.; Theodorakopoulos, G.; Nicolaides, C. A. Theory and Ab-Initio Calculations of Partial Widths And Interchannel Coupling in Predissociation Diatomic States. Application to HeF. *Chem. Phys. Lett.* **1991**, *182*, 561.
90. Komninos, Y.; Nicolaides, C. A. Electron Correlation, Geometry and Energy Spectrum of Quadruply Excited States. *Phys. Rev. A* **1994**, *50*, 3782.
91. Bacalis, N. C.; Komninos, Y.; Nicolaides, C. A. Toward the Understanding of the He_2^- Excited States. *Chem. Phys. Lett.* **1995**, *240*, 172.
92. Peierls, R. E. Complex Eigenvalues in Scattering Theory. *Proc. R. Soc. Lond.* **1959**, *A253*, 16.
93. Dykhne, A. M.; Chaplik, A. V. Normalization of the Wavefunctions of Quasistationary States. *Sov. Phys. JETP* **1961**, *13*, 1002.
94. Nicolaides, C. A.; Gotsis, H. J.; Chrysos, M.; Komninos, Y. Resonances and Exterior Complex Scaling. *Chem. Phys. Lett.* **1990**, *168*, 570.
95. Bylicki, M.; Themelis, S. I.; Nicolaides, C. A. State-Specific Theory and Computation of a Polyelectronic Atomic State in a Magnetic Field. Application to Doubly Excited States of H^-. *J. Phys. B* **1994**, *27*, 2741.
96. Nicolaides, C. A.; Haritos, C.; Mercouris, Th. Theory and Computation of Electron Correlation in the Continuous Spectrum: Double Photoionization Cross-Section of H^- and He Near and Far from Threshold. *Phys. Rev. A* **1997**, *55*, 2830.

97. Haritos, C.; Mercouris, Th.; Nicolaides, C. A. Single and Double Photoionization Cross-Sections of the He^- 1s2s2p $^4P^o$ State at and Far from Threshold, from the Many-Electron, Many-Photon Theory. *J. Phys. B* **1998**, *31*, L783.
98. Silverstone, H. J. High-order perturbation theory and its application to atoms in strong fields. *NATO ASI Ser. B Phys.* **1990**, *212*, 295. in ref. [100].
99. Silverman, J. N.; Nicolaides, C. A. Energies and Widths of the Ground and Excited States of Hydrogen in a dc Field via Variationally-Based Large-Order Perturbation Theory. *NATO ASI Ser. B Phys.* **1990**, *212*, 309. in ref. [100].
100. Nicolaides, C. A.; Clark, C. W.; Nayfeh, M. H., Eds.; *NATO ASI Series B: Physics*, Vol. 212; Plenum Press: New York, **1990**.
101. Reinhardt, W. P. Padé Summations for the Real and Imaginary Parts of Atomic Stark Eigenvalues. *Int. J. Quantum Chem.* **1982**, *21*, 133.
102. Silverman, J. N.; Nicolaides, C. A. Complex Stark Eigenvalues via Analytic Continuation of Real High-Order Perturbation Series. *Chem. Phys. Lett.* **1988**, *153*, 61.
103. Silverman, J. N.; Nicolaides, C. A. Complex Stark Eigenvalues for Excited States of Hydrogenic Ions from Analytic Continuation of Real Variationally Based Large-Order Perturbation Theory. *Chem. Phys. Lett.* **1991**, *184*, 321.
104. Schultze, M.; Fiess, M.; Karpowicz, N.; Gagnon, J.; Korbam, M.; Hofstetter, M.; Neppl, S.; Cavalieri, A. L.; Komninos, Y.; Mercouris, Th.; Nicolaides, C. A.; Pazourek, R.; Nagele, S.; Feist, J.; Burgdörfer, J.; Azzeer, A. M.; Ernstorfer, R.; Kienberger, R.; Kleineberg, U.; Goulielmakis, E.; Krausz, F.; Yakovlev, V. S. Delay in Photoemission. *Science* **2010**, *328*, 1658.

CHAPTER NINE

High-Temperature Superconductivity in Strongly Correlated Electronic Systems

Lawrence J. Dunne[*,†,‡,1], **Erkki J. Brändas**[§], **Hazel Cox**[‡]
[*]School of Engineering, London South Bank University, London, United Kingdom
[†]Imperial College London, London, United Kingdom
[‡]University of Sussex, Brighton, United Kingdom
[§]Institute of Theoretical Chemistry, Ångström Laboratory, University of Uppsala, Uppsala, Sweden
[1]Corresponding author: e-mail address: dunnel@lsbu.ac.uk

Contents

Abstract

In this chapter we give a selective review of our work on the role of electron correlation in the theory of high-temperature superconductivity (HTSC). The question of how electronic repulsions might give rise to off-diagonal long-range order (ODLRO) in high-temperature superconductors is currently one of the key questions in the theory of condensed matter. This chapter argues that the key to understanding the occurrence of HTSC in cuprates is to be found in the Bohm–Pines Hamiltonian, modified to include a polarizable dielectric background. The approach uses reduced electronic density matrices and discusses how these can be used to understand whether ODLRO giving rise to superconductivity might arise from a Bohm–Pines-type potential which is comprised of a weak long-range attractive tail and a much stronger short-range repulsive Coulomb interaction. This allows time-reversed electron pairs to undergo a

Advances in Quantum Chemistry, Volume 74
ISSN 0065-3276
http://dx.doi.org/10.1016/bs.aiq.2016.06.003

superconducting condensation on alternant cuprate lattices. Thus, a detailed summary is given of the arguments that such interacting electrons can cooperate to produce a superconducting state in which time-reversed pairs of electrons effectively avoid the repulsive hard-core of the interelectronic Coulomb interaction but reside on average in the attractive well of the effective potential. In a superconductor the plasma wave function becomes the longitudinal component of a massive photon by the Anderson–Higgs mechanism. The alternant cuprate lattice structure is the key to achieving HTSC in cuprates with $d_{x^2-y^2}$ symmetry condensate symmetry.

1. PER OLOV LÖWDIN

In celebrating the centenary of Per Olov Löwdin's birth and reflecting upon his contributions to science, we cannot help being in awe at the breadth and depth of his insights. The review we present here is in a real sense a small tribute to Per Olov Löwdin's enormous contribution to theoretical chemistry and physics. Configuration interaction, electron correlation, density matrices, eigenvalues of density matrices, natural orbitals and geminals, variational problems, and effective Hamiltonians all have a prominent part and Per Olov played a key role in their development. In a series of seminars on correlation and density matrix theory given by Per Olov in Uppsala, during the very cold Swedish winter of 1978/1979, he reviewed the conditions for superconductivity stressing the necessity of a macroscopically large eigenvalue of the second-order reduced electronic density matrix analogous to Bose–Einstein condensation. It is not widely known that complementing the famous work of Yang, such a condition was also discovered at about the same time by Sasaki working in Per Olov's group in Uppsala as discussed here. It was during these seminars in 1978/1979 and inspired by Per Olov that the ideas discussed in this chapter were born.

2. INTRODUCTION TO SUPERCONDUCTIVITY

Superconductors are materials which transport electric charge without resistance[1] and with the display of associated macroscopic quantum phenomena such as persistent electrical currents and magnetic flux quantization. These quantum phenomena are associated with macroscopic wave functions characteristic of off-diagonal long-range order (ODLRO), a name first introduced by Yang.[2] This order is characterized by a long-range coherence of the quantum mechanical phase which demonstrates itself as macroscopic quantum phenomena. The focus of this chapter is to give a very selective review of the role of electron correlation in high-temperature superconductivity (HTSC) where we put forward the view that this is very largely an electron correlation effect due to electronic repulsions but with a prominent role played by the plasma modes in the energetics and electrodynamics of superconductors. This review follows earlier work on aspects of these topics.[3–7]

Bednorz and Müller[8] first made the discovery of HTSC in the cuprates which set off a very well-known avalanche of theoretical and experimental activity resulting in the discovery of many hole-doped and electron-doped superconducting cuprates and their related properties. Sometime after the discovery of the cuprates (see Fig. 1 for chemical structures), iron-based high-temperature superconducting compounds[9,10] were discovered in Japan. These two types of chemical structures highlighted the role of alternant lattices which exist in these materials and shown specifically for the cuprates below in Fig. 1 and discussed in Refs. 1–7. For structures of the iron-based compounds see, eg, Ref. 11.

The microscopic origin driving the superconducting condensation in these solids is the focus of very many theories and speculations which are too numerous to review here. However, there is a widely held view that HTSC materials are "electronic superconductors" in which the participation of phonons is at best secondary. No theory has yet been widely accepted despite numerous studies and many notable and highly interesting

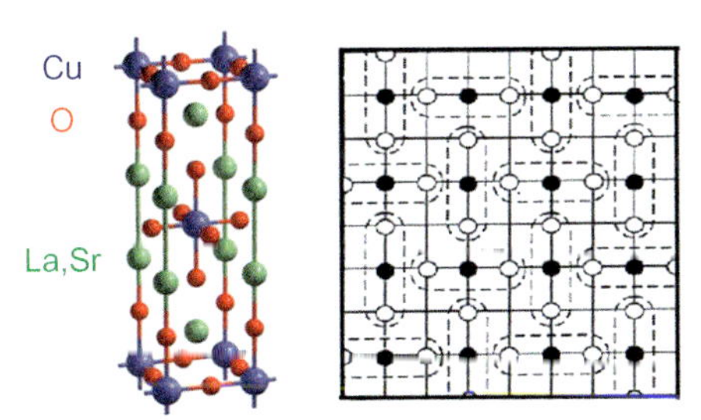

Fig. 1 *Left side*: Structure of lanthanum strontium cuprate. *Right side*: Alternant cuprate layer. •, copper; ∘, oxygen. Left side: *Reproduced from Barišić, N.; Chan, M. K.; Li, Y.; Yu, G.; Zhao, X.; Dressel, M.; Smontara, A.; Greven, M. Universal Sheet Resistance and Revised Phase Diagram of the Cuprate High-Temperature Superconductors.* Proc. Natl. Acad. Sci. U.S.A. ***2013,*** 110 *(30), 12235–12240. DOI: 10.1073/pnas.1301989110 with permission of National Academy of Sciences of the United States of America.* Right side: *Reproduced from Dunne, L. J.; Murrell, J. N.; Brändas, E. J. Off-Diagonal Long Range Order from Repulsive Electronic Correlations in a Localised Model of a High Tc Cuprate Superconductor.* Physica C ***1990,*** 169, *501–507 with permission of Elsevier.*

proposals.[12–14] There are many aspects and properties of HTSC materials which we do not consider here such as charge density waves and the origin of the pseudogap which have both attracted wide attention. Hence here we will focus in a limited way on our view of the role of Coulomb repulsions in high-temperature cuprate superconductivity.

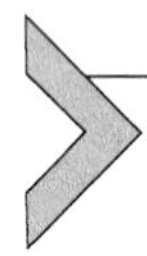

3. THE PAIRING INSTABILITY IN CLASSICAL AND HIGH-TEMPERATURE SUPERCONDUCTORS

The theory of conventional low-temperature metallic superconductors is due to Bardeen–Cooper–Schrieffer (BCS) theory[15] where there is an attractive phonon-induced electron–electron attraction which causes a Cooper pair instability.[12] In a conventional metallic superconductor[12] the effective electron–electron interaction V_{eff} is made up of the contributing interactions $V_{eff} = V_{sc} + V_{el-ph-el}$. Here V_{sc} is the screened Coulomb repulsion and $V_{el\text{–}ph\text{–}el}$ is the phonon-induced electron–electron interaction.[12,13,15] The BCS Hamiltonian or variants of it have been widely discussed, but some original points will be revisited later. It does not seem likely that the BCS Hamiltonian applies to the cuprates but the study of this points to the features which will give rise to superconductivity in the cuprates.

There is a well-known symmetry in the doping phase diagram for HTSC cuprates where electron doping or hole doping of the cuprate layer both lead to similar magnetic and superconducting features.[5] To give our point of view, in alternant lattices (see Fig. 1) unit cells may be given a sign with opposite signs for nearest neighbors. Electron pairs in time-reversed states on alternant cuprate lattices interact with a short-range Coulomb repulsion and a longer range attraction. Electrons in the coherent ground state can avoid each other at short range. It is suggested here that the cause of the long-range attractive part of the effective interaction is to be found in the short-range part of the Bohm–Pines[16–20] potential to be discussed later and shown in Fig. 2.

The symmetry of the condensate wave function has been experimentally established in hole-doped cuprates, and there is a broadly held opinion that the condensate wave function has $d_{x^2-y^2}$ symmetry in superconducting hole-doped cuprates[12,13,21,22] and this seems probable too in electron-doped cuprates.[23]

The model describes in detail a mechanism in which the Coulombic interactions allow a superconducting state on a cuprate layer to appear. Pairs

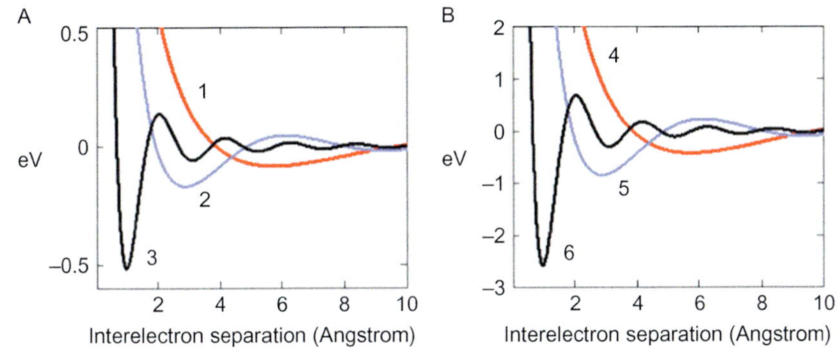

Fig. 2 Bohm–Pines effective electron–electron potential, $v_{sr}(r)$ calculated from Eq. (20) for (A) 1, $k_c = 0.5$ Å^{-1}. 2, $k_c = 1.0$ Å^{-1}. 3, $k_c = 3$ Å^{-1}, each with $\varepsilon = 5$, (B) 4, $k_c = 0.5$ Å^{-1}. 5, $k_c = 1.0$ Å^{-1}. 6, $k_c = 3$ Å^{-1}, each with $\varepsilon = 1$. Note that for the same k_c value the potential is inversely proportional to ε.

of electrons in the superconducting condensate correlate over longer distances in the attractive well in the Bohm–Pines potential[5,17,18] and avoid each other at very short separations. This gives a lower Coulomb repulsion than in the uncorrelated normal state with which it competes. It is a mechanism for electrons to stay out of the hard Coulombic core but reside in the attractive region of the potential at longer range. A conventional BCS singlet-type s-wave condensate wave function does not have such embedded correlations. Yet, a d-wave or sign-alternating s-wave does have such in built electronic correlations.

BCS used an independent particle model whose success was a surprise. In an important paper which is seldom referred to in recent discussions of superconductivity, Bardeen and Pines[16] give a careful justification for such a theoretical line of attack on the problem of superconductivity. As is well known the long-range Coulomb interaction between electrons in a good metal is screened out. The basis for the success of the apparent neglect of the long-range Coulomb interaction arises from the role of the zero-point energies of the plasmon modes in a metal. This is intimately tied up with currently hotly discussed questions about the massive longitudinal photons and the Anderson–Higgs mode in superconductors. Historically, the idea of the Higgs particle and mass acquisition by elementary particles finds its origin in Anderson's famous paper[24] written following an idea put forward by Schwinger.[25] Anderson's idea was taken up by Higgs[26] (see Ref. 27 for a more complete historical perspective) and others and the rest, including the Large Hadron Collider at CERN, as they say—is history.

In a superconductor the system undergoes a spontaneously broken symmetry and chooses a phase. The low-lying phase excitations or Goldstone mode becomes the longitudinal degree of freedom of a massive vector Boson via the Anderson–Higgs mechanism.[28,29] Much more work remains to be done to clarify the interconnection between the Thomas–Fermi screening and the role of the Anderson–Higgs mechanism.

The plasma modes are comprised of the long-wavelength Fourier components of the Coulomb interaction leaving behind a residual screened electron–electron interaction. The plasmon energies are high enough that they are not thermally populated at room temperature and therefore they do in a sense drop out of the problem. So the BCS wave function is actually a wave function for the quasiparticles.

In conventional superconductors exchange of virtual phonons induces an attractive effective interaction between electrons of opposite spin and momentum whereby the Fermi sea becomes unstable to electron pair formation. This is the so-called Cooper pair instability[5,13,15] which is widely discussed as an n-fold stabilization effect. Consider n-degenerate states which are Slater determinants in a basis of pairwise occupied states required to obtain a coherence of the sign of the matrix elements. The diagonal elements are at an energy U above some reference energy with off-diagonal elements matrix element $-V$ due to the attractive phonon-induced electron–electron coupling. One of the eigenvalues at energy $U-nV$ splits off from the rest giving the well-known Cooper pair instability (see Refs. 3–7,12,13 for discussion).

However as has been discussed in detail elsewhere beginning in 1979[3–7,30] (see Section 1 for the historical context) other scenarios with repulsive matrix elements are possible which give a similar type of n-fold stabilization effect. We consider a real symmetric $2k$-dimensional configurational interaction Hamiltonian matrix composed of two blocks. The first block is a k-dimensional diagonal submatrix with diagonal elements equal to U. The second is a k-dimensional full off-diagonal block with a repulsive matrix element, $V>0$ in all sub-block elements as indicated later.[3–7]

The following eigenvalue/eigenvector relationships given below are easily verified:

$$\left(\begin{array}{cc|cc} U & 0 & V & \cdots \\ 0 & \ddots & V & V \\ \hline V & V & U & 0 \\ \cdots & V & 0 & \ddots \end{array}\right)\begin{pmatrix} 1 \\ \cdots \\ -1 \\ \cdots \end{pmatrix} = (U-kV)\begin{pmatrix} 1 \\ \cdots \\ -1 \\ \cdots \end{pmatrix} \qquad \left(\begin{array}{cc|cc} U & 0 & V & \cdots \\ 0 & \ddots & V & V \\ \hline V & V & U & 0 \\ \cdots & V & 0 & \ddots \end{array}\right)\begin{pmatrix} 1 \\ \cdots \\ 1 \\ \cdots \end{pmatrix} = (U+kV)\begin{pmatrix} 1 \\ \cdots \\ 1 \\ \cdots \end{pmatrix} \tag{1}$$

The lowest energy (U–kV) eigenvector is a $2k$-dimensional vector with elements ± 1. Real-space pairwise occupation of the single-particle localized states making up the Slater determinant basis functions is essential to obtain the coherence of the matrix element signs and give the block structure to the Hamiltonian matrix shown earlier.

Subject to the magnitudes of U and kV the lowest eigenvalue at U–kV can thus cross over into a new low energy ground state just as in Cooper's problem.

In the condensed matter literature there has been extensive discussion[12,31] of the BCS gap equation given by

$$\Delta_k = -\sum_{k'} V_{k,k'} \frac{\Delta_{k'}}{2E_{k'}} \tag{2}$$

for the zero temperature case where E_k and Δ_k are, respectively, the BCS excitation energy and the gap parameter. Superconducting low energy solutions exist for repulsive matrix elements $V_{k,k'}$ if the variational coefficients Δ_k change sign coherently across the Fermi surface. Although expressed somewhat differently, such a result is a special case of the more general situation discussed for the first time in Ref. 30 and discussed in Section 1 and followed up in other publications.[3–7,32] We will return to this question where Eq. (2) is derived by the assumption of an extreme state and the existence of Yang's ODLRO is to be discussed later.

4. ODLRO IN SUPERCONDUCTORS

A key insight into the ordered nature of the superconducting state was suggested by London[33] who proposed the centrally important idea of momentum space ordering of the electrons in superconductors and the rigidity or stiffness of the superconducting wave function. It has taken a very long time since this and other early insights[34] to understand even conventional superconductivity. It was some while after BCS theory[15] appeared that the nature of the ordering in superconductivity became clearer. Ginzburg and Landau[35] had created a very successful model of a superconductor near to the transition temperature by invoking a superconducting order parameter with an associated wave function. Following earlier work of Onsager and Penrose[36] on the condensate in liquid ^{4}He, and other important work by Gorkov on a nondiagonal order parameter,[37] Yang[2] introduced the concept of ODLRO. Yang's analysis gives a method for deciding whether a many-electron wave function has the ODLRO property which characterizes the superconducting state whereby an electron pair population analysis is undertaken using the second-order reduced electronic density matrix $\rho_2(\mathbf{x}_1, \mathbf{x}_2; \mathbf{x}_1', \mathbf{x}_2')$[3–7,38,39] (see also Sasaki[40] referred to in Section 1). Evaluation of the eigenvalues of a pair-space sub-block P of $\rho_2(\mathbf{x}_1, \mathbf{x}_2; \mathbf{x}_1', \mathbf{x}_2')$ can demonstrate the existence of ODLRO in an electronic wave function. The second-order reduced electronic density matrix for a many-electron wave function $\Psi(\mathbf{x}_1, \mathbf{x}_2, \ldots, \mathbf{x}_{2M})$ is defined in the Yang normalization as

$$\rho_2\left(\mathbf{x}_1, \mathbf{x}_2; \mathbf{x}_1', \mathbf{x}_2'\right) = 2M(2M-1)\int \Psi(\mathbf{x}_1, \mathbf{x}_2, \mathbf{x}_3, \ldots, \mathbf{x}_{2M})\Psi^* \times \left(\mathbf{x}_1', \mathbf{x}_2', \mathbf{x}_3, \ldots, \mathbf{x}_{2M}\right) d\mathbf{x}_3 \cdots d\mathbf{x}_{2M} \tag{3}$$

In order to undertake a pair population analysis $\rho_2(\mathbf{x}_1, \mathbf{x}_2; \mathbf{x}_1', \mathbf{x}_2')$ may be expressed as

$$\rho_2\left(\mathbf{x}_1, \mathbf{x}_2; \mathbf{x}_1', \mathbf{x}_2'\right) = \sum_{ij,kl} g_{ij}(\mathbf{x}_1, \mathbf{x}_2) g_{kl}^*\left(\mathbf{x}_1'\mathbf{x}_2'\right) P_{ij,kl} = \mathbf{gPg}^\dagger \tag{4}$$

$P_{ij,kl}$ is an element of the pair subspace population coefficient matrix $\mathbf{P}$. In Eq. (4) $g_{ij}(\mathbf{x}_1, \mathbf{x}_2)$ is a two-electron Slater determinant and $\mathbf{x}_1, \mathbf{x}_2, \ldots$ are spin–space variables. The normalization of $g_{ij}(\mathbf{x}_1, \mathbf{x}_2)$ is such that the density matrix eigenvalues correspond to populations of electron pairs in a particular germinal state. ODLRO is present and characterizes a superconducting condensate for a many-electron wave function $\Psi(\mathbf{x}_1, \mathbf{x}_2, \ldots, \mathbf{x}_{2M})$ when one of the eigenvalues λ_L of the matrix $\mathbf{P}$ is macroscopically large. This macroscopically large number of electron pairs populating the same pair state is analogous to Bose–Einstein condensation.[2,5,6] Leggett[12] has given that an extensive discussion of the eigenvector of the density matrix associated with the large eigenvalue is superconducting condensate wave function $\psi(\mathbf{x}_1, \mathbf{x}_2)$ and the relation to superconductivity.

Diagonalization of $\mathbf{P}$ in $\mathbf{gPg}^\dagger$ gives

$$\mathbf{gPg}^\dagger = \mathbf{gSS}^\dagger\mathbf{PSS}^\dagger\mathbf{g}^\dagger = \left(g_{11}' \;\; g_{22}', \;\; \ldots\right)\begin{pmatrix} \lambda_1 & & & & \\ & \lambda_2 & & & \\ & & \ddots & & \\ & & & \ddots & \\ & & & & \ddots \\ & & & & & \lambda_N \end{pmatrix}\begin{pmatrix} g_{11}'^* \\ g_{22}'^* \\ \vdots \\ g_{NN}'^* \end{pmatrix} \tag{5}$$

The unitary transformation $\mathbf{S}$ which is the matrix of orthonormal eigenvectors of the matrix $\mathbf{P}$ relates the bases $\mathbf{g}$ and $\mathbf{g}'$.

Hence in diagonal form

$$\rho_2\left(\mathbf{x}_1, \mathbf{x}_2; \mathbf{x}_1', \mathbf{x}_2'\right) = \sum_i^n g_{ii}'(\mathbf{x}_1, \mathbf{x}_2) g_{11}'^{*}\left(\mathbf{x}_1', \mathbf{x}_2'\right)\lambda_i \tag{6}$$

The mean of the eigenvectors associated with the small eigenvalues are assumed to be negligible in the limit when the couple at $(\mathbf{x}_1 \sim \mathbf{x}_2)$ is well separated from $\left(\mathbf{x}_1' \sim \mathbf{x}_2'\right)$. The electron pair density matrix then factorizes to give a Ginzburg–Landau-type macroscopic wave function[35] of the form $\Psi(\mathbf{x}_1, \mathbf{x}_2) = \sqrt{\lambda_L} g_L'(\mathbf{x}_1, \mathbf{x}_2)$. To connect with the n-fold stabilization problems above consider the simplest Coleman[41,42] extreme state where we have M electron pairs distributed in a pairwise fashion over possible permutations of pair states occupying N time-reversed pair states. We will consider both attractive and repulsive cases where each Slater determinant basis function has the same absolute weight and so this extreme state is close in form to a projected BCS wave function. For both attractive and repulsive cases the matrix $\mathbf{P}$ has a macroscopically large eigenvalue. First, as shown below for Cooper's problem with attractive matrix elements we have

$$\begin{pmatrix} \frac{M}{N} & M\frac{(N-M)}{N(N-1)} & M\frac{(N-M)}{N(N-1)} & \cdots \\ M\frac{(N-M)}{N(N-1)} & \ddots & \vdots & \vdots \\ M\frac{(N-M)}{N(N-1)} & \vdots & \frac{M}{N} & \vdots \\ \vdots & \vdots & \vdots & \ddots \end{pmatrix} \begin{pmatrix} 1 \\ 1 \\ 1 \\ \vdots \\ \vdots \\ 1 \end{pmatrix} = \left(M\left(1-\frac{M}{N}\right)+\frac{M}{N}\right)\begin{pmatrix} 1 \\ 1 \\ 1 \\ \vdots \\ \vdots \\ 1 \end{pmatrix} \tag{7}$$

and in Eq. (35) for repulsive matrix elements. The details of how these matrix elements are obtained are given in full in Refs. 3,7. It can be seen that both scenarios for attractive and repulsive matrix elements in these degenerate systems can produce low energy states exhibiting ODLRO with superconducting properties. The point of view presented here is that the cuprates and possibly the iron-based superconductors[9–11] are examples of the latter scenario. The reader is referred to recent developments[43] relating to the large eigenvalue of the cumulant part of the two-electron reduced density matrix as a measure of ODLRO relevant to HTSC. In this chapter we will consider the reduced electronic density matrix of the quasiparticle wave function which comes from the Bohm–Pines method.[17–19]

Before moving on we will make a connection between BCS theory[15] and Yang's concept of ODLRO[2] and Coleman's extreme state given in Ref. 32. We consider a many-electron Hamiltonian for n electrons given by

$$H = \sum_{i=1}^{n} h_i + \sum_{i\langle j}^{n} h_{ij} \tag{8}$$

where the terms on the right are the sums of one-body and two-body interactions.

We focus only on that part of the energy (the pairing energy) which changes in a BCS condensation.

Eq. (8) can be decomposed into a sum of electron pair Hamiltonians using center of mass (R) and internal coordinates (r):

$$H_2 = \left(\frac{p_r^2}{2\mu_r} + \frac{P_R^2}{2M_R}\right) + V(r) \tag{9}$$

where $\mu_r = m_e/2 \text{and} M_R = 2m_e$. In an extreme state the total pairing energy $\langle E \rangle$ can be expressed as

$$\langle E \rangle = \lambda_L \langle g_L | H_2 | g_L \rangle + \lambda_s \sum_{k=2}^{N} \langle g_k | H_2 | g_k \rangle \tag{10}$$

where λ_L and λ_s are the large and small eigenvalues of the box P defined earlier.

Expanding $\sqrt{\lambda_L} |g_L\rangle = \sum_k F_k^L \exp(i\mathbf{k}.\mathbf{r})$ and $\sqrt{\lambda_s} |g_s\rangle = \sum_k F_k^s \exp(i\mathbf{k}.\mathbf{r})$

Following Refs. 12,32 the BCS pairing energy is

$$\langle E \rangle = \sum_k \left| F_k^L \right|^2 \frac{\hbar k^2}{m} + \sum_{s=2}^{N} \sum_k \left| F_k^s \right|^2 \frac{\hbar k^2}{m} + \sum_{k,k'} F_k^* F_{k'} V_{k,k'} \tag{11}$$

If we now assume that the magnitude of the Fourier coefficients is essentially constant over the Fermi surface, we obtain

$$\langle E\rangle = \frac{1}{N}(\lambda_L + (N-1)\lambda_s)\sum_k \left(\frac{\hbar k^2}{m}\right) + \sum_{k,k'} F_k^* F_{k'} V_{k,k'} \tag{12}$$

which for an extreme state can be written

$$\langle E\rangle = \frac{M}{N}\sum_k \left(\frac{\hbar k^2}{m}\right) + \sum_{k,k'} F_k^* F_{k'} V_{k,k'} \tag{13}$$

Following Ref. 32 we also have

$$\left|F_k^L\right|^2 = \lambda_L/N = \frac{M}{N} - \left(\frac{M}{N}\right)^2 \tag{14}$$

which can be solved to give

$$2\frac{M}{N} = 1 - \sqrt{1 - 4|F_k|^2} \tag{15}$$

Insertion into Eq. (14) gives

$$\langle E\rangle = \sum_k \left(1 - \sqrt{1 - 4|F_k|^2}\right)\left(\frac{\hbar k^2}{2m}\right) + \sum_{k,k'} F_k^* F_{k'} V_{k,k'} \tag{16}$$

Minimization of $\langle E\rangle$ with respect to F_k^* and appropriate identification of F_k^* with E_k and Δ_k which are, respectively, the BCS excitation energy and the gap parameter gives the BCS gap equation given by Eq. (2) and in detail in Refs. 12,32, for example. It is appropriate to remark that Eq. (15) has another root which may be appropriate to hole doping of a nearly full band.

5. THE BOHM PINES HAMILTONIAN

The existence of plasma oscillations in metals was suggested many years ago, yet the understanding of this continues to develop. A plasmon[44,45] may be described as a quantized excitation of a neutral system made up of positive and negative charges. The longitudinal oscillations of such a system (like sound waves) give rise to unique behavior and here it is argued that it is intimately related to the occurrence of HTSC in cuprates and iron-based materials.

The importance of plasma oscillations in metals was not appreciated until the work of Bohm and Pines was published during the early 1950s in a series of groundbreaking papers. Although initially controversial, the physical picture which has emerged has been very significant in theoretical condensed matter physics. At that time the theory of metals was still in an unsatisfactory state (for an early review, see Ref. 45 and for a more recent review by David Pines, see Ref. 13, p. 85).

Bohm and Pines[16] demonstrated that the Coulomb interaction in a metal can be split into two regions. These are a long-range (*lr*) part due to the collective plasmon modes and a short-range (*sr*) part which is the screened Coulomb interaction. It is this "screened" interaction which will interest us here. Plasmons have been described as "longitudinal photons."[17] In a superconductor following the Anderson–Higgs mechanism a Plasmon is a longitudinal component of a massive vector Boson which also gives mass to the transverse photons.[24,27–29]

We consider a collection of charges $\{\mu\}$ with masses m_μ, charges Z_μ, and coordinates $\mathbf{r}_\mu$ in a cuprate superconductor. The system has an overall electrical neutrality so that $\sum_\mu Z_\mu = 0$.

The system of charges interacts in the presence of polarizable atomic cores which give rise to a high-frequency background dielectric constant ε into which the set of charges $\{Z_\mu\}$ is introduced. The charge carriers associated with the cuprate layers give rise to screening of the type appropriate to an electron gas, yet the polarizable atomic cores give rise to the screening associated with a classical dielectric medium which we represent by the high-frequency dielectric constant ε. The standard Hamiltonian for all the particles is given by

$$H = \sum_\mu \frac{1}{2m_\mu}\left(\mathbf{p}_\mu - \frac{Z_\mu}{c}\mathbf{A}(\mathbf{r}_\mu)\right)^2 + \frac{1}{2\varepsilon}\sum_{\mu,\nu}{}' \frac{Z_\mu Z_\nu}{r_{\mu\nu}} \tag{17}$$

where $\mathbf{A}$ is an irrotational magnetic vector potential. The electric scalar potential is zero and does not appear in the Hamiltonian in the temporal gauge.

We will split the total Hamiltonian into $H = H_{sr} + H_{lr}$ with $\mathbf{A}.\mathbf{p}$, $\mathbf{p}.\mathbf{A}$-type coupling terms neglected. Thus, H is split into two decoupled Hamiltonians given by

$$H = H_{sr} + H_{lr} \tag{18}$$

which corresponds to decomposition into short-range H_{sr} and long-range H_{lr} Hamiltonians. Henceforth in the short-range Hamiltonian H_{sr} the nuclei will be regarded as infinitely heavy so that the Born–Oppenheimer approximation is effectively made so that only electronic motions will be considered.

5.1 The Short-Range Hamiltonian H_{sr}

The short-range Hamiltonian is given by

$$H_{sr} = \sum_{\mu} \frac{\mathbf{p}_{\mu}^{2}}{2m_{\mu}} + \frac{2\pi}{\varepsilon\Omega} \sum_{\mu,\nu}{}' Z_{\mu} Z_{\nu} \sum_{k>k_c} \frac{1}{k^2} \exp\left(i\mathbf{k}.\mathbf{r}_{\mu\nu}\right) \tag{19}$$

The last term will now be evaluated by replacing the sum by integration over k-space.

Hence the short-range Hamiltonian is given by

$$H_{sr} = \sum_{\mu} \frac{\mathbf{p}_{\mu}^{2}}{2m_{\mu}} + \frac{1}{2} \sum_{\mu,\nu}{}' \frac{Z_{\mu} Z_{\nu}}{\varepsilon r_{\mu\nu}} \left(1 - \frac{2}{\pi} \int_{0}^{k_c r_{\mu\nu}} \frac{\sin(kr)}{kr} d(kr) \right) \tag{20}$$

The short-range potential itself, $v_{sr}(r) = Z_{\mu} Z_{\nu}/\varepsilon r \left(1 - 2/\pi \int_{0}^{k_c r} \sin(kr)/kr\, d(kr) \right)$, between two charges Z_{μ}, Z_{ν} at separation r is the second term in Eq. (20). This is easily evaluated numerically and is extremely interesting and will be discussed in detail next.

k_c is the inverse screening length. We will estimate k_c using the inverse Thomas–Fermi screening length given by $k_c^2 = 16\pi^2 me^2 (3\eta_c/\pi)^{1/3}/\varepsilon h^2$, where η_c is the carrier density and ε the high-frequency dielectric constant of the polarizable background taken to be about 5 as used in Refs. 3–7. We are forced by the difficulty of the problem to use the Thomas–Fermi approximation[46] to estimate screening lengths. For a hole-doped valence band or an electron-doped conduction band as in cuprate superconductors the screening length is readily related to the carrier concentration η_c by $k_c^2 = 16\pi^2 me^2 (3\eta c/\pi)^{1/3}/\varepsilon h^2$. If the bands are anisotropic, the details change but the qualitative picture remains.

In Fig. 2 the potential $v_{sr}(r)$ calculated from Eq. (20) for two electrons is plotted. The potential has a long-range oscillatory tail.

At short range the Bohm–Pines effective potential is very close numerically with the Thomas–Fermi potential $v(r)_{\mathrm{TF}} = 1/\varepsilon r \exp\left(-k_c^r\right)$.

It is the long-range oscillations in such a potential which will be the focus for the occurrence of HTSC. These are also well known as Friedel oscillations[47] and discussed by March and Murray[48] and Langer and Vosko.[49]

5.2 Long-Range Hamiltonian and Plasma Oscillations

The long-range Hamiltonian is given by

$$H_{lr} = \frac{1}{2\varepsilon}\sum_{k<k_c} P_k^* P_k + \frac{\omega^2}{2}\sum_{k<k_c} Q_k^* Q_k - \left(\sum_{\mu}\frac{2\pi Z_{\mu}^2}{\varepsilon\Omega}\right)\sum_{k<k_c}\frac{1}{k^2} \tag{21}$$

where

$$\omega^2 = \frac{4\pi}{\Omega}\sum_{\mu}\frac{Z_{\mu}^2}{m_{\mu}}$$

The last term in H_{lr} is a constant self-energy term.

The above readily gives the energies of the plasmons as

$$E_k = \left(n_k + \frac{1}{2}\right)\hbar\omega_p, \text{ where } \omega_p^2 = \frac{4\pi}{\varepsilon\Omega}\sum_{\mu}\frac{Z_{\mu}^2}{m_{\mu}} \tag{22}$$

where $n_k = 0, 1, 2, \ldots$. The coefficients $\{Q_k\}$ are the Fourier components of the longitudinal magnetic vector potential discussed in detail in Ref. 18. The plasmon energies are large compared to thermal energies. Hence a large part of the correlation energy is accounted for by the plasmon zero-point energies namely $1/2\hbar\omega_p$ for each mode k.

The eigen functions of H_{lr} are the plasmon wave functions. The plasmons are sometimes regarded as longitudinal photons. A longitudinal electromagnetic wave can propagate in the electron gas at ω_p. The plasmon energies are usually regarded as too high to be easily excited and thus may be regarded as remaining in their ground states. Thus the ground state plasmon wave function is simply the product of the ground state wave functions of all the plasmonic oscillators which we will denote by X_{plasmon}.

6. TOTAL HAMILTONIAN

We study an effective short-range Bohm–Pines Hamiltonian H_{sr} (the random phase approximation) for the electrons on the cells of a square alternant arrangement of unit cells with local C_{4v} point group symmetry:

$$H_{sr} = \sum_i h(i) + \frac{1}{2}\sum_{i,j}{}' v_{sr}\left(r_{ij}\right) \tag{23}$$

The many-electron wave function may be expanded in a basis of Slater determinants $\{\phi_k\}$ where $\Psi(\mathbf{x}_1, \mathbf{x}_2, \ldots) = \Sigma c_k \phi_k$, where $\{c_k\}$ are the set of expansion coefficients obtained as an eigenvector of the Hamiltonian matrix. The total wave function will be a product of the form

$$X_{plasmon}\Psi(\mathbf{x_1}, \mathbf{x_2}, \ldots) = X_{plasmon}\sum_k c_k\phi_k \quad (24)$$

The total energy is then

$$\langle E\rangle = \langle H_{sr}\rangle + \sum_{k<k_c}\frac{1}{2}\hbar\omega_p - \frac{2\pi}{\varepsilon\Omega}\left(\sum_\mu Z_\mu^2\sum_{k<k_c}\frac{1}{k^2}\right) \quad (25)$$

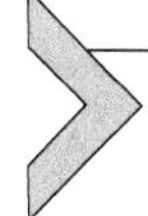

7. CHOICE OF LOCALIZED BASIS FUNCTIONS AND SUMMARY OF GROUP THEORETICAL ANALYSIS OF CUPRATE SUPERCONDUCTOR REAL-SPACE CONDENSATE WAVE FUNCTION

The cuprate lattice is an alternant layer structure with local C_{4v} local point group symmetry as shown in Fig. 1. The lattice may be partitioned into + or − sublattices where each unit cell has nearest neighbors with opposite signs. It is established experimentally that the superconducting condensate pair function in the hole-doped cuprates has the singlet $^1B_1$ symmetry under the operations of the C_{4v} point group.[50] This information about the symmetry of the cuprate two-point function condensate wave function permits some very fundamental deductions to be made about the localized electronic orbitals which can be used to construct the condensate wave function. The distance over which the condensate wave function $\psi(\mathbf{x}_1, \mathbf{x}_2)$ stays finite as $|\mathbf{r}_1 - \mathbf{r}_2| \to \infty$ is a measure of the superconducting coherence length ξ_0 or the pair size. Experiments point to a very short ξ_0 of only a few Ångstroms in cuprate superconductors[12] and this characteristic is commonly believed to indicate real-space pairing occurs in cuprate superconductors. Hence we have chosen to work in a localized basis. A widely held view is that the active electronic orbitals are derived from oxygen (2p) or copper (3d) bands or some hybrid of these. We are seeking to identify a single or group of localized orbitals which allows pairs of electrons to evade the short-range Coulomb repulsion and yet to exploit any longer range attractive region of the effective electron–electron interaction. Early on after the discovery of superconducting cuprates Sawatzky et al.[51] made a group theoretical analysis of the charge transfer in CuO but which was before the discovery of the d-wave condensate in superconducting cuprates. Following this, some years ago[50] we identified a pair of localized Wannier-type orbitals which we labeled as ($\boldsymbol{px}$, $\boldsymbol{py}$) with e-symmetry in the C_{4v} point group, which seem on balance likely to play a significant role in cuprate superconductivity. In the cuprate layers oxygen $2p_x$, $2p_y$, $2p_z$ orbitals hybridize with the Cu d-orbitals to form a set of symmetry-adapted orthogonalized localized basis functions $\{\phi_i(\mathbf{r})\}$ which make up the condensate wave function. The shapes of these is given in Fig. 3A and B.

Decomposing $\psi(\mathbf{x}_1, \mathbf{x}_2)$ leads to

$$\Psi(\mathbf{x}_1, \mathbf{x}_2) = \chi(\Gamma) + \sum_k \varphi_k \tag{26}$$

The first term on the right-hand side χ (Γ) indicates the sum of the pair functions centered on the principal axis at the center of the unit cell transforming as the irreducible representation Γ in the C_{4v} point group. The second term on the right represents a linear combination of all the terms which transform collectively as the irreducible representation Γ and most significantly these are derived from the surrounding crystal. If the condensate wave function $\psi(\mathbf{x}_1, \mathbf{x}_2)$ has $^1B_1$ symmetry under the operations of the C_{4v} group, then all the terms in Eq. (26) are required to individually have $^1B_1$ symmetry.

If the Wannier functions are assumed real, the $(\boldsymbol{px}(1)\boldsymbol{px}(2) - \boldsymbol{py}(1)\boldsymbol{py}(2))$ pairing is composed of electrons in time-reversed states where such a choice of

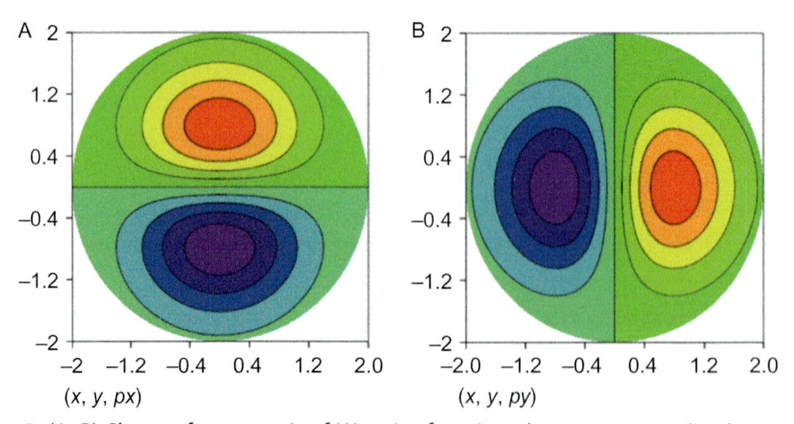

Fig. 3 (A, B) Shape of *px–py* pair of Wannier functions (or *e*-representation in square symmetry) for cuprate superconductors. These orbitals are expected to be largely out-of-phase combinations of ligand O(2p) orbitals as discussed in the text. The lobes with *red* and *blue* centers have opposite signs. *Reproduced from Dunne, L. J.; Murrell, J. N.; Brändas, E. J. Off-Diagonal Long Range Order from Repulsive Electronic Correlations in a Localised Model of a High Tc Cuprate Superconductor.* Physica C ***1990,*** 169, *501–507 with permission of John Wiley.*

pairing is considered the best candidate for producing robust HTSC. We disregard other pairing candidates as unlikely for reasons given elsewhere.[3–50,52] These are shown on a cuprate lattice below and we will now discus how we can obtain a low energy superconducting state from electronic correlations where the ($\boldsymbol{px}, \boldsymbol{py}$) choice of active localized basis functions leads straightforwardly to low energy ground state exhibiting ODLRO.

8. THE SUPERCONDUCTING GROUND STATE

We consider a square cuprate lattice composed of $N/2$ cells as shown in Fig. 1 (right). A pair $\{\phi_{l,px}(\mathbf{x})\}$, $\{\phi_{l,py}(\mathbf{x})\}$ of Wannier-type functions are localized at the center of each cell with index l.

Each member of the pair $(l,px\uparrow,\ l,px\downarrow)$ of Wannier orbitals is assigned a signature $(-1)^l$ and each pair $(l,py\uparrow,\ l,py\downarrow)$ a signature $(-1)^{l+1}$ as shown in Fig. 4. The layer is decorated with Wannier pair functions which show an alternant pattern. Most importantly, each Wannier function has nearest and next-nearest neighbors with opposite signs as shown in the blue/red crosses in Fig. 4.

We consider a basis of Slater determinants generated by populating M singlet-coupled electron pairs randomly over the N Wannier orbitals so that each pair is either occupied by two electrons or vacant as depicted in Fig. 5 with a filling fraction $\rho = M/N$. The number of ways of arranging M

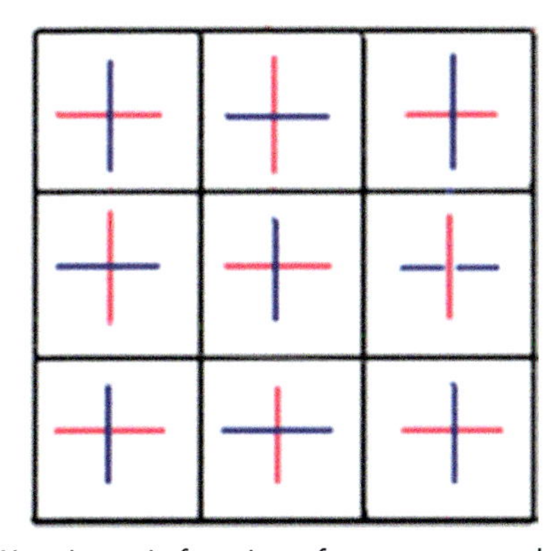

Fig. 4 Signatures of Wannier pair functions from a cuprate layer showing nine unit cells. The Wannier pair functions are labeled with a positive (*red* color) or negative sign (*blue* color). *Reproduced from Dunne, L. J.; Murrell, J. N.; Brändas, E. J. Off-Diagonal Long Range Order from Repulsive Electronic Correlations in a Localised Model of a High Tc Cuprate Superconductor.* Physica C ***1990,*** 169, *501–507 with permission of John Wiley.*

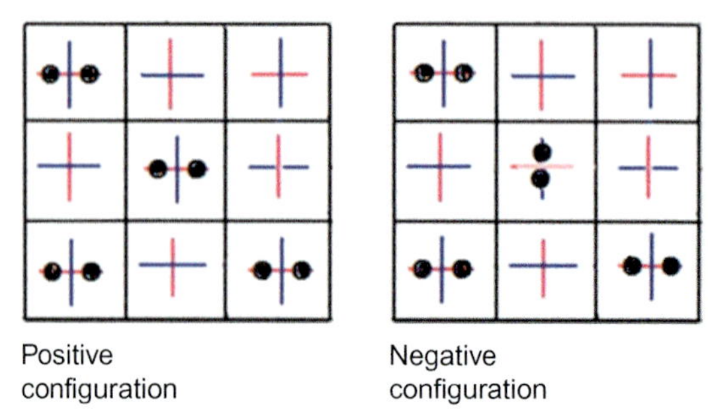

Fig. 5 Two typical configurations of singlet-coupled time-reversed electron pairs (*black balls* referred to as a "dimer" in the text) on the lattice with opposite signatures. *Reproduced from Dunne, L. J.; Murrell, J. N.; Brändas, E. J. Off-Diagonal Long Range Order from Repulsive Electronic Correlations in a Localised Model of a High Tc Cuprate Superconductor.* Physica C ***1990,*** 169, *501–507 with permission of John Wiley.*

electron singlet-coupled pairs randomly over N occupied orbitals is $N!/M!(N-M)!$. Each configuration is given an overall signature given by the product of the signs of the occupied pairs of orbitals such that each Slater determinant basis function may be grouped into one of two classes according to the overall signature.

Expansion in a basis of Slater determinants $\{\phi_k\}$ gives the many-electron wave function as $\Psi(\mathbf{x}_1, \mathbf{x}_2, \ldots) = \sum_k c_k \phi_k$. The set of variationally determined coefficients $\{c_k\}$ for a given state is obtained by calculation of the associated eigenvector of the Hamiltonian matrix where for the ground state $\{c_k\}$ is determined remarkably simply from the product $c_k = \prod_{i \in k} \sigma_i$ which runs over all occupied pairs of orbitals in the configuration k.

In second quantized form the ground state wave function is

$$\Psi(\mathbf{x}_1, \mathbf{x}_2, \ldots) = \left(\sum_l (-1)^l \left(a^\dagger_{lpx\uparrow} a^\dagger_{lpx\downarrow} - a^\dagger_{lpy\uparrow} a^\dagger_{lpy\downarrow} \right) \right)^M |0\rangle \qquad (27)$$

On average each configuration has $5M(1-\rho)$ nearest or next-nearest neighbor interactions with basis functions of the opposite sign where there is one Wannier pair difference in the occupation numbers. The matrix elements between these states populate off-diagonal blocks of the Hamiltonian matrix.

The matrix element intracell two-electron integrals for px–py pair transfers inside cell is given by

$$v = \left\langle \phi_{i,px}(\mathbf{r_1}) \phi_{ipx}(\mathbf{r_2}) | v_{sr}(r_{12}) | \phi_{i,py}(\mathbf{r_1}) \phi_{i,py}(\mathbf{r_2}) \right\rangle \qquad (28)$$

whereas next-nearest neighbor pair transfer matrix elements between px–px (or py–py) orbitals, between cells i and are given by

$$V = \left\langle \phi_{i,px}(\mathbf{r}_1)\phi_{i,px}(\mathbf{r}_2)|v_{sr}(r_{12})|\phi_{j,px}(\mathbf{r}_1)\phi_{j,px}(\mathbf{r}_2)\right\rangle \tag{29}$$

The Coulomb repulsion between a pair of electrons in the same pair of px–px (or py–py) Wannier orbitals is

$$u = \left\langle \phi_{i,px}(\mathbf{r}_1)\phi_{ipx}(\mathbf{r}_2)|v_{sr}(r_{12})|\phi_{i,px}(\mathbf{r}_1)\phi_{i,px}(\mathbf{r}_2)\right\rangle \tag{30}$$

Significant values for V and v demand significant localized orbital overlap. Only nearest- and next-nearest-neighbor interactions are important and can be neglected unless there is significant overlap. This happens most readily for Slater determinants with opposite signature and this feature renders the Hamiltonian matrix with block structure shown in Eq. (1) and discussed earlier.

If the minimum in $v_{sr}(r)$ falls outside the overlap region of two Wannier orbitals, then the contribution of the attractive parts of $v_{sr}(r)$ to the matrix elements v and V will be small. At very short range screening is not really effective and so we may use the approximation

$$v \approx \left\langle \phi_{i,px}(\mathbf{r}_1)\phi_{ipx}(\mathbf{r}_2)\left|\frac{e^2}{\varepsilon r_{12}}\right|\phi_{i,py}(\mathbf{r}_1)\phi_{i,py}(\mathbf{r}_2)\right\rangle \tag{31}$$

and similarly for V. Yet u defined in Eq. (30) can be effectively screened where the minimum in $v_{sr}(r)$ can make a significant contribution to lowering u so that it becomes less energetically demanding to bring a pair of electrons into the same Wannier orbital.

Let us assume for a very simple model in which the superconducting ground state energetically competes with a normal ground state whose wave function is a single Slater determinant Ψ_{normal}. In a localized basis with screened but locally strong Coulomb repulsive interactions and maximally unpaired electrons in the normal phase the energy is given by the expectation value of

$$\langle E\rangle = \langle H_{sr}\rangle + \sum_{k<k_c}\frac{1}{2}\hbar\omega_p - \frac{2\pi}{\varepsilon\Omega}\left(\sum_{\mu} Z_{\mu}^2 \sum_{k<k_c}\frac{1}{k^2}\right) \tag{32}$$

But the last two terms do not contribute to the energy difference between states as long as the use of the same screening constant k_c is a reasonable approximation so that only $\langle H_{sr}\rangle$ is important. An *estimate* of the energy of the short-range Hamiltonian $\langle H_{sr}\rangle$ for such a normal state with the long-range part removed is

$$\langle H_{sr}\rangle = 2M\langle h\rangle + M_D u \tag{33}$$

M_D is the number of paired electrons in the configuration. $2M\langle h\rangle$ represents the one-body terms. These energy differences between the superconducting state and such a normal state cancel out each other. Thus assuming that the sum of the Plasmon zero-point energies which describe long-range energy differences are the same in both the normal and superconducting state, then we estimate of the energy difference/orbital between the energetically competing normal and superconducting phases as

$$\begin{array}{ll} (1-\rho)u-(4V+v)\rho(1-\rho) & \text{if } \rho>1/2 \\ \rho u-(4V+v)\rho(1-\rho) & \text{if } \rho<1/2 \end{array} \tag{34}$$

The energy expression given in Eq. (34) can become negative as the screening increases with doping and this allows the superconducting state to become the ground state, as depicted in Fig. 6.

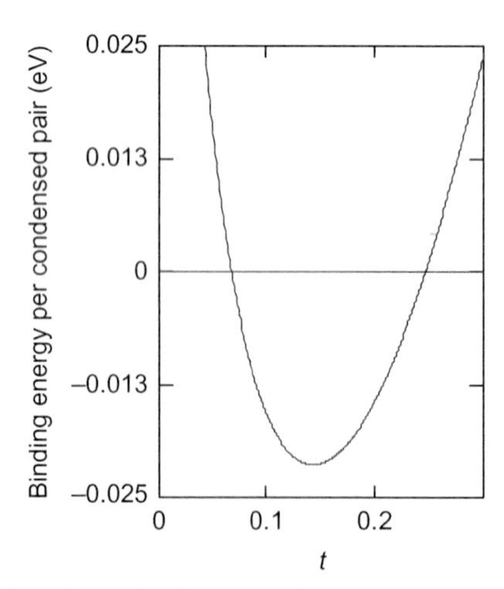

Fig. 6 Binding energy/condensed pair against dopant concentration for the parameters discussed in the text. For electron doping $t=\rho$ and for hole doping $t=(1-\rho)$. The parameterization $u\approx e^2/\varepsilon L/2\exp(-k_c L/2)$, $(4V+v)=0.64\text{eV}$ has been used. L is the unit cell length ($\approx$4 Å). *Reproduced from Dunne, L. J. High-Temperature Superconductivity and Long-Range Order in Strongly Correlated Electronic Systems.* Int. J. Quant. Chem. ***2015,*** 115, *1443–1458 with permission of John Wiley.*

The behavior of the appearance/disappearance of superconductivity observed on doping is reproduced along with the well-known electron-/hole-doping symmetry.

9. CONDENSATE WAVE FUNCTIONS

For the wave function given in Eq. (35) the pair population density coefficient pair submatrix **P** is given by

$$\begin{pmatrix} \frac{M}{N} & -M\frac{(N-M)}{N(N-1)} & M\frac{(N-M)}{N(N-1)} & \cdots \\ -M\frac{(N-M)}{N(N-1)} & \ddots & \vdots & \vdots \\ M\frac{(N-M)}{N(N-1)} & \vdots & \frac{M}{N} & \vdots \\ \vdots & \vdots & \vdots & \ddots \end{pmatrix} \begin{pmatrix} 1 \\ -1 \\ 1 \\ \vdots \\ \vdots \\ 1 \end{pmatrix} = \left(M\left(1-\frac{M}{N}\right)+\frac{M}{N}\right) \begin{pmatrix} 1 \\ -1 \\ 1 \\ \vdots \\ \vdots \\ 1 \end{pmatrix} \tag{35}$$

Such a macroscopically large eigenvalue shown above which indicates a superconducting condensation also occurs for the ground state wave function given in Eq. (1). Hence, n-fold stabilization with ODLRO can occur with both the attractive and repulsive matrix elements in the Hamiltonian matrix of the type discussed earlier.

The matrix **P** has a macroscopically large eigenvalue λ_L given by

$$\lambda_L = M\left(1-\frac{M}{N}\right)+\frac{M}{N} \tag{36}$$

The associated eigenvector as given in Eq. (9) or in normalized form

$$\psi(\mathbf{x}_1,\mathbf{x}_2) = \left(\rho-\rho^2\right)^{1/2}\sum_l (-1)^l \left(a^\dagger_{lpx\uparrow}a^\dagger_{lpx\downarrow} - a^\dagger_{lpy\uparrow}a^\dagger_{lpy\downarrow}\right)|0\rangle \tag{37}$$

The macroscopically large eigenvalue represents a condensate of electron pairs and numerically equals

Yang's upper bound[2] which was also derived by Sasaki[40] in Uppsala at about the same time as Yang did his work but Sasaki's work is rarely discussed.

The superconducting condensate density n_s is proportional to $\lambda_L/N=\rho(1-\rho)$ and is an example of Coleman's extreme case[41] discussed by Weiner and Ortiz.[42]

For hole doping when $\rho \rightarrow 1$ and electron doping when $\rho \rightarrow 0$ and then n_s is practically linear in the dopant density. This prediction has been discussed by Dunne[5] who reworked the analysis made by Dunne and Spiller[53] for two bands relevant to the case here and to which the reader is referred. Valence counting was used to calculate the hole densities used in Ref. 53. The result given in Ref. 5 is that the broad features of the experimental observations[54] are satisfactorily described by the theory. A model of thermal properties of the model is also given in Ref. 5. The singlet condensate wave function has d-wave symmetry in **k**-space and in real space and as shown below in Fig. 7A and B.

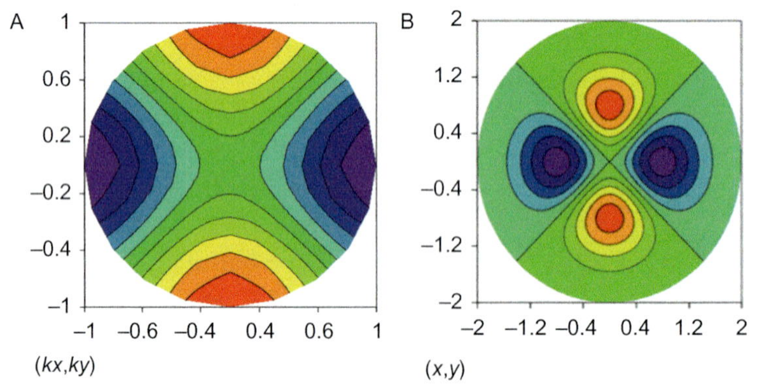

Fig. 7 (A, B) Shape of cuprate condensate wave function in *k*-space (F_k) and real-space ($\psi(r)$). The relative coordinate is $r=r_1-r_2$. The range over which $\psi(r)$ is significant measures the pair size or the coherence length. In the above figures the range shown is 4 Å but the condensate wave function remains significant out to about 10 Å. The lobes with *red* and *blue centers* have opposite signs. The *green areas* are close to zero. *Reproduced from Dunne, L. J.; Murrell, J. N.; Brändas, E. J. Off-Diagonal Long Range Order from Repulsive Electronic Correlations in a Localised Model of a High Tc Cuprate Superconductor.* Physica C ***1990,*** 169, *501–507 with permission of John Wiley.*

10. ROLE OF ELECTRON CORRELATION

An examination of the energy difference between the normal state and the superconducting state shows that there is a coherent lowering of the electronic energy on the cuprate lattice.

Let us focus simply on the electron-doped case where we have the energy difference as $\rho u - (4V + v)\rho(1 - \rho)$ from Eq. (34). The first term is the increase in Coulomb repulsion between the paired electrons. There is a long-range contribution to the integral u which is due to the minimum in v_{sr}. The latter term is a correlation term which arises from keeping pairs of electrons apart at very short range. It is the utmost importance to appreciate that the last term is made up from contributions which act collectively to lower the energy. In Fig. 4 for any unit cell the region in between red/blue lines for nearest and next-nearest neighbors are regions of space with a reduced probability of finding a pair of electrons with opposite spin close together. This collective reduction in the Coulomb repulsion allows the attractive part of the long-range Bohm–Pines potential to play a significant role in the energetics.

The sign alternation in the condensate wave function indicates that a "hole" develops in regions of space around each electron keeping pairs of electrons out of the hard-core Coulombic repulsion. This allows them to reside with higher probability at the minimum of $v_{sr}(\mathbf{r})$.

A weakness in the current approach is the use of a uniform high-frequency dielectric constant ε.

It has been shown recently[55,56] that a two-electron system can hold a bound state with a mean nuclear-electron distance $\langle r \rangle = 3.5\,\text{Å}$ for the outer electron at the critical nuclear charge for binding. It would be highly interesting to study a two-electron atomic problem with the nucleus replaced by a polarizable body to mimic the Cu atom in the situation shown in Fig. 7.

Finally we remark that electron correlation was central to Per Olov's work and in Refs. 39,57,58 the ground work for key aspects of the discussion in this contribution was laid. Also, in his book[59] on linear algebra, he discusses the "mirror theorem." In principle, it goes beyond the Born–Oppenheimer (BO) approximation. A superconductor would be divided into two parts: (a) the light fermions and (b) the heavy nuclear framework which is linked through the "mirror theorem." Investigating the consequences of this linkage for superconductors, possibly by tracing over the nuclear coordinates, might be of interest.

REFERENCES

1. Annett, J. F. *Superconductivity, Superfluidity and Condensates*; Oxford University Press: Oxford, 2004.
2. Yang, C. N. Concept of Off-Diagonal Long-Range Order and the Quantum Phases of Liquid He and of Superconductors. *Rev. Mod. Phys.* **1962**, *34*, 694–704.
3. Dunne, L. J.; Brändas, E. J. Superconductivity from Repulsive Electronic Correlations on Alternant Cuprate and Iron-Based Lattice. *Int. J. Quantum Chem.* **2013**, *113*, 2053–2059.
4. Dunne, L. J.; Brändas, E. J. Review of Off-Diagonal Long-Range Order and High-Temperature Superconductivity from Repulsive Electronic Correlations. *Adv. Quant. Chem.* **2013**, *66*, 1–30.
5. Dunne, L. J. High-Temperature Superconductivity and Long-Range Order in Strongly Correlated Electronic Systems. *Int. J. Quant. Chem.* **2015**, *115*, 1443–1458.
6. Dunne, L. J.; Murrell, J. N.; Brändas, E. J. Off-Diagonal Long Range Order from Repulsive Electronic Correlations in a Localised Model of a High Tc Cuprate Superconductor. *Physica C* **1990**, *169*, 501–507.
7. Dunne, L. J. Off-Diagonal Long-Range Order and Essential Phenomenological Description of Some Properties of High Tc Cuprate Superconductors. *Physica C* **1994**, *223*, 291–312.
8. Bednorz, J. G.; Müller, K. A. Possible High Tc Superconductivity in the Ba-La-Cu-O System. *Z. Phys. B* **1986**, *64*, 189–193.
9. Kamihara, Y.; Watanabe, T.; Hirano, M.; Hosono, H. Iron-Based Layered Superconductor La[O1-xFx]FeAs (x=0.05–0.12) with Tc=26 K. *J. Am. Chem. Soc.* **2008**, *130*, 3296–3297.
10. Takahashi, H.; Igawa, K.; Arii, K.; Kamihara, Y.; Hirano, M.; Hosono, H. Superconductivity at 43 K in an Iron-Based Layered Compound LaO1-xFxFeAs. *Nature* **2008**, *453*, 376–378.
11. Andersen, O. K.; Boeri, L. On the Multi-Orbital Band Structure and Itinerant Magnetism of Iron-Based Superconductors. *Ann. Phys.* **2011**, *523*, 8–50.
12. For a good review of early work see Leggett, A. J. *Quantum Liquids*; Oxford University Press: New York, 2006.
13. See, eg, Anderson, P. W. BCS: The Scientific "Love of My Life". In: Cooper, L. N.; Feldman, D., Eds.; BCS: 50 Years. World Scientific Publishing Co. Pte. Ltd, 2011; pp 127–142. http://dx.doi.org/10.1142/9789814304665_0008.
14. Mazumdar, S.; Clay, R. T. The Chemical Physics of Unconventional Superconductivity. *Int. J. Quantum Chem.* **2014**, *114*(16), 1053–1059.
15. Bardeen, J.; Cooper, L. N.; Schrieffer, J. R. Theory of Superconductivity. *Phys. Rev.* **1957**, *108*, 1175–1204.
16. Bardeen, J.; Pines, D. Electron-Phonon Interaction in Metals. *Phys. Rev.* **1955**, *99*, 1140–1150.
17. Bohm, D.; Pines, D. A Collective Description of Electron Interactions: III. Coulomb Interactions in a Degenerate Electron Gas. *Phys. Rev.* **1953**, *92*, 609–625.
18. Dunne, L. J. Order in Superconductors and the Repulsive Electronic Correlation Model of High Tc Cuprate Superconductivity. *J. Mol. Struct. (THEOCHEM.)* **1995**, *341*, 101–114.
19. Huag, A. *Theoretical Solid State Physics*; Pergamon Press: Oxford, 1972.
20. Donovan, B. *Elementary Theory of Metals*; Pergamon Press: Oxford, 1967.
21. Wollman, D. A.; Van Harlingen, D. J.; Lee, W. C.; Ginsberg, D. M.; Leggett, A. J. Experimental-Determination of the Superconducting Pairing State in YBCO from the Phase Coherence of YBCO-Pb DC Squids. *Phys. Rev. Lett.* **1993**, *71*, 2134–2137.
22. Tsuei, C. C.; Kirtley, J. R. Phase-Sensitive Tests of Pairing Symmetry in Cuprate Superconductors. *Physica C* **1997**, *282–287*, 4–11.

23. Tsuei, C. C.; Kirtley, J. R. Pairing Symmetry in Cuprate Superconductors. *Rev. Mod. Phys.* **2000**, *72*, 969–1016.
24. Anderson, P. W. Plasmons, Gauge Invariance, and Mass. *Phys. Rev.* **1962**, *130*, 439–442.
25. Schwinger, J. Gauge Invariance, and Mass. *Phys. Rev.* **1962**, *125*, 397–398.
26. Higgs, P. Broken Symmetries and the Masses of Gauge Bosons. *Phys. Rev. Lett.* **1964**, *13*, 508–509.
27. Allen, R. E. The London–Anderson–Englert–Brout–Higgs–Guralnik–Hagen–Kibble–Weinberg Mechanism and Higgs Boson Reveal the Unity and Future Excitement of Physics. *J. Mod. Opt.* **2014**, *61*(1). http://dx.doi.org/10.1080/09500340.2013.818170.
28. Greiter, M. Is Electromagnetic Gauge Invariance Spontaneously Violated in Superconductors?*Ann. Phys.* **2005**, *319*, 217–219.
29. Negele, J. W.; Orland, H. *Quantum Many-Particle Systems Frontiers in Physics*; Addison-Wesley: Redwood City, CA, 1988.
30. Dunne, L. J.; Brändas, E. J. Possible New Coherence Effects in Condensed Systems with Repulsive Interactions. *Phys. Lett. A* **1979**, *71*, 377–381. See also misprint correction. *Phys. Lett.* **1990,** *A149* (488), 377–381.
31. Scalapino, D. J. The Case for dx2 − y2 Pairing in the Cuprate Superconductors. *Phys. Rep.* **1995**, *250*, 329–365.
32. Brändas, E.; Dunne, L. J. Bardeen-Cooper-Schrieffer BCS Theory and Yang's Concept of Off-Diagonal Long-Range Order ODLRO. *Mol. Phys.* **2014**, *112*(5–6), 694–699.
33. London, F. *Superfluids*; Wiley: New York, 1950.
34. Gorter, C. J.; Casimir, H. B. G. On supraconductivity I. *Physica* **1934**, *1*, 306–320.
35. Ginzburg, V. L.; Landau, L. D. On Theory of Superconductivity. *Zh. Eksperim. I. Teor. Fiz.* **1950**, *20*, 1064–1082.
36. Penrose, O.; Onsager, L. Bose-Einstein Condensation and Liquid Helium. *Phys. Rev.* **1956**, *104*, 576–584.
37. Abrikosov, A. A.; Gorkov, L. P.; Dzyaloshinski, I. E., Eds.; *Methods of Quantum Field Theory in Statistical Physics*; Dover Publications New York, 1963. Translated and edited by R.A. Silverman.
38. Mc Weeney, R.; Sutcliffe, B. T. *Methods of Molecular Quantum Mechanics*; Academic Press: New York, 1969.
39. Löwdin, P. O. Quantum Theory of Many-Particle Systems. I. Physical Interpretations by Means of Density Matrices, Natural Spin-Orbitals, and Convergence Problems in the Method of Configurational Interaction. *Phys. Rev.* **1955**, *97*, 1474–1489.
40. Sasaki, F. Eigenvalues of Fermion Density Matrices. *Phys. Rev.* **1965**, *138B*, 1338–1342. Technical Note Quantum Chemistry Group, University of Uppsala, Sweden, 1962 no. 77.
41. Coleman, A. J. Structure of Fermion Density Matrices. *Rev. Mod. Phys.* **1963**, *35*, 668–689.
42. Weiner, B.; Ortiz, J. V. Construction of Unique Canonical Coefficients for Antisymmeterized Geminal Power States. *Int. J. Quant. Chem.* **2004**, *97*, 896–907.
43. Raeber, A. E.; Mazziotti, D. A. Large Eigenvalue of the Cumulant Part of the Two-Electron Reduced Density Matrix as a Measure of Off-Diagonal Long-Range Order. *Phys. Rev. A* **2015**, *92*, 052502. http://dx.doi.org/10.1103/PhysRevA.92.052502.
44. Pines, D.; Nozières, P. *Theory of Quantum Liquids*; Vol. I; W.A. Benjamin, Inc.: New York, 1966.
45. Raimes, S. Theory of Plasma Oscillations in Metals. *Rep. Prog. Phys.* **1957**, *20*, 1–37.
46. Mott, N. F.; Davis, E. A. *Electronic Processes in Non-crystalline Materials*; Clarendon Press: Oxford, UK, 1971.
47. Blandin, A.; Daniel, E.; Friedel, J. On the Knight Shift of Alloys. *Philos. Mag.* **1959**, *4*, 180–182.

48. March, N. H.; Murray, A. M. Electronic Wave Functions Round a Vacancy in a Metal. *Proc. R. Soc. Lond. A Math. Phys. Sci.* **1960**, *256*(1286), 400–415.
49. Langer, S.; Vosko, S. H. The Shielding of a Fixed Charge in a High-Density Electron Gas. *J. Phys. Chem. Solids* **1960**, *12*(2), 196–205.
50. Dunne, L. J.; Brändas, E. J.; Murrell, J. N.; Coropceanu, V. Group Theoretical Identification of Active Localised Orbital Space in High Tc Cuprate Superconductors. *Solid State Commun.* **1998**, *108*, 619–623.
51. Eskes, H.; Tjeng, L. H.; Sawatzky, G. A. Cluster-Model Calculation of the Electronic Structure of CuO: A Model Material for the High-Tc Superconductors. *Phys. Rev.* **1990**, *41B*, 288–299.
52. Barišić, N.; Chan, M. K.; Li, Y.; Yu, G.; Zhao, X.; Dressel, M.; Smontara, A.; Greven, M. Universal Sheet Resistance and Revised Phase Diagram of the Cuprate High-Temperature Superconductors. *Proc. Natl. Acad. Sci. U.S.A.* **2013**, *110*(30), 12235–12240. http://dx.doi.org/10.1073/pnas.1301989110.
53. Dunne, L. J.; Spiller, T. P. Condensate Fraction in High Tc Cuprate Superconductors. *J. Phys. Condens. Matter* **1993**, *4*, L563.
54. Uemura, Y. J.; Luke, G. M.; Sternlieb, B. J.; Brewer, J. H.; Carolan, J. F.; Hardy, W. N.; Kadono, R.; Kempton, J. R.; Kiefl, R. F.; Kreitzman, S. R.; Mulhern, P.; Riseman, T. M.; Williams, D. L.; Yang, B. X.; Uchida, S.; Takagi, H.; Gopalakrishnan, J.; Sleight, A. W.; Subramanian, M. A.; Chien, C. L.; Cieplak, M. Z.; Xiao, G.; Lee, V. Y.; Statt, B. W.; Stronach, C. E.; Kossler, W. J.; Yu, X. H. Universal Correlations Between Tc and n_s/m^* (Carrier Density over Effective Mass) in High-Tc Cuprate Superconductors. *Phys. Rev. Lett.* **1989**, *62*, 2317–2320.
55. King, A. W.; Rhodes, L. C.; Cox, H. Inner and Outer Radial Density Functions in Correlated Two-Electron Systems. *Phys. Rev. A* **2016**, *93*, 022509.
56. King, A. W.; Rhodes, L. C.; Readman, C. A.; Cox, H. Effect of Nuclear Motion on the Critical Nuclear Charge for Two-Electron Systems. *Phys. Rev. A* **2015**, *91*, 042512.
57. Löwdin, P.-O. Some Aspects of the Correlation Problem and Possible Extensions of the Independent-Particle Model. *Adv. Chem. Phys.* **1969**, *14*, 283–340.
58. Löwdin, P.-O. Band Theory, Valence Band Tight-Binding Calculations. *J. Appl. Phys.* **1962**, *33*, 251–280.
59. Löwdin, P.-O. *Linear Algebra for Quantum Theory*; Wiley-Interscience: New York, 1998.

CHAPTER TEN

Quantum Chemistry and Superconductors

Sven Larsson[1]
Chalmers University of Technology, Göteborg, Sweden
[1]Corresponding author: e-mail address: slarsson@chalmers.se

Contents

Abstract

Thirty years after the discovery of high temperature (HT) superconductivity (SC), no by all accepted theory exists. The Bardeen, Cooper, Schrieffer (BCS) model, hewed into the Bloch theory for metals, is unfit for local systems such as cuprates and organic superconductors. In this chapter, we will use a theory that dates back to Landau and Pekar, but we will avoid the effective mass approach by using a total free energy model, as designed for electron transfer problems by Marcus and Jortner. A diffusion equation is used to derive the resistivity in the local case. The original definition of Hubbard U by Mott as a metal-to-metal (or molecule-to-molecule) charge transfer energy will be updated by including the neglected negative terms. It will be shown that the absorption at 2 eV in the cuprates is indeed due to Cu–Cu charge transfer, identical to the Hubbard U or Mott transition. The model accounts for bond-length fluctuations due to occupancy of d-orbitals (extended over the ligands), or in the molecular case the π orbitals, and this makes it necessary to make a distinction between adiabatic and vertical Hubbard U. $U_{vert} = 1.5$–3 eV while U_{ad} may be a few hundred times smaller. Organic SC in aromatic hydrocarbons will be shortly reviewed and found consistent with the general model. Finally, we will discuss SC in tungsten bronzes discovered in 1964 by Matthias.

Advances in Quantum Chemistry, Volume 74
ISSN 0065-3276
http://dx.doi.org/10.1016/bs.aiq.2016.06.005

1. INTRODUCTION

In 1964 superconductivity (SC) was discovered in transition metal oxides by Matthias et al. in tungsten bronzes (alkali + WO_3) ($T_C = 0.6$ K)[1] and by Cohen et al. in reductively doped and black $SrTiO_3$ ($T_C = 0.3$ K).[2] $SrTiO_3$ is essentially a crystal built on the insulating oxides SrO and TiO_2. Not surprisingly pure WO_3, SrO, and TiO_2 form transparent, insulating compounds in their pure crystalline form. The gap to the conduction band is >3 eV. Both WO_3 and $SrTiO_3$ become superconductors or conductors after reductive doping and also acquire black or bronze colors. The connection between SC and color is surprising since the energy scales for the two phenomena are very different. On the other hand ordinary metals are known which are fully transparent for visible light in the bulk. Is there a connection between SC and the absorption in the visible region? The theory for SC given below suggests that this is the case.

In 1973 a superconducting oxide was found ($LiTi_2O_4$) at a quite high critical temperature, $T_C = 12.4$ K.[3] The Li2s lightly bound valence electrons transfer to the Ti3d band. The ligand field absorption spectrum is covered by strong absorption in the visible region. Since the band gap in TiO_2 corresponds to UV absorption (>3 eV), there are immediate difficulties to explain this absorption as due to O2p → Ti3d. As we will see in the following section, there is another explanation.

In 1975 Sleight et al. found SC with $T_C = 13$ K in $BaPb_{1-x}Bi_xO_3$.[4] Sleight pointed out that the undoped compound consists of half Bi^{3+} and half Bi^{5+} sites,[5] suggesting that there is a local $6s^2$ pair replacing the Cooper pair of the BCS model. A remarkable connection to the spectrum was also noticed. The SC region was always "black."

In the case of cuprates Bednorz and Müller used oxidative doping to remove electrons from an already unfilled Cu3d valence band in $La_{2-x}Sr_xCuO_4$. The critical temperature is high ($T_C = 35$ K), higher than in any previously known SC,[6,7] in spite of the fact that the undoped system is an insulator. Chu et al. managed to raise the critical temperature to 40 K by applying pressure.[8] Other cuprates were also found to be superconducting. $YBa_2Cu_3O_{7-x}$ (YBCO) that has a critical temperature as high as 93 K.[9] Particularly remarkable is that SC is found among transition metal oxides, systems where great efforts have been made in the past to explain why they are *insulators* at $T = 0$. This "Mott problem" will be further discussed in Section 2.

At the same time several physicists with a background in chemistry pointed out that there may be local models to understand SC. The crystal properties of the YBCO compound were studied by David et al.[10] In this connection a mechanism was suggested, equivalent to electron pair formation by disproportionation (very likely by Day; one of the coauthors). Sleight also pointed out that disproportionation [2Cu(II) → Cu(I) + Cu(III)][5] is a possible model for explanation of SC.

The present chapter commemorates the birthday 100 years ago of the late Prof. Per-Olov Löwdin. His papers on many-electron methods, including electron correlation methods, became the basis of electron structure theory of local systems (Quantum Chemistry),[11] including a rational treatment of electron correlation.

There is a need for conductivity models for local systems. Hopping models have been suggested by, for example, Pohl and Pollak.[12] Incidentally both authors were members of the Quantum Chemistry Group in Uppsala, created by Per-Olov Löwdin, when the present author joined the Group in 1964. The advantage with hopping models is that the probability of electron transfer from one metal site to the next may be calculated with the help of experimental quantities or, alternatively, using the methods of Quantum Chemistry. Conductivity may be obtained from the hopping rate. In 1989 the present author suggested to calculate conductivity and SC[13] on the basis the Marcus model for electron and electron pair transfer.[14–16]

Landau and Pekar formulated an effective mass polaron model already in 1948, and this model may be regarded as the origin of current SC models used in solid-state physics.[17] They carefully pointed out an "internal contradiction" of the theory. Nuclear coordinates are held fixed while in reality they are moving. The electrons adiabatically follow the ion motion. This "internal contradiction" is solved and forms a cornerstone in Marcus theory.

Effective mass single-electron methods are difficult to handle. On the other hand hopping model based on Marcus theory may be written as a conductivity model for electrons and electron pairs. The resistivity as a function of temperature is monotonically increasing for metals and monotonically decreasing for semiconductors. However, for cuprates $\rho(T)$ has a minimum at about 100 K[18] (see later). In other words cuprates are neither insulators nor ordinary metals.

The two-electron transfer model involves two electrons in two orbitals, centered on adjacent metal sites; in the cuprates two Cu($x^2 - y^2$) orbitals (with ligand admixtures) on adjacent sites, A and B. Generally sites can be molecular sites or even "nanoparticles." Each site can be occupied by two electrons [Cu(I)], one electron [Cu(II)], or no electron [Cu(III)]. Three sequential oxidation states are indeed typical for superconductors[4,5] and is required for high mobility. The coupling is due to a correlation effect. In a Heitler–London model SC requires active participation of the ionic states (charged states).[13]

The free energy Marcus model[14,15] has been of great importance for the understanding of ET between metal ions in solution. The model explains very well differences in rate constants of ET reactions, varying over more than 10 orders of magnitude. The dependence on nuclear coordinates[17] explains the variation of rate constants. The electron–nuclear coupling occurs by the vibration that accomplishes the geometry change. Theories for electron conductivity can be relevant only if the small geometry differences in bond distances are accounted for. Activation energies in the normal region are one important result.

Bloch theory should not be applied on local systems (~strongly correlated systems). The metal-to-metal (or molecule-to-molecule) charge transfer (MMCT) transition is polarized in the CuO_2 plane, and this is the case for the absorption at 2 eV.[19] This transition is the reason why most superconductors are black, bronze, or generally with strong absorption. TiO_2 is white since the band gap to the empty 3d band is >3.2 eV, and $CuSO_4 \cdot 5H_2O$ is blue due to crystal field transitions, In this case the MMCT transition is in the UV region since the distance between the Cu ions is quite large. The energy scale of SC is much smaller, and the main reason is the order of magnitude difference between spectroscopic and adiabatic U (see later).

2. COUPLING AND LOCALIZATION

In the Marcus model the total free energy is a function of nuclear coordinates and the nuclei move according to classical mechanics.[14–16] For low temperatures this motion is quantized and the model has to be replaced by the Jortner model,[20,21] in which the vibrational levels are introduced. The "nuclear tunneling" results in rates that decrease less than exponentially with $1/T$. The model replaces the "variable range hopping model."

A collective nuclear coordinate is placed on the x-axis and the total free energy on the y-axis (Fig. 1).[14–16] There are energy minima in the equilibrium points and an approximate parabolic dependence along collective coordinates between the equilibrium points. This introduces a coupling

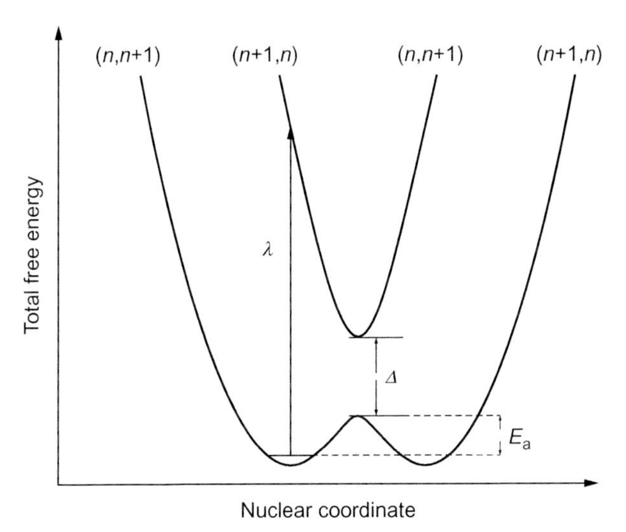

Fig. 1 Potential free energy surfaces in the Marcus model for one-electron transfer; λ is reorganization energy, $\Delta = 2H_{12}$ is coupling, and E_a is activation energy. $(n,n+1)$, etc., refer to site orbital occupation.

between the nuclear and electronic coordinates which replaces the Cooper coupling in the itinerant case, and which also controls the width of spectral absorptions. A critical quantity is the reorganization energy (λ), the energy gap between the ground state and charge transfer state at constant geometry. λ is equal to spectroscopic Hubbard U (U_{vert}).

CuO and cuprates are local systems, except in certain doping regions and temperatures when they may become itinerant.[18] The methods of Quantum Chemistry allow calculation of geometry as a function of local occupancy, as well as coupling.[22] This in turn allows barriers for ET (Fig. 1) to be calculated. The condition for an itinerant state is that the activation energies become zero.

In the case of excitations the different equilibrium geometry of the excited state leads to a Stokes shift[23] that is in fact equal to the reorganization energy[24] for excitation, responsible for the redshift at emission. Localization in the Marcus model has been discussed by Creutz,[25] by Allen and Hush,[26] by Reimers et al.,[27] and in Refs. 13,16,18,22,24 Localization is directly connected to the structural reorganizations. In Fig. 1 the activation energy E_a may be written as[16]:

$$E_a = \frac{\lambda}{4}\left(1 - \frac{\Delta}{\lambda}\right)^2 \tag{1}$$

Thus $E_a = \lambda/4$ for $\Delta = 0$ and $E_a = 0$ for $\Delta > \lambda$. One criterion for SC is thus that the coupling between the sites is large. The coupling $H_{12} = \Delta/2$ between different ET states is well defined and may be calculated accurately.[27,28] Coupling decreases exponentially with distance roughly in the same way as the orbital overlap, ie, exponentially.

In the physics literature $H_{12}/U_{vert} = \Delta/2\lambda$ is considered as the critical quantity for localization,[29] whereas Δ/λ is the one used of ET theory,[25–27] because of Eq. (1). It is understood that the transition then has to be the one suggested by Mott, ie, the MMCT transition. λ is the reorganization energy and strongly dependent on small modifications of the geometry when the number of electrons is changed at a site (but the total number of electrons kept constant). For copper the ML distance increases by 0.1 Å from Cu(III) to Cu(II) and another 0.1 Å from Cu(II) to Cu(I). Such a difference leads to $\lambda \approx 2$ eV. For Ni(II), $\lambda \approx 3$. Earlier a number of calculations have been performed where the equation $U = I - A$ is used (ionization energy minus electron affinity). U is then overestimated. Taking I and A from the free copper atom leads to a disastrous $U = 6.3$ eV (Eq. 2 below leads to even larger values).

For the case of electron pair transfer, three potential energy surfaces are necessary. The coupling between sites is an interaction between a spin-coupled state and a charged state (Fig. 2).[22]

The Bloch band model can be used for repetitive and infinite systems which are delocalized in the sense that removal of a hole, electron, or excitation can be done with fixed nuclear coordinates. This is usually not the case for transition metal oxides, for example, NiO or nickelates, or for CuO and cuprates. The band model does not produce a band gap where one is expected, at 2–4 eV. This is the historical reason for the introduction of "Hubbard U".[30] In the interpretation by Mott, U is the excitation energy for charge transfer to the adjacent site.

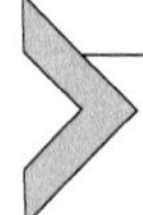

3. THE HUBBARD GAP AND THE ABSORPTION SPECTRUM OF TRANSITION METAL OXIDES

In the case of the cuprates Hubbard U_{vert} is simply the gap from a spin-coupled $ab + ba$ state to the ionic aa and bb states (or vice versa if the ground state is a charged state). a and b are orbitals on adjacent metal sites.

Mott focused on the repulsion between two electrons in the same atomic orbital $\phi = a$ and defined

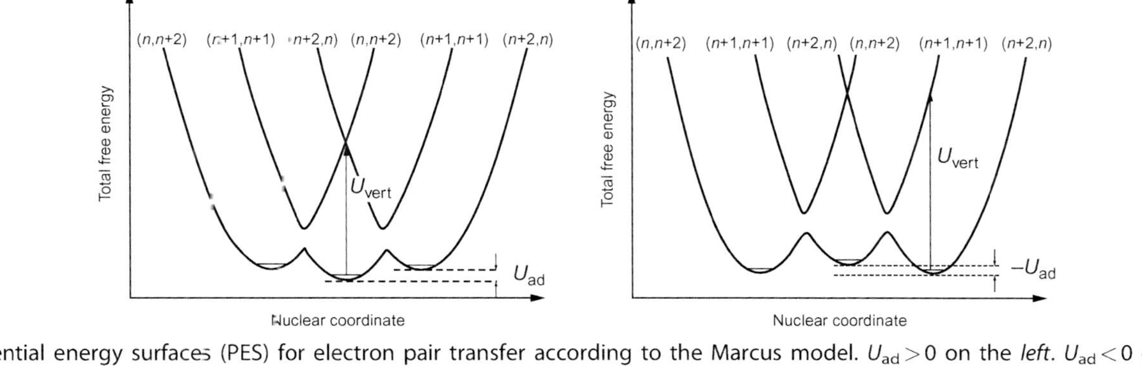

Fig. 2 Potential energy surfaces (PES) for electron pair transfer according to the Marcus model. $U_{ad} > 0$ on the *left*. $U_{ad} < 0$ on the *right*. $(n,n+2)$, etc. refers to site orbital occupation.

$$U = \int \phi^*(1)\phi^*(1)\frac{1}{r_{12}}\phi(2)\phi(2)\mathrm{d}V_1\mathrm{d}V_2 \tag{2}$$

to be the gap by which the Bloch model must be amended.[30] Mott neglects attractive terms, such as between electron and hole, and therefore enormously large U are calculated using Eq. (2). The purpose of Eq. (2) may be to simplify the discussion, but nevertheless we cannot escape the conclusion that Eq. (2) is not meaningful as a definition of an excitation energy (it may perhaps be used as a *symbol* for electronic repulsion). A quick look into the Born–Haber cycle for calculating free energy contribution in ionic crystals[24] shows us that, for example, Na^+Cl^- is a considerably better approximation to the wave function of table salt than NaCl with neutral Na and Cl atoms. To move one electron from Na to Cl does not lead to extremely large repulsion, but to attraction. The same is the case in transition metal oxides. Bi(IV) ions are unstable relative to Bi(III) and Bi(V) disproportionation.[4,5,13,31] Another case is the Tl(II) positive ion, disproportionating into Tl(I) and Tl(III). Bi(IV) and Tl(II) are said to be missing oxidation states, but in reality some valencies only have unfavorable reduction potentials, rather than being missing.

Nevertheless there have been suggestions to predict SC on the basis of electron repulsion only.[32] In fact the idea that U is large and has to be reduced by some novel mechanism, has been a favorite concept in theoretical papers after 1986. Actually structural fluctuations depending on the number of electrons at a site lead to such a reduction. $\lambda = U = 2$ eV is still far above the energy scale of the SC gap and the reduction often results in an adiabatic U (U_{ad}) only slightly above zero for many transition metal oxides.[16] $U = 0$ is required for SC.

The structural fluctuation idea is old and was stated as early as 1948, by Landau and Pekar.[17] Unfortunately the authors suggested an effective mass treatment, which, although extremely difficult to materialize, is still dominating the field.

Fig. 2 shows the Marcus model for electron pair transfer. As in the case of ET we assume two adjacent sites. On the x-axis are plotted collective nuclear coordinates for a breathing or half-breathing motion. In other words when the ligand coordinates around one site shrink, they open up on the other site. The active electrons are highly sensitive to the nuclear positions and localize in regions with the lowest energy, which means that the electron transfer at the curve crossing or slightly afterward.

In the case of two electrons and two sites there are three possible ground states, a priori (Fig. 2). The leftmost minimum in total free energy is the equilibrium point, if there are two additional electrons on metal ion A (or molecule A). The rightmost minimum is the total free energy when the electron pair is on the adjacent site (B). If there is no interaction with the other metal site, the energy curves may be approximated as perfect parabolas. When we move to the right on the x-axis, we first encounter another state, however, with a completely different wave function. This is the central parabola corresponding to the spin-coupled wave function with one electron on each site and symmetric geometry. This parabola couples the two others strongly.

Adiabatic U, U_{ad}, is the negative of the difference in free energy for the minimum of the central parabola, compared to the minimum of the two outer parabolas. Clearly if U_{ad} is large, the interaction cannot open up a SC gap. The same is the case if U_{ad} is negative (right picture). In cuprates $U_{ad} > 0$ since the ground state is the spin-coupled state represented by the central parabola. This is the case for La_2CuO_4 and other cuprates. There is no possibility for pairing. Fig. 3 explains both absorption (U_{vert}) and

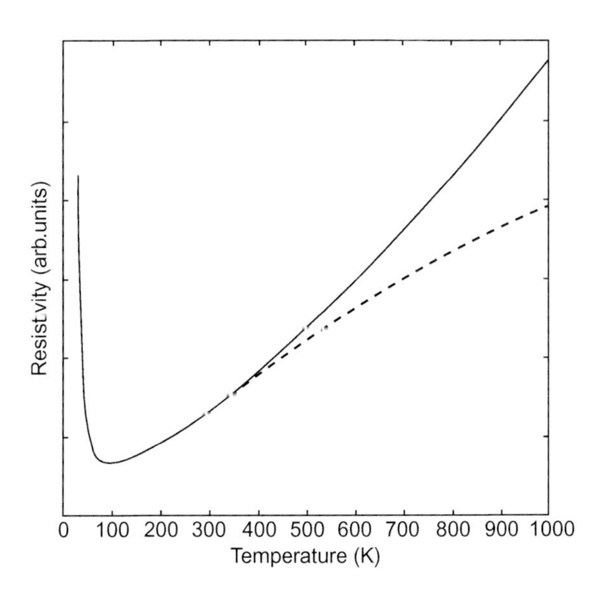

Fig. 3 Resistivity for $La_{2-x}Sr_xCuO_4$ with $x = 0.06$. The *full-drawn curve* shows the contribution from one-electron exchange alone. Contribution to conductivity from the pseudogap is added in the *dashed curve*.

SC (in the case of disappearing activation barrier, $U_{ad}=0$). A quantum mechanical calculation using the CASSCF method, containing four Cu ions, confirms the picture in Fig. 2.[33]

In the Marcus model conductivity is possible in mixed-valence systems as ET with vanishing energy barrier (E_a):

$$Cu_A(III) + Cu_B(II) \leftrightarrow Cu_B(II) + Cu_A(III) \tag{3}$$

Eq. (3) has been rewritten as a diffusion model for electron mobility[34] which has been applied to the normal state cuprate resistivity.[18] SC corresponds to $U_{ad} \leq 0$ (Fig. 2) and vanishing activation energy. On the other hand, Eq. (2) can never become negative, of course. It is therefore unlikely that any predictions can be made on the basis of Eq. (2).

The transition U_{vert} has high intensity in the optical conductivity spectrum. Obviously MMCP absorptions lead to conductivity contrary to O2p → Cu3d (LMCT) transitions. Another piece of evidence for concluding that the transition at 2 eV is a MMCT is that it is polarized in the CuO_2 plane,[18,19] which is typical for a MMCT transition, but not for a O2p → Cu3d transition. The 2 eV absorption leads to persistent conductivity.[18]

Finally we may consider the variation in energy among different copper compounds. MMCT transitions are extremely sensitive to the distance for CT. For example, in $CuSO_4$, with a larger distance between the copper ions than in a cuprate, the MMCT transition is in the UV region and the blue crystal field spectrum can be seen. There is no reason to believe that the O2p → Cu3d transition has a large variation; however, since it is a transition within the same complex. The absorption at 2 eV is thus a MMCT transition.

In the normal region, conductivity occurs primarily by the mechanism of Eq. (3), ie, for the cuprates due to Cu(II)/Cu(III) hopping. This mechanism has nothing to do with pair transfer, but leaves a signature in the absorption spectrum; for cuprates below 1 eV.[18,19] For dopings between 7% and 20% the resistivity suddenly drops to zero below T_C. A superconducting phase can only appear, if the ground state is a superposition between spin-coupled and site-paired electrons.[13,22] Then $U_{ad} \rightarrow 0$ and disproportionation takes place

$$2Cu(II) \rightarrow Cu(I) + Cu(III) \quad (\Delta G_0 = U_{ad}) \tag{4}$$

Conductivity also appears due to electron pair hopping[18] of the following type

$$Cu(I) + Cu(III) \rightarrow Cu(III) + Cu(I) \tag{5}$$

When U_{ad} tends to zero in cuprates the antiferromagnetic phase changes to a charged phase, where Cu(III) and Cu(I) sites take over from Cu(II) sites. In between these two phases the SC phase may appear, provided the activation energy disappears by Eq. (1).

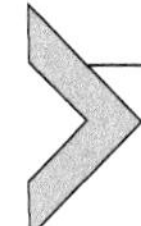

4. WAVE FUNCTIONS AND CONDUCTIVITY IN THE LOCAL STATE

A theory for conductivity local systems is given in another paper.[18] The probability for jumping from one site to the next is given in Marcus theory. By applying an electric field and subtracting the probabilities in either field directions we obtain an expression for the conductivity and, by inverting, the resistivity. The latter is obtained as a function of temperature (T):

$$\rho = \text{const.}^{-1}\,\frac{1}{n}\frac{4\pi\hbar}{\Delta^2}\left(\frac{\lambda k_B}{\pi^3}\right)T^{3/2}\exp\left(\frac{\lambda}{4k_B T}\right) \tag{6}$$

The $T^{3/2}$-factor enters because of the differentiation (T), and because of kinetic energy ($T^{1/2}$). For high kinetic energies the Landau–Zener avoided crossing tends to be passed without hopping. The probability is calculated in Ref. 18 High temperature therefore causes a decrease of the ET rate[18] and increase of the resistivity (Fig. 3).

In electron exchange of one-electron type (ET) (Eqs. 3 and 6), only a single electron is involved. This mechanism actually provides most of the conductivity in the normal state of the doped system. Doping creates a hole and an electron can be exchanged with a Cu(II) site which leads to conductivity according to the standard Marcus model. The agreement with the results of Takagi et al.[35] is satisfactory. The resistivity is typical for stepwise electron transfer from site to site, when the sites are weakly coupled.

For low temperatures the result should be corrected using Jortner theory. The resistivity then tends to a finite value for $T \rightarrow 0$.[18,20,21]

In two-electron exchange systems adjacent sites have equivalent orbitals, a and b, occupied with either one electron in each, with different spin (spin-coupled state), or with two electrons in one and none in the other (charged state). The wave functions are $ab+ba$, and aa and bb, respectively (or $aa+bb$ and $aa-bb$), with $\alpha\beta-\beta\alpha$ as spin function. If the electrons have equally directed spin (the spin functions $\alpha\alpha$, $\alpha\beta+\beta\alpha$, or $\beta\beta$), we get a triplet state with the spatial function $ab-ba$, otherwise a singlet state $ab+ba$. We call the latter state the spin-coupled state. The charged state and the spin-coupled state are both spin-singlet states. There is interaction between $ab + ba$ and $aa + bb$ as a part of the electronic correlation effects. Two states are formed with an energy gap. Finally there is coupling of all electronic states to the vibrational breathing or half-breathing modes, with a vibronic SC gap as the result.

In conclusion, we find that the minimum in Fig. 3 depends on one-electron exchange according to Eq. (3), unrelated to SC. We may compare to two other transition metal systems, $BaPb_{1-x}Bi_xO_3$ and $LiTi_2O_4$. For $T > T_C$ the latter behaves as $\rho = \rho_0 + aT^2$ as some superconducting cuprates. We may expect a similar disproportionation for Ti(III):

$$2\mathrm{Ti(III)} \rightarrow \mathrm{Ti(IV)} + \mathrm{Ti(II)} \quad (\Delta G_0 = U_{ad}) \tag{7}$$

Conductivity then arises with the help of electron pairs. The Ti3d band is populated by electrons. Since O2p$\rightarrow$Ti3d is in the UV region the black color arises from MMCT transitions as in the case of cuprates. Only the spinel structure is superconducting. Possibly the absence of SC in the other alkali titanates depends on the high spin of Ti(II) in octahedral geometry.

In $BaPb_{1-x}Bi_xO_3$ the ground state is of the type Bi(III) and Bi(V), which is strongly suggested by metal–ligand bond lengths, meaning that $U_{ad} < 0$.[4,36] The disproportionated state is thus the ground state for $0.3 < x < 1$.[4] Absorption in the visible region is due to:

$$h\nu + \mathrm{Bi(IV)} + \mathrm{Bi(II)} \rightarrow 2\mathrm{Bi(III)} \tag{8}$$

Spectrum and resistivity have been measured by Tajima et al.[37] The optical conductivity spectrum has a maximum at $10^4\ \mathrm{cm}^{-1}$ (1.3 eV). The excited state is the spin-coupled $Bi6s_A$–$Bi6s_B$ state. There is no one-electron exchange of the type seen in Eq. (3). All conductivity arises from activated electron pair transport, Eq. (5). The resistivity tends to infinity as $T \rightarrow 0$ and seems to tend to zero for $T \rightarrow \infty$[37] as expected for a semiconductor.

5. ORGANIC SC

Organic SC was first suggested by London[38] and by Little.[39] London was considering ring currents in single aromatic molecules. Little considered a polymer chain. We will show that SC in organic systems is in accordance with the total free energy model.

Organic SC was not realized until 1982, in the work by Jerome and Bechgaard.[40] Each molecule corresponds to a metal site in the case of superconducting transition metal ions. Recently aromatic molecules have been found to be superconducting,[41] if doped with alkali. SC in the fullerenes C_{60} molecules was discovered already in 1991[42–44] in K_3C_{60} and Rb_3C_{60}. In Cs_3C_{60} SC appears only after applied pressure. In both cases the Marcus model provides good explanations.[45]

The C_{60} molecule accepts three electrons on the average. Disproportionation: $2K_3C_{60}{}^{-3} \rightarrow K_3C_{60}{}^{-2} + K_3C_{60}{}^{-4}$ emphasizes the appearance of three consecutive charges, corresponding to three different oxidation states.[45,46] The negative counter ions in oxidatively doped Bechgaard salts[40] correspond to the positive alkali ions in A_3C_{60}.

Typical for A_3C_{60} SC is the closeness of antiferromagnetic, charged, and SC phases. In K_3C_{60} and Rb_3C_{60} disproportionation has taken place:

$$2C_{60}{}^{3-} \rightarrow C_{60}{}^{2-} + C_{60}{}^{4-} \tag{9}$$

The charged phase is the ground state above T_C: On the other hand the spin-coupled phase is the ground state in Cs_3C_{60}. In Rb_3C_{60} the distance between the C_{60} molecules is larger than in K_3C_{60}, hence U less negative, and configuration interaction between the charged state and the spin-coupled state more important. Not surprisingly SC has been found in K_3C_{60} and Rb_3C_{60} with $T_C = 19.3$ and $T_C = 30$ K, respectively. In Cs_3C_{60} some pressure is necessary to make U to negative.[45,46] Applying pressure makes Cs_3C_{60} superconducting at $T_C = 38$ K,[44] which apparently corresponds to the situation with $U = 0$ and thus with the highest T_C.

We conclude that organic SC in the A_3C_{60} compounds can be explained in the same way as SC in the cuprates. The color of A_3C_{60} is dark, demonstrating that a charge transfer transition exists in the visible region of the disproportionation type.

SC in polycyclic aromatic molecules, doped by alkali, was discovered in recent years.[47–49] As in the case of A_3C_{60} the doping with three alkalis per molecule gives a SC with high critical temperature and, as in the case of A_3C_{60}, the system becomes diamagnetic, hinting that the molecules have alternating charges −2 and −4. The aromatic alkali doped superconductors are also dark.[47] Phenanthrene (Ph),[48] picene (Pi),[48] and dibenzopentacene (DBP)[49] are three of about 10 such polycyclic aromatic molecules (Fig. 4). K_3Ph with $T_C = 4.95$ K, K_3Pi with $T_C = 6.9$ and 14 K, and K_3DBP with $T_C = 7.4$, 28.2, 33 K are found to be superconducting. Different critical temperatures are obtained for different crystal structures.

The Raman spectrum for neutral and negative ions has been calculated and measured by Kambe et al.[50] for picene. The bond stretch vibrations in

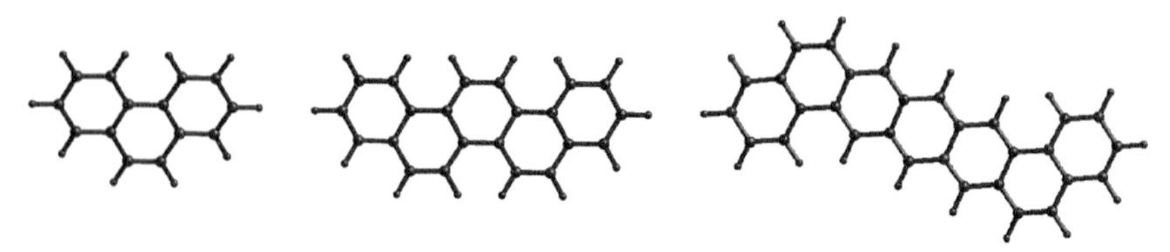

Fig. 4 (*Left*) phenanthrene (Ph), (*middle*) picene (Pi), and (*right*) dibenzopentacene (DPB).

the region 1300–1400 cm^{-1} form breathing modes (along the x-axis in Fig. 1) for the crystal, coupled to superconductivity. The energy for the high-intensity vibronic states is set by the intersection of the parabolas, agreeing well with the theory used in this paper.

The resistivity curve for K_3-picene shows no resemblance to the cuprate result.[35] As for $BaPb_{1-x}Bi_xO_3$,[37] it tends to zero for $T \rightarrow 0$, goes through a maximum, and finally tends to zero for $T \rightarrow \infty$.[51] This can be explained as due to absence of single-electron transfer.

6. TUNGSTEN AND MOLYBDENUM BRONZES

In 1823 Wöhler prepared the first tungsten bronze, Na_xWO_3.[52] Pure W(VI) oxide ($x=0$) is yellow and has an empty 5d-subshell. The yellow color is due to O2p → W5d with a 2.5–3 eV band gap. When doped with Na, W(V) is formed, meaning that the W5d band of orbitals becomes occupied, first in the t_{2g} orbitals. The 5d electrons interact between sites and create CT states. The MMCT transitions give rise to wide absorption of high intensity at 1–2 eV.

Molybdenum bronzes were discovered in 1895.[53] The properties are similar to the corresponding tungsten compounds. K_xMoO_3, for example, is strongly red or blue colored, depending on x and crystal structure. For $x=0$, ie, in WO_3 and MoO_3 the color is lemon yellow and white, respectively. For increased x, ie, for increased population of the 4d and 5d subshells, respectively, the systems become deeply blue ($x \approx 0.3$) and then goes through red, yellow, and golden.[54,55] The oxides Mo_2O_5 and W_2O_5 are unstable.[56] For $x=1$, sodium tungsten bronze adopts approximately a perovskite crystal structure. In this form, the structure is corner-sharing WO_6 octahedra with sodium ions in the interstitial regions.

Another way to populate the d-shells is to use fluorine doping, for example, $WO_{3-x}F_x$. $x=0.4$ gives a superconducting compound.[57] Tungsten bronzes belongs to the first transition metal oxides proven to be superconducting.[1,57–60]

For $x=0$ the excitation is across a band gap which in the case of WO_3 is just below the border to the ultraviolet. For x small there are transitions of the type:

$$h\nu + W_A(V) + W_B(VI) \rightarrow W_B(VI) + W_A(V) \quad (10)$$

of the same type as for the cuprates. Since for small x there is a high concentration of W(VI) sites the energy of the transition is quite small. Red light is absorbed and blue light emitted. For increasing x the number of W(VI) sites is decreasing and hence the energy of the charge transfer transition higher. In the case of cuprates, on the other hand, the number of high oxidation state sites is increasing, and the charge transfer transitions lower in energy.[19]

The Mo(V) and W(V) oxides may disproportionate:

$$2W(V) \rightarrow W(VI) + W(IV) \quad (11)$$

The spectral process causes the black or bronze color and the thermal process SC. Na_xWO_3 crystals do not form a fully repetitive structure[59,60] and the SC critical temperature therefore varies. T_C may be as high as 80 K on the surface.

The mechanism for SC is almost the same as for the cuprates, titanates, and organic superconductors, although cuprates can be both electron and hole doped.

7. CONCLUSION

The basic principles for a local, stochastic model that can explain SC in cuprates and organic superconductors are now at hand. Ordinary conductivity and SC may be considered as the itinerant limit of two different mobility phenomena in localized systems. Ordinary conductivity corresponds to a mixed-valence system with two successive oxidation states, say Cu(II)/Cu (III), where two transition metal ions or a molecular system are exchanging one single electron. Only two-electron processes can lead to SC. Adiabatic Hubbard U needs to be ≤ 0.

The superconducting phase is a quantum mechanically mixed state, due to configuration interaction between Cu(II)/Cu(II) and Cu(III)/Cu(I). Each of the two states is insulating, but after interaction they become conducting.[24] In the configuration interaction only states with an even number of electrons are included. Hence Bose–Einstein statistics becomes visible for low T. The total electronic wave function is antisymmetric, of course.

The total free energy description of SC, based on electron and electron pair mobility, leads to satisfactory results. The cumbersome particle description in terms of effective electron mass can be avoided. Directly related to the difference in magnitude between U_{vert} and U_{ad} is the remarkable fact that superconductors are black or have a metallic luster.

ACKNOWLEDGMENTS

This chapter is dedicated to the memory of Prof. Per-Olov Löwdin, my supervisor, on his 100th birthday, and to the memory of Prof. Berndt Matthias, who used to give the last lecture on the Sanibel Conference, organized by Prof. Löwdin. We, students of Prof. Löwdin, had an exciting life thanks to Per and Berndt.

REFERENCES

1. Raub, C. J.; Sweedler, A. R.; Jensen, M. A.; Broadston, S.; Matthias, B. T. Superconductivity of Sodium Tungsten Bronzes. *Phys. Rev. Lett.* **1964**, *13*, 746.
2. Schooley, J. F.; Hosler, W. R.; Cohen, M. L. Superconductivity in Semiconducting $SrTiO_3$. *Phys. Rev. Lett.* **1964**, *12*, 474–475.
3. Johnston, D. C.; Prakash, H.; Zachariasen, W. H.; Viswanathan, R. High Temperature Superconductivity in the Li-Ti-O Ternary System. *Mater. Res. Bull.* **1973**, *8*, 777–784.
4. Sleight, A. W.; Gillson, J. L.; Bierstedt, P. E. High-Temperature Superconductivity in the $BaPb_{1-x}Bi_xO_3$ System. *Solid State Commun.* **1975**, *17*, 27–28.
5. Sleight, A. W. Oxide Superconductors: A Chemist's View. *MRS Proc.* **1987**, *99*, 3. http://dx.doi.org/10.1557/PROC-12-3.
6. Bednorz, J. G.; Müller, K. A. Perovskite Type Oxides—The New Approach to High-T_C Superconductivity. *Nobel Lect.* **1987**, (Dec. 8).
7. Bednorz, J. G.; Müller, K. A. Possible High T_c Superconductivity in the Ba–La–Cu–O System. *Z. Phys. B* **1986**, *64*, 189–193.
8. Chu, C. W.; Hor, P. H.; Meng, R. L.; Gao, L.; Huang, Z. J.; Wang, Y. Q. Evidence for Superconductivity Above 40 K in the La-Ba-Cu-O Compound System. *Phys. Rev. Lett.* **1987**, *58*, 405–407.
9. Wu, M. K.; Ashburn, J. R.; Torng, C. J.; Hor, P. H.; Meng, R. L.; Gao, L.; Huang, Z. J.; Wang, Y. Q.; Chu, C. W. Superconductivity at 93 K in a New Mixed-Phase Y-Ba-Cu-O Compound System at Ambient Pressure. *Phys. Rev. Lett.* **1987**, *58*, 908–910.
10. David, W. I. F.; Harrison, W. T. A.; Gunn, J. M. F.; Moze, O.; Soper, A. K.; Day, P.; Jorgensen, J. D.; Hinks, D. G.; Beno, M. A.; Soderholm, L.; Capone, D. W., II; Schuller, I. K.; Segre, C. U.; Zhang, K.; Grace, J. D. Structure and Crystal Chemistry of the High-T_c Superconductor $YBa_2Cu_3O_{7-x}$. *Nature* **1987**, *327*, 310–312.
11. Löwdin, P.-O. Quantum Theory of Many-Particle Systems. I. Physical Interpretations by Means of Density Matrices, Natural Spin-Orbitals, and Convergence Problems in the Method of Configurational Interaction. *Phys. Rev.* **1955**, *97*, 1474; Quantum Theory of Many-Particle Systems. II. Study of the Ordinary Hartree-Fock Approximation. *Phys. Rev.* **1955**, *97*, 1490; Quantum Theory of Many-Particle Systems. III. Extension of the Hartree-Fock Scheme to Include Degenerate Systems and Correlation Effects. *Phys. Rev.* **1955**, *97*, 1509.
12. Pohl, H. A.; Pollak, M. Nomadic Polarization in Quasi-One-Dimensional Solids. *J. Chem. Phys.* **1975**, *66*, 4031–4040.
13. Larsson, S. Localization Condition for Metallic Conductivity and Superconductivity. *Chem. Phys. Lett.* **1989**, *157*, 403–408.
14. Marcus, R. A. On the Theory of Oxidation-Reduction Reactions Involving Electron Transfer. *J. Chem. Phys.* **1956**, *24*, 966–978.

15. Marcus, R. A. Chemical and Electrochemical Electron-Transfer Theory. *Annu. Rev. Phys. Chem.* **1964**, *15*, 155.
16. Brunschwig, B. S.; Logan, J.; Newton, M. D.; Sutin, N. A Semiclassical Treatment of Electron-Exchange Reactions. Application to the Hexaaquoiron(II)-Hexaaquoiron(III) System. *J. Am. Chem. Soc.* **1980**, *102*, 5798–5809.
17. Landau, L. D.; Pekar, S. I. Effective Mass of a Polaron. *Zh. Eksp. Teor. Fiz.* **1948**, *18*, 419–423. (in Russian), English translation: *Ukr. J. Phys.* **2008**, *53*, 71–74 (special issue).
18. S. Larsson, Conductivity in Cuprates Arises from Two Different Sources: One-Electron Exchange and Disproportionation. *J. Supercond. Nov. Magn.* accepted for publication.
19. Uchida, S.; Ido, T.; Takagi, H.; Arima, T.; Tokura, Y.; Tajima, S. Optical Spectra of $La_{2-x}Sr_xCuO_4$: Effect of Carrier Doping on the Electronic Structure of the CuO_2 Plane. *Phys. Rev. B* **1991**, *43*, 7942–7954.
20. Jortner, J. Temperature Dependent Activation Energy for Electron Transfer Between Biological Molecules. *J. Chem. Phys.* **1976**, *64*, 4860–4867.
21. Bixon, M.; Jortner, J. Intramolecular Vibrational Excitations Accompanying Solvent Controlled Electron Transfer Reactions. *J. Chem. Phys.* **1988**, *89*, 3392–3393.
22. Larsson, S. Electronic Structure of Planar Superconducting Systems. From Finite to Extended Model. *Chem. Phys.* **1998**, *236*, 133–150.
23. Stokes, G. G. On the Change of Refrangibility of Light. *Philos. Trans. R. Soc. London* **1852**, *142*, 463–562.
24. Larsson, S. *Chemical Physics. Electrons and Excitations*; CRC Press, Taylor & Francis Group, 2012. Boca Raton, London, New York, chapter 12.
25. Creutz, C. Mixed-Valence Complexes of d5–d6 Metal Centers. *Prog. Inorg. Chem.* **1983**, *30*, 1–73.
26. Allen, G. C.; Hush, N. S. Intervalence-Transfer Absorption. Part 1. Qualitative Evidence for Intervalence-Transfer Absorption in Inorganic Systems in Solution and in the Solid State. *Prog. Inorg. Chem.* **1967**, *8*, 357–389.
27. Reimers, J. R.; Wallace, B. B.; Hush, N. S. Towards a Comprehensive Model for the Electronic and Vibrational Structure of the Creutz–Taube Ion. *Philos. Trans. R. Soc. A* **2008**, *366*, 15–31.
28. Braga, M.; Larsson, S. Correlation Effects in the Electronic Coupling Between π Electrons Through a Cyclohexane Spacer. *Chem. Phys. Lett.* **1992**, *200*, 573–579.
29. Gunnarsson, O.; Koch, E.; Martin, R. M. Mott Transition in Degenerate Hubbard Models: Application to Doped Fullerenes. *Phys. Rev. B* **1996**, *54*, R11026–R11029.
30. Mott, N. F. Metal-Insulator Transition. *Rev. Mod. Phys.* **1968**, *40*, 677–683.
31. Larsson, S. Lattice Enthalpy Drives Hubbard U to Zero. *J. Mod. Phys.* **2013**, *4*, 29–32.
32. Dunne, L. J.; Brändas, E. Superconductivity from Repulsive Electronic Correlations on Alternant Cuprate and Iron-Based Lattices. *Int. J. Quantum. Chem.* **2013**, *113*, 2053.
33. Klimkāns, A. *Quantum Mechanical Studies of Electron Transport Processes Including Superconductivity*; thesis, Chalmers University of Technology, Department of Chemistry, 2001.
34. Johansson, E.; Larsson, S. Electronic Structure and Mechanism for Conductivity in Thiophene Oligomers and Regioregular Polymer. *Synth. Met.* **2004**, *144*, 183–191.
35. Takagi, H.; Batlogg, B.; Kao, H. L.; Kwo, J.; Cava, R. J.; Krajewski, J. J.; Peck, W. F., Jr. Systematic Evolution of Temperature-Dependent Resistivity in $La_{2-x}Sr_xCuO_4$. *Phys. Rev. Lett.* **1992**, *69*, 2975–2978.
36. Sleight, A. W. Bismuthates: $BaBiO_3$ and Related Superconducting Phases. *Physica C Supercond.* **2015**, *514*, 152–165.
37. Tajima, S.; Uchida, S.; Masaki, A.; Takagi, H.; Kitazawa, K.; Tanaka, S. Electronic States of $BaPb_{1-x}Bi_xO_3$ in the Semiconducting Phase Investigated by Optical Measurements. *Phys. Rev. B* **1987**, *35*, 696–703.
38. London, F. Supraconductivity in Aromatic Compounds. *J. Chem. Phys.* **1937**, *5*, 837–838.
39. Little, W. A. Superconductivity at Room Temperature. *Sci. Am.* **1965**, *212* (2), 21–27.

40. Jerome, D.; Mazaud, A.; Ribault, M.; Bechgaard, K. Superconductivity in a Synthetic Organic Conductor (TMTSF)2PF_6. *J. Phys. (Paris) Lett.* **1980**, *41*, L95–L98.
41. Kubozono, Y.; Goto, H.; Jabuchi, T.; Yokoya, T.; Kambe, T.; Sakai, Y.; Izumi, M.; Zheng, L.; Hamao, S.; Nguyen, H. L. T.; Sakata, M.; Kagayama, T.; Shimizu, K. Superconductivity in Aromatic Hydrocarbons. *Physica C Supercond.* **2015**, *514*, 199–205.
42. Hebard, A. F.; Rosseinsky, M. J.; Haddon, R. C.; Murphy, D. W.; Glarum, S. H.; Palstra, T. T. M.; Ramirez, A. P.; Kortan, A. R. Superconductivity at 18 K in Potassium-Doped C_{60}. *Nature (London)* **1991**, *350*, 600–601.
43. Rosseinsky, M. J.; Ramirez, A. P.; Glarum, S. H.; Murphy, D. W.; Haddon, R. C.; Hebard, A. F.; Palstra, T. T. M.; Kortan, A. R.; Zahurak, S. M.; Makhija, A. V. Superconductivity at 28 K in RbxC60. *Phys. Rev. Lett.* **1991**, *66*, 2830–2832.
44. Ganin, A. Y.; Takabayashi, Y.; Jeglič, P.; Arčon, D.; Potočnik, A.; Baker, P. J.; Ohishi, Y.; McDonald, M. T.; Tzirakis, M. D.; McLennan, A.; Darling, G. R.; Takata, M.; Rosseinsky, M. J.; Prassides, K. Polymorphism Control of Superconductivity and Magnetism in Cs_3C_{60} Close to the Mott Transition. *Nature* **2010**, *466*, 221–225.
45. Larsson, S. T_C Dependence on Hubbard U in the Fullerenes. *J. Supercond. Nov. Magn.* **2015**, *28*, 315–317.
46. Larsson, S. Effect of Pressure on Superconducting Properties. *J. Supercond. Nov. Magn.* **2015**, *28*, 1693–1698.
47. Mitsuhashi, R.; Suzuki, Y.; Yamanari, Y.; Mitamura, H.; Kambe, T.; Ikeda, N.; Okamoto, H.; Fujiwara, A.; Yamaji, M.; Kawasaki, N.; Maniwa, Y.; Kubozono, Y. Superconductivity in Alkali-Metal-Doped Picene. *Nature* **2010**, *464*, 76–79.
48. Wang, X. F.; Liu, R. H.; Gui, Z.; Xie, Y. L.; Yan, Y. J.; Ying, J. J.; Luo, X. G.; Chen, X. H. Superconductivity at 5 K in Alkali-Metal-Doped Phenanthrene. *Nat. Commun.* **2011**, *2. article 507.*
49. Xue, M.; Cao, T.; Wang, D.; Wu, Y.; Yang, H.; Dong, X.; He, J.; Li, F.; Chen, G. F. Superconductivity Above 30 K in Alkali-Metal-Doped Hydrocarbon. *Sci. Rep.* **2012**, *2*, 389.
50. Kambe, T.; He, X.; Takahashi, Y.; Yamanari, Y.; Teranishi, K.; Mitamura, H.; Shibasaki, S.; Tomita, K.; Eguchi, R.; Goto, H.; Takabayashi, Y.; Kato, T.; Fujiwara, A.; Kariyado, T.; Aoki, H.; Kubozono, Y. Synthesis and Physical Properties of Metal-Doped Picene Solids. *Phys. Rev. B* **2012**, *86*, 214507.
51. Teranishi, K.; He, X.; Sakai, Y.; Izumi, M.; Goto, H.; Eguchi, R.; Takabayashi, Y.; Kambe, T.; Kubozono, Y. Observation of Zero Resistivity in K-Doped Picene. *Phys. Rev. B* **2013**, *87*, 060505(R).
52. Wöhler, F. Natriumwolframbronzen. *Ann. Phys. Chem. (Poggendorf)* **1824**, *2*, 350.
53. Stavenhagen, A.; Engels, E. Über Molybdänbronzen. *Ber. Deut. Chem. Gesell.* **1895**, *28*, 2280–2281.
54. Hägg, G. The Spinels and the Cubic Sodium-Tungsten Bronzes as New Examples of Structures with Vacant Lattice Points. *Nature* **1935**, *135*, 874.
55. Magneli, A. Studies on the Hexagonal Tungsten Bronzes of Potassium, Rubidium, and Cesium. *Acta Chem. Scand.* **1953**, 7, 315–324.
56. van Liempt, J. A. M. Die Darstellung von reinem WO_2 und W_2O_5. *Zeits. Anorg. Allgem. Chem.* **1923**, *126*, 183–184.
57. Hirai, D.; Climent-Pascual, E.; Cava, R. J.; Hirai, D.; Climent-Pascual, E.; Cava, R. J. Superconductivity in $WO_{2.6}F_{0.4}$ synthesized by reaction of WO_3 with teflon. *Phys. Rev. B* **2011**, *84*, 174519.
58. Reich, S.; Tsabba, Y. Possible Nucleation of a 2D Superconducting Phase on WO_3 Single Crystals Surface Doped with Na^+. *Eur. Phys. J. B* **1999**, *9*, 1–4.
59. Reich, S.; Leitus, G.; Popovitz-Biro, R.; Goldbourt, A.; Vega, S.; Possible, A. A Possible 2D H_xWO_3 Superconductor with a T_C of 120 K. *J. Supercond. Nov. Magn.* **2009**, *22*, 343–346.
60. Remeika, J. P.; Geballe, T. H.; Matthias, B. T.; Cooper, A. S.; Hull, G. W.; Kelly, E. M. Superconductivity in Hexagonal Tungsten Bronzes. *Phys. Lett. A* **1967**, *24*, 565–566.

CHAPTER ELEVEN

State-Quantum-Chemistry Set in a Photonic Framework

Orlando Tapia[1]
Chemistry Ångström, Uppsala University, Uppsala, Sweden
[1]Corresponding author: e-mail address: orlando.tapia@fki.uu.se

Contents

Abstract

The photonic scheme provides an abstract perspective to describing chemical and physical processes; it is well adapted for biologically sustained processes too. The scheme is used to help analyze semiclassic pictures in order for a deeper understanding of natural processes to arise.

A q-state is not an object (eg, a molecule) but convoluted with a photon field, it hangs somehow on sensitive surfaces revealing an image constructed from q-events: these q-events are joint q-energy and angular momentum bridging probe-to-probing systems. Exchanges between physical states and probing ones establish a reality for a q-state.

Thus, in the photonic scheme, a q-state may emerge as an image if appropriately recorded via q-events. Initially collected q-events seem to indicate a random process.

Advances in Quantum Chemistry, Volume 74
ISSN 0065-3276
http://dx.doi.org/10.1016/bs.aiq.2016.07.001

However, after gathering these q-events in sufficient numbers, as in a two-slit example, a supportive *image* develops corresponding more and more to what is known as an interference pattern.

Moreover, the unlocking of a spin-triplet state is used to illustrate applications: for instance, the opening required a path starting from a parent spin-singlet excited electronic state. A *low-frequency* multiphoton mechanism regulated by conservation laws permits the description of a triplet state activation.

Of course, the materiality sustaining a q-state must transfer *information* that is richer than that a classical particle impact would convey. The use we make of quantum mechanics is basically the same that everyone does though without current interpretations; inclusion of photon fields makes the difference by providing quantum mechanisms to accomplish measurements.

Per-Olov Löwdin (POL), scientist, research leader, and teacher, in his memorable 1955 trilogy[1] provided mathematical foundations to the quantum theory of many-electrons systems; these and many other fields were extended and completed with many more publications besides summer and winter schools in Scandinavia and Florida (USA). POL contributed to an extraordinary expansion of mathematical physics within chemistry, thereby creating a sound basis for the development of quantum mathematical chemistry. This chapter is intended to be a tribute to his revered memory.

State-quantum chemistry refers to those fundamental elements not being representable with semiclassic theory. We are living in laser times. Coherent light is a tool to address theoretically and experimentally not only the chemical processes that might be assisted and/or prompted by photon fields; photonics combine matter and electromagnetic (EM) energy aspects. Both coherence and entanglement take center stage to help describe processes ranging from physics to chemistry; extended to biology; they provide a framework to examine photobiological situations beyond standard interpretations of q-mechanics.

The photonic quantum chemistry (PQC) framework[2,3] requires electronic quantum states independent from classic degrees of freedom, eg, instantaneous nuclear positions.[4] Elementary constituents *sustain* quantum states; electrons and nuclei support electronuclear (EN) q-states. The photonic mode of relating these levels takes away the representational role that early theory assigned to q-states.

On the other hand, difficulties encountered in developing quantum electrodynamics (QED)[5,6] prevented earlier construction of useful schemes wedding matter q-states and quantum EM radiation, eg, Fock photon number base states. The PQC scheme[2,3] makes it possible to give presence to Planck's fundamental quantum event (q-event) derived from his 1900 seminal work. Q-interaction and q-entanglement are implemented via *nonseparable* base elements.

Semiclassic models in quantum physics and chemistry are pervading and certainly most useful; they help develop computing replicas in QED and molecular quantum mechanics, including quantum technology.[6] However, this class of models sidesteps q-events that find expression in real space interactions, in particular between measuring (probing) laboratory devices and the q-system under probe (measurement). Interaction between q-systems and measuring (probing) is embodied by q-events opening one space to the other.

Moreover, the concept of quantum state has been changing under impulses originated from developments in quantum technology; papers in Refs. 7–9 address the theme from initially opposite directions, while with the present approach they tend to converge; this is possible once a physical framework allowed abstract and laboratory fields to be relationally bridged. To achieve this, EM and material purviews must both be quantized (q-fields); Fock space elements $|n_{\omega j}\rangle$ for EM q-states and $\{|ik(i)\rangle\}_{i,k}$ for matter sustained basis,[3,4] respectively. These sets open a way to handle q-events, thus targeting the energy/angular momentum (AM) exchange epitomized by Planck's discovery; this quantum possibility was not available to earlier approaches of molecular wavefunctions.[10,11] Importantly, simultaneous variation in both q-fields implies nonseparability at the level of base states; this calls for entanglement/disentanglement as a resource to bridge these ends, thereby forcing the introduction of entanglement events.

Considering this new situation, it is necessary to move the theory's grounds beyond semiclassic levels.[9,12–14] Moreover, as R.G. Newton clearly asserts,[13] a state vector does not describe the system itself; the idea of representation falls off as remarked by us in Refs 9,14. Amplitudes not only serve to calculate probabilities but then also intervene in q-interaction between matter sustained q-state and those associated to probing space.[14] Classical physics is not central to the measurement event to the extent that local realism can be safely set aside.[15,16] A graded scheme permits linking abstract to laboratory spaces, as will be illustrated later.

The chapter takes as ground QED ideas from Dirac, Fock, and Podolsky,[17] although these are used to rather illustrate connections and subtleties of the relativistic limit and help to relate with nonrelativistic models.

1. PHOTON Q-STATES: FOCK SPACE

In Fock space, photon number base states: $\{|n_{\omega}\rangle\}$ prompts for coherent states that come out naturally by using the basis set $|F-\omega \text{ basis}\rangle$:

$$(|0_{\omega}\rangle|1_{\omega}\rangle|2_{\omega}\rangle\ldots|n_{\omega}\rangle\ldots)^{\mathrm{T}} \rightarrow |F-\omega \text{ basis}\rangle \tag{1}$$

Connections are set up by creation $\left(\widehat{a}\dagger|n\rangle = \sqrt{n+1}\cdot|n+1\rangle\right)$ and annihilation $\left(\widehat{a}\,|n\rangle = \sqrt{n}\cdot|n-1\rangle\right)$ operators acting on Fock base states; no particles are "created" or annihilated in a physical sense; in fact, it is a reckoning device for abstract spaces. Note that the "colored" vacuum base in (1) indicates it is a label (subindex).

A particular q-state comes out as infinite dimensional row vector with complex numbers (amplitudes) ordered according to the sequence found in (1)-|basis⟩ and using basis labels as:

$$(C_0\,C_1\,C_2\ldots C_n\ldots)^{\mathrm{T}} \rightarrow |\text{q-state}\rangle \tag{2}$$

Families of photon q-states can be obtained as follows. Select a complex number α as eigenvalue of equation: $\widehat{a}\,|\alpha\rangle = \alpha|\alpha\rangle$, where $\langle\alpha|\hat{a}\dagger = \alpha^*\langle\alpha|$, so that one obtains amplitudes $C_0 = \alpha^0/\sqrt{0}!$, $C_1 = \alpha^1/\sqrt{1}!$, $C_2 = \alpha^2/\sqrt{2}!, \ldots,$ $C_n = \alpha^n/\sqrt{n}!$, with normalization factor, $A_0 = \exp\left(-|\alpha|^2/2\right)$; inserting these values in (2) a coherent quantum state comes out as the special scalar product the bra of (1) is a simple transposition:

$$\langle F-\omega \text{ basis}|\text{q-state}(\alpha)\rangle := \sum_{k=0,1,\ldots\infty} C_k(\alpha)|n_{\omega,k}\rangle \tag{3}$$

Shining coherent light say at frequency ω only base states with nonzero amplitude value can originate a transition response. The commutator $[H, |i\rangle\otimes\langle j|]$ with base state as eigenstates of H leads to $\left(\varepsilon_i - \varepsilon_j\right)|i\rangle\otimes\langle j|$. Applying this to a q-state, one gets $\left(\varepsilon_i - \varepsilon_j\right)|i\rangle\otimes\langle j|\text{q-state}\rangle$ and, therefore, it picks up amplitude at the *j*th base state to modulate a response (interaction).

The energy quantum connects two energy levels and Bohr's map: $\left(\varepsilon_i - \varepsilon_j\right) \rightarrow$ "quantity of EM energy measured by the value of radiation frequency." Note that the absolute value for energy levels is not meaningful; Bohr's map relates to energy level differences, and in this sense becomes measurable as the quantity of energy: a resource offered by the quantum formalism.

For a given q-state at the moment of probe, it is apparent that engaging excitation from a given base state (root state) requires particular amplitude to be different from zero. This fact may act as a sort of "selection rule."

2. BASIC PHOTONIC SCHEME

Direct products of EN-base $\{|ik(i)\rangle\}$ with photon-base states enter as elements of an extended basis set. A basis column infinite dimensional vector for a one-photon case reads:[2,3]

$$(\ldots|ik(i)\rangle\otimes|1_\omega\rangle\ldots|ik(i);1_\omega\rangle\otimes|0_\omega\rangle\ldots|i'k'(i');0_\omega\rangle\otimes|0_\omega\rangle\ldots \\ |i'k'(i')\rangle\otimes|0_\omega\rangle\ldots)\rightarrow|\text{basis}\rangle \tag{4}$$

A particular q-state comes out as row vectors with complex numbers (amplitudes) ordered according to the sequence found in (4)-$|\text{basis}\rangle$ and using basis labels as:

$$\left(\ldots C_{ik(i)\otimes 1\omega}\ldots C_{ik(i);1\omega\otimes 0\omega}\ldots C_{i'k'(i');0\omega\otimes 0\omega'}\ldots C_{i'k'(i')\otimes 0\omega'}\ldots\right) \\ \rightarrow|\text{q-state}\rangle \tag{5}$$

Nonzero amplitudes control interactions with external probing devices, while zero-valued ones keep the set ordered.[3,4] A particular q-state corresponds to an implicit scalar product: $\langle\text{basis}|\aleph|\text{q-state}\rangle$, where $\aleph$-sign prevents us taking a simple (finite) sum; thus $\langle\text{basis}|\text{q-state}\rangle$ stands for the appropriate scalar product, $\langle(4)|(5)\rangle$. Besides *basis vectors* being an information resource, they remain fixed once a particular model is chosen.

All terms in (4) form the nonseparable basis as they stand as possibilities associated to the quantum system; state changes can happen via amplitude transformations in (5). No partitioning is yet included in the matter scheme that would allow for constituting subsystems; inclusion will be assured later on, if necessary.[2,3,14]

The element $|ik(i)\rangle\otimes|1_\omega\rangle$ from (4) shares a common origin with remaining basis elements; otherwise, it would stand for independent photon source and matter location. Finding this in (4) implies a constraint eliciting an opening to q-interaction. On the other hand, the term $|ik(i);1_\omega\rangle\otimes|0_\omega\rangle$ indicates the photon number depletion and entanglement of photon and matter field states, eg, $|ik(i);1_\omega\rangle$; note that for this case there is no free photon (this latter one is the name for a quantum of EM field energy) available to be given back. It is a photon-dressed base state (*photon–matter entangled state*). Next, the model uses a type of base state, eg, $|i'k'(i');0_\omega\rangle\otimes|0_\omega\rangle$ to imply the excited state entangled to the photon field vacuum, while the simple direct product form $|i'k'(i')\rangle \otimes |0_\omega\rangle$ opens matter sustained excited state interacting with a photon field vacuum. The actual amplitude value controls interaction process.[2,3]

If all four amplitudes in (2) were different from zero, a coherent photon–matter q-state would be obtained.

2.1 Reading q-States

When, eg, a q-state amplitude changes say from $C_{ik(i)\otimes 1\omega} = 1$ & $C_{i'k'(i');0\omega\otimes 0\omega} = 0$ to $C_{ik(i)\otimes 1\omega} = 0$ & $C_{i'k'(i');0\omega} = 1$, one gets a state of photon/matter entanglement relating not to objects but q-state mutation prompted by a q-process (not explicitly given) leading to relocated amplitudes. Consider a change:

$$\text{From}: \; \left(\ldots 1_{ik(i)\otimes 1\omega} \ldots 0_{ik(i);1\omega\otimes 0\omega} \ldots 0_{i'k'(i');0\omega} \ldots 0_{i'k'(i')\otimes 0\omega} \ldots\right) \tag{6a}$$

$$\text{To}: \; \left(\ldots 0_{ik(i)\otimes 1\omega} \ldots 1_{ik(i);1\omega\otimes 0\omega} \ldots 0_{i'k'(i');0\omega} \ldots 0_{i'k'(i')\otimes 0\omega} \ldots\right) \tag{6b}$$

Taken together, (6a) and (6b) define a one-photon entrance channel seen from the matter element's viewpoint. Materiality sustaining these q-states is *conserved* and never engaged in "filling" any energy eigenvalue.

Finite model linear superpositions, such as $\left(C_{6a}|(6a)\rangle + C_{6b}|(6b)\rangle\right)$, and normalized $|C_{6a}|^2 + |C_{6b}|^2 = 1$, correspond to projected coherent states that, apprehended globally, may show possible finite lifetimes. This situation may connect to q-events that would result in displacements (transfers) of energy and/or AM. In fact, shifting amplitudes from (6a) to (6b) displaces a unit of AM from a photon field to a matter sustained field; and consequently, selection rules apply to couplings' base states sustained by materiality. These rules arise from conservation principles.

Taken in the opposite direction, namely, from (6b) to (6a), the emission mode opens as a possibility. The emitted state belongs to a limit that does not have its proper place in the space used to introduce $\langle(1)|(2)\rangle$. This simply means that the abstract theory has to be supplemented to describe events as one would expect. Old completeness claims are not granted.

Once this caveat is understood, one may go on. The amplitude at the entangled base state in (6b′) defeats propagation; this may act as the initial state so that an external action may yield, for instance, new steps such as amplitude change from (6b′) to (6c):

$$\left(\ldots 0_{ik(i)\otimes 1\omega} \ldots 1_{ik(i);1\omega\otimes 0\omega} \ldots 0_{i'k'(i');0\omega\otimes 0\omega} \ldots 0_{i'k'(i')\otimes 0\omega} \ldots\right) \tag{6b'}$$

In Eq. (6b′), the entrance (emission) channel is closed, ie, $0_{ik(i)\otimes 1\omega}=0$. From state (6b′), state (6c) may result via coherent states, though here we show a pure state:

$$\left(\ldots 0_{ik(i)\otimes 1\omega}\ldots 0_{ik(i);1\omega\otimes 0\omega}\ldots 1_{i'k'(i');0\omega\otimes 0\omega}\ldots 0_{i'k'(i')\otimes 0\omega}\ldots\right) \quad (6c)$$

This latter stands as an excited state entangled to a *ω-vacuum*.

Possibilities for having registered histories are brought in. This might be helped by the use of laser sources; for physical examples analyzed, see Ref. 14.

It is important to remember that elementary materiality does not occupy individual base states (energy levels) so that all basis positions remain open as possibilities modulated by amplitudes; these ones turn out operative whenever a time evolution process starts up or a laboratory event prepared by an experimenter is flashed.[14]

Why do we not use the propensity idea? Simply because PQC is not a representational scheme; nothing is explicitly said on the elementary material constituents except that they must always be present (sustaining as a configuration space would do). This is a much weaker "metaphysical" assumption concerning material reality than the one found in standard QM.

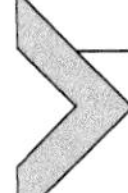

3. ABSTRACT AND LABORATORY SPACES: LINKING/RELATING SYSTEMS

In terms of information data, the bridge is clearly identified (see earlier). The special relativity theory (SRT) framework (I-frame) opens a connection between abstract and laboratory domains. It helps introduce configuration spaces with the dimension defined by *the number* of classical degrees of freedom, ie, a dimensionless number. This includes two I-frames to help define relative origins and orientation required by SRT. There is no absolute space and time.

3.1 Role of Configuration Space

For abstract configuration space, $\mathbf{x}\rightarrow(\mathbf{x}_1, \ldots, \mathbf{x}_n)$ collects information on the number $3n$ of classic degrees of freedom; these n-tuples support a linear vector space over real numbers. In the abstract space, these numbers do not refer to "particle" properties, and they enter as labels in rigged Hilbert spaces either in configuration $\{|\mathbf{x}\rangle\}$ or in reciprocal $\{|\mathbf{k}\rangle\}$ spaces. These base states would bridge abstract q-states $\{|\psi\rangle\}$ to laboratory sets via projected states $\langle \mathbf{x}|\psi\rangle$ or $\langle \mathbf{k}|\phi\rangle$; or simply *wavefunctions*: $\psi(\mathbf{x})$ or $\phi(\mathbf{k})$ so the link is via I-frames that can be located in the laboratory space (real space). $\psi(\mathbf{x})$ or $\phi(\mathbf{k})$ are complex functions over real numbers support. Bases functions such as $\langle \mathbf{x}|\mathbf{k}\rangle$ or $\langle \mathbf{k}|\mathbf{x}\rangle$ have the form $\exp(i\mathbf{k.x})$ or $\exp(-i\mathbf{k.x})$ with measures $\mathbf{dx}$ or $\mathbf{dk}$, respectively. In $\mathbf{x}$-space, $\mathbf{k}$ plays the role of a quantum number, albeit a continuous one; $A(\mathbf{x})$ gives the wave packet state in $\mathbf{x}$-space: $\int A(\mathbf{x})\exp(i\mathbf{kx})\mathbf{dx} \Rightarrow B(\mathbf{k})$.

In $\mathbf{k}$-space, wave packets read: $\int B(\mathbf{k})\exp(-i\mathbf{kx})\mathbf{dk} \Rightarrow A(\mathbf{x})$ maps to a wave packet state in $\mathbf{k}$-space. Configuration spaces are just collections of real numbers, there is no special assigned meaning, and they correspond to a type of abstract space.

3.2 Frequency–Time Regime

Bases functions such as $\langle t|\omega\rangle$ and $\langle \omega|t\rangle$ have the form $\exp(i\omega|t)$ or $\exp(-i\omega t)$ with measures $d\omega$ or dt, respectively. In t-space, ω plays the role of a quantum number, albeit a continuous one covering all possibilities; $A(t)$ gives the wave packet state in t-space: $\int A(t)\exp(i\omega t)dt \Rightarrow B(\omega)$. Additionally, in ω-space, wave packets read: $\int B(\omega)\exp(-i\omega t)d\omega \Rightarrow A(t)$ maps to a wave packet state in ω-space.

To address the energy-time regime, Planck's constant $h/2\pi$ ($\hbar$) is required: It is a dimension-based condition, not a quantization constraint.

A warning should be noted: a primal nonclassical attribute of q-states unfortunately may acquire a pseudo-classical gloss if one imagines the superposition terms as a kind of classical interference of wave functions. Let us avoid this type of theorizing.

For semiclassic models, coordinates refer to particle positions, eg, electrons and/or nuclei. The numbers $\{\mathbf{x}\}$ are thus loaded with a meaning (that is, to be classical particle positions). It goes without saying that these are not required by the mathematical structure, as no classical representation is sought. Labels 1 to n can only be traced to particular classical material elements; this choice allows the use of invariant ordered configuration space; for identical elements a second subindex facilitates handling of permutation symmetries (see POL).[1] Quantum degrees of freedom would enter as new quantum numbers (labels, eg, 2- and 4-spinors). Note the possibility to label the wavefunction with a global spin quantum number.

With respect to I-frames, one can distinguish internal and external q-states. In what follows, we focus on quantum numbers for the internal as well as for quantized EM system where a source defines an origin or signals a target for emission. Including quantum numbers covering spectral products and intermediate excited (transition) states in Eq. (1), schemes were obtained where the chemical processes become explicit via amplitude changes of states type-(2).[2,4–8] In practice—and it is the stance of the paper's author—semiclassic and full quantum physical schemes complement each other.[2,5–8] The photonic approach projected with the help of configuration space leads to a different view of chemical processes.

To help in the appraisal of the q-state notion, consider an experimental double-slit experiment[16] viewed from the present framework.

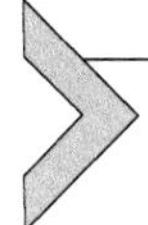

4. QUANTUM PHYSICS OF TONOMURA DOUBLE-SLIT EXPERIMENT

This experiment, using electrons as carriers, touches on key aspects of quantum physical foundations: its interpretation provides a new awareness on quantum states.

Electron states are basic elements supplemented by detection devices where q-events are registered (clicks). Entangled states mediate possible q-energy transfers to or from one subsystem to the other; location is not determined by the electron q-state, only changes can be given some space-time characters. Tonomura developed technology so that the equivalent of one electron at a time was present in the region supporting a device comparable to a double slit.[16] Actually, the statement "only *one electron* at a time came" has now changed: only one q-state sustained by the elementary materiality was present at a time. This latter element is the relevant component for the quantum theoretical analysis.

In other words, a q-state *interacts* with the double-slit technically given by an interaction operator transporting geometric information. It is neither a particle nor a path that is the relevant feature. It is the interaction that is explored first from an abstract standpoint, thereby skimming all possibilities. Quantum states are first handed in Hilbert–Fock space, not in real (laboratory) space.

In abstract space, all possibilities would be included, so that the result (in this case) is a q-state signaling an interference pattern. This situation (in abstract space) can be set in correspondence (linked or related) to simultaneous interaction at the double-slit device with *one and the same incident q-state*. The experimental setup grants this sameness. Thus, interposing a detecting surface, and in agreement to (abstract) calculations, an interference pattern emerges, as predicted. Experimentally, the case can be presented as follows.

Detection of electron states one by one apparently generates a random image during initial collection of spots (clicks). However, according to the present perspective (sustainment), any two q-events will be correlated via the final quantum state containing information on the q-interactions with the double slit. Any sustaining material would be a "carrier" for the *same* q-state. No independent particles can do it; we are thus in the presence of q-interactions at the double-slit (or its equivalent) yielding q-states in an interference scheme. The abstract interference pattern will be there, first as possibility, and until enough incoming q-events impinge with the recording device, an *image* of that interference would slowly emerge.

It is a q-event that can generate a "click." This one would correspond to quantum energy transfer; a sensitive surface records the event that initially appears *as if they were random in location*. Actually, early events do *appear* as being quite randomly distributed (noticed by Tonomura[16]); this is the impression at least and suggests that a too-positivistic interpretation misses the physics encapsulated in a q-event.

For, according to the present view, each one of these registered spots is the expression of the amplitude of scattered quantum states plus incident q-state taken in the intensity regime (square modulus of the total wavefunction at the sensitized surface).

4.1 A Century Later

Almost a century passed before we acknowledged that in quantum physics, what is measured corresponds to quantum states sustained by material carriers and these latter express particle states (not classical particles). There is no such a thing as wave-particle duality either.[15] This latter is present in our way of speaking only.[17]

Naturally, it happens that scattering and interference are present as well in classical wave phenomena; however, *discovering interference in a quantum setting does not logically imply a wave property in a classical sense*. Such a conclusion would not be logically granted and is quantum physically a nonsense, as the matter involves physics without objects.[9,14,18]

Both measured and measuring devices comprise quantum elements that might prompt for q-events. The quantum system can be subjected to q-interactions that do not imply a detectable energy swap; this is the case (at least in part) with the double-slit device.

Thus, quantum mechanics is not a representational scheme. It does not describe either particles or waves in a classical physics manner. This can be taken as the "message of the quantum": do not mix up the levels.

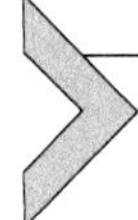

5. ONE-PHOTON INITIATED QUANTUM PHYSICAL PROCESSES

What does one-photon case achieve? Consider: (i) Base state $|i=0\rangle\otimes|1_\omega\rangle$ features spin 1 from the photon-base state and possible interaction with electronic ground state; (ii) entanglement base $|i=0;1_\omega\rangle\otimes|0_\omega\rangle$ displays spin 1 this time sustained by the entangled photon–matter term; (iii) first electronic excited state $|i=1, S=0, L=1;0_\omega\rangle\otimes|0_\omega\rangle$ AM lies at $L=1$; and (iv) lowest spin triplet $|i=2, T\rangle\otimes|0_\omega\rangle$; it shows angular momentum 2, namely $L=1$, $S=1$. The updated base vector looks as (6):

$$(|i=0\rangle\otimes|1_\omega\rangle\ldots|i=0;1_\omega\rangle\otimes|0_\omega\rangle\ldots|i=1,S=0;0_\omega\rangle\otimes|0_\omega\rangle\ldots|i=1, S=0\rangle\otimes|0_\omega\rangle\ldots|i=2,T\rangle\otimes|0_\omega\rangle\ldots) \quad (6)$$

These elements open to states sustained by the elementary materiality. Q-state (6) is a unity, a one-partite state, where the first entry would mediate the connection with a bi-partite base associated to, eg, free photon and elementary materiality states. It is beyond a q-theory to represent (as it were) the link, and only a q-interaction would *relate* these two worlds; this underlines a sort of relational approach.

The energy associated to the first four terms in (6) is the same; so, there is no jump when one sees (6) as a unity; only if one focuses on one partner (materiality sustained states for instance) is there a change of both energy and AM. A photon–matter entangled basis cannot be separated; it is a "unit," though not an object. A way out to the left side (propagation) would change amplitudes so that the response from the entangled basis changes into a response from amplitudes at $|i=1, S=0\rangle\otimes|0_\omega\rangle$ that is an isolated excited state. If one eliminates all entanglements of this kind, then a standard basis set for one EN Hilbert space is obtained. But even in this case, the system can only display coherent states. The inclusion of photon basis, in the way shown so far, corresponds to a physical Hilbert space prepared to bridge both systems. Let us explore some possibilities that the framework offers.

5.1 Opening Access to Spin-Triplet States: Optically Assisted Way

Amplitude at S_1 that can be directly open from a ground closed shell state (S_0), while spin-triplet base (T_1) state features zero amplitude value because $S_0 \rightarrow T_1$ and $S_1 \rightarrow T_1$ are both forbidden transitions by AM conservation.

A possible activation takes a path similar to the optically activated zero field magnetic resonance phenomena.[19,20]

Thus, $\left(C_{i=0\otimes 1\omega}\ldots C_{i=0;1\omega\otimes 0\omega}\ldots 0_{1;0\omega}\ldots 0_{1\otimes 0\omega}\ldots 0_{2,\otimes 0\omega}\ldots\right)$ is a coherent state that might open an entrance channel to q-states covering other spectral sectors. This heralds a photon state entangled with ground state basis. Both amplitudes are different from zero simultaneously; energy conserves if $\left|C_{i=0\otimes 1\omega}\right|^2 + \left|C_{i=0;1\omega\otimes 0\omega}\right|^2 = 1$. The very activation process is sustained by entanglement of one-photon state and the ground state.

This situation permits apprehending grounds for elastic scattering: targeting state $\left(1_{i=0\otimes 1\omega}\ldots 0_{i=0;1\omega\otimes 0\omega}\ldots\right)$, entanglement with $\left(0_{i=0\otimes 1\omega}\ldots 1_{i=0;1\omega\otimes 0\omega}\ldots\right)$, and scattering via q-state $\left(\exp(-i\mathbf{k}'.\mathbf{r})1_{i=0\otimes 1\omega}\ldots 0_{i=0;1\omega\otimes 0\omega}\ldots\right)$, where $\mathbf{k}'$ signals a direction away from the I-frame of the targeted state. Note that the term $\exp(-i\mathbf{k}'.\mathbf{r})1_{i=0\otimes 1\omega}$ cannot be taken in isolation; otherwise, one reintroduces a particle concept. The coherent state speaks of possible cases, not of representing (whatever the case might be).

It is useful to examine propagation from state (6′) over base states made accessible by a second one-photon interaction. We focus our attention on fluorescent-like states:

$$\left(0_{i=0\otimes 1\omega}\ldots 0_{i=0;1\omega\otimes 0\omega}\ldots 1_{1;0\omega}\ldots 0_{1\otimes 0\omega}\ldots 0_{2,\otimes 0\omega}\ldots 0\ldots\right) \tag{6'}$$

From (6′), triplet state activation may follow a path, eg: (1) information–injection of S_1–T_1 energy gap via a supplementary photon state $|1_{\omega S_1-T_1}\rangle$; and (2) entanglement suggested by state (7) identified by $C_{1^*;0\omega\otimes 1\omega_{ST}}$:

$$\left(0_{i=0\otimes 1\omega}\ldots 0_{i=0;1\omega\otimes 0\omega\otimes 1\omega_{ST}}\ldots C_{1;0\omega\otimes 0\omega_{ST}}\ldots C_{1^*;0\omega\otimes 1\omega_{ST}}\ldots 0_{2,\otimes 0\omega\otimes 0\omega_{ST}}\ldots\right) \tag{7}$$

Note the addition of the excited state base state $|1^*;0\omega\otimes 1\omega_{ST}\rangle\otimes|0_\omega\rangle$; by construction it has two pieces of information, ie, energy level above the first excited state (S_1) and the energy level when measured from ground state (T_0) displays the equivalent of $2\hbar\omega_{ST}$ so that the triplet base state label now gains new information on the S_1–T_1 channel. Energy conservation demands a second external photon state $|1\omega_{ST}\rangle$; this one would "harvest" an assisted consecutive two-photon emission; it would result in the triplet state opening with a nonzero value for the amplitude, ie:

$$\left(0_{i=0\otimes 1\omega}\ldots 0_{i=0;1\omega\otimes 0\omega\otimes 1\omega_{ST}}\ldots 0_{1;0\omega\otimes 0\omega_{ST}}\ldots 0_{1^*;0\omega\otimes 1\omega_{ST}}\ldots 1_{2,\otimes 0\omega\otimes 0\omega_{ST}}\ldots 0\ldots\right) \tag{7'}$$

The triplet base state shows $L=1$, $M_L=0$ sustained by space antisymmetric term and spin state with q-number equal $S=1$, $M_S=0$. The effective production would consume two units of AM measured as two photons; the required AM was taken from external photon fields. This is the equivalent of spin–orbit coupling in the semiclassic scheme.

Observe that a direct one-photon activation from singlet ground state S_0 to the triplet is not possible, due to AM conservation rules, unless the photon state also displays space AM.

Once the activation channel displays state (7), only the first excited state S_1 would act as root state for supplementary electronic excitation events. A nonrotating wave model supplies energy at $2\hbar\omega_{ST}$ that adds label information at $0_{2,T\otimes 0\omega}$ so that it can play the role of a dressed vacuum, ie, $0_{2,T\otimes 0\omega\otimes 0\omega_{ST}}$.

Shine a *second* ST-gap photon state $|1\omega_{ST}\rangle$ at (7), to prompt for the cascade process noted above, and that would end up with nonzero amplitude at $C_{|i=2,S=1\rangle\otimes|0\omega\rangle\otimes|0\omega_{ST}\rangle}$. These states remain implicit in (7) that now we change into the explicit end result:

$$\left(\ldots 0_{i\otimes 1\omega}\ldots 0_{i;1\omega\otimes 0\omega}\ldots 0_{i+1;0\omega\otimes}\ldots 1_{i=2,T\otimes 0\omega 0\omega_{ST}}\ldots\right) \tag{8}$$

The information brought up by injection is registered by the second ST-photon gap state that adds to amplitude $1_{i+2,T\otimes 0\omega 0\omega_{ST}}$; the operation has the flavor of information supplement to the label. In other words, chemical and photophysical processes starting at a triplet as their root state can now begin. The ST-gap two photons pay for two units of AM required by the "transfer" from the spin-singlet electronic excited state. The space part corresponds to changing from $L=0$ to $L=1$.

Note that one ST-gap photon state could be radiated, thereby acting as a catalyst to activation of the triplet state reflected in (8).

Observe that the elementary event that may lead to a transfer of one energy quantum can take (6′) as portal state. *Mutatis mutandi* this type of state may also act as possible portal for photon emission. At any rate, information circulates expressed via q-state amplitudes.

5.2 Photon Upconversion: Triplet–Triplet Two-Photon Interactions

Upconversion processes involving (spin) triplet states play key roles in organic light-emitting diodes. Here, we present a brief photonic description in the style of optically assisted opening of spin-triplet states.

Consider a bi-partite (dimer) with two separate sets of elementary materials sustaining equivalent spectra that can be controlled independently. Each partite can be monitored following its particular I-frame.

We apply the procedure of the preceding section so that each dimer element (partite) is set at a triplet + triplet state (cf. Eq. 8). The model assumes that each partite crystal site (or molecule), for instance, sustains the respective I-frame, thereby allowing for control of relative real space distance between internal states. To fuse both quantum internal spaces, interaction would ask for (setup) entanglement. Thus the weakly interacting pair would appear as a linear combination of site states:

$$\begin{aligned} &1/\sqrt{2}\left[(\ldots 1_{2,\otimes 0\omega\otimes 0\omega_{ST}}\cdots)_1 \pm (\ldots 1_{2,\otimes 0\omega\otimes 0\omega_{ST}}\cdots)_2\right] \\ &\to 1/\sqrt{2}(|T_1\rangle_1 \pm |T_1\rangle_2) = |\pm\rangle \end{aligned} \tag{9}$$

This equation looks like LCAO, but here via entanglement the elementary material is expanded and the local q-states $(\ldots 1_{2,\otimes 0\omega\otimes 0\omega_{ST}}\cdots)_{i=1,2}$ are disposed in the extended Hilbert–Fock space.

The energy level for $|T_0\rangle$ defines the zero energy level (triplet ground state) measured from S_0 ground state. Thus, under resonance condition elicited by (9), even a weak interaction would mix $|+\rangle$ and $|-\rangle$ q-states, thereby leading to $2E_{T_1}$ at one site $|2T_1\rangle$ and a triplet ground state $|T_0\rangle$ at the other site. The interesting result: characteristic frequency $\omega_{T_0T_1}$ is then doubled at one site $2\omega_{T_0T_1}$ and formally zeroed (vacuum triplet state).

From the base state $|2T_1\rangle$, there are now several possibilities. Remember that two triplets combine to give: $S=0$, 1, or 2 spin manifolds. Among the new possibilities, there will be an excited singlet state with energy level about $2E_{T_1} \to E_{S_1^*}$. The energy gap w.r.t. the singlet ground state presents frequency ω' much larger than the one for the first singlet excited state. This is then one of the sources for the so-called photon upconversion.

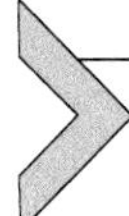

6. PROBE THROUGH X-RAY AND HIGHER FREQUENCY PHOTON STATES

Recently, the extension of laser techniques to coherent beams into the X-ray region of the EM spectrum to tabletop equipment opens opportunities for applications in exciting research fields.[21] The X-ray q-states generated represent a quantum coherent mode of the Roentgen X-ray tube in the soft X-ray region.

The energy quanta associated to this type of probe might be found well above the first or second ionization limit of materiality sustaining the processes. Consider a generic circumstance: the first step in the interaction with matter shares the same pattern, which corresponds always to entanglement, thence a number of possibilities (nonradiative changes) are possible, including reaching at an ionization channel threshold indicated with q-number n^*:

$$\begin{aligned}(|i=0\ k(0)\rangle\otimes|1_\omega\rangle\ldots|i=0\ k(0);1_\omega\rangle\otimes|0_\omega\rangle\ldots|i\\ =n^*\ k(n^*);0_\omega\rangle\otimes|0_\omega\rangle\ldots|n\ k(n)\rangle\otimes|0_\omega\rangle\ldots)^{\mathrm{T}}\end{aligned} \quad (10)$$

The energy labels n are always larger than n^*; they belong to a continuum. A generic quantum state takes the form:

$$\left(C_{|i=0\ k(0)\rangle\otimes|1\omega\rangle}\cdots C_{|i=0\ k(0);1\omega\rangle\otimes|0\omega\rangle}\cdots C_{|i=n^*\ k(n^*);0\omega\rangle\otimes|0\omega\rangle}\ldots C_{|n\ k(n)\rangle\otimes|0\omega\rangle}\cdots\right)$$

The first interaction slot opens the material system via (12), namely photon–matter field entanglement:

$$\left(C_{|i=0\ k(0)\rangle\otimes|1\omega\rangle}\cdots C_{|i=0\ k(0);1\omega\rangle\otimes|0\omega\rangle}\cdots 0_{|i=n^*\ k(n^*);0\omega\rangle\otimes|0\omega\rangle}\cdots 0_{|n\ k(n)\rangle\otimes|0\omega\rangle}\cdots\right) \quad (11)$$

These are the entangled states with energy equivalent to the incoming photon field. The initial slot is: $\left(1_{|i=0\ k(0)\rangle\otimes|1\omega\rangle}\cdots 0_{|i=0\ k(0);1\omega\rangle\otimes|0\omega\rangle}\cdots 0_{|i=n^*\ k(n^*);0\omega\rangle\otimes|0\omega\rangle}\cdots 0_{|n\ k(n)\rangle\otimes|0\omega\rangle}\cdots\right)$.

The form (11) of this q-state yet entangled via coherence with the photon vacuum requires flowing in, so to speak, the matter field space to become the special q-state:

$$\left(0_{|i=0\ k(0)\rangle\otimes|1\omega\rangle}\cdots 1_{|i=0\ k(0);1\omega\rangle\otimes|0\omega\rangle}\cdots 0_{|i=n^*\ k(n^*);0\omega\rangle\otimes|0\omega\rangle}\cdots 0_{|n\ k(n)\rangle\otimes|0\omega\rangle}\cdots\right) \quad (12)$$

The initial probing channel $0_{|i=0\ k(0)\rangle\otimes|1\omega\rangle}$ is closed. Now, if a *low-frequency field* (eg, microwave or radio waves) acts on the system, one can expect time dependence first at entrance channel amplitudes imposing time dependence to (11):

$$\left(C_{|i=0\ k(0)\rangle\otimes|1\omega\rangle}(t)\ldots C_{|i=0\ k(0);1\omega\rangle\otimes|0\omega\rangle}(t)\ldots 0_{|i=n^*\ k(n^*);0\omega\rangle\otimes|0\omega\rangle}\ldots 0_{|n\ k(n)\rangle\otimes|0\omega\rangle}\ldots\right)$$

since the energy level $\varepsilon_{n|k(n)\rangle\otimes|0\omega\rangle}$ features the same energy of $\varepsilon_{|i=0\ k(0);1\omega\rangle\otimes||0\omega\rangle}$.

There is a large energy gap between activated states and the rest; observe that the lowest energy level shifts so that the coherent state forbids direct access to any external radiation. Thus, there are no free-electron states in the continuum for the present model. The ionization event would be mediated by an entangled state linking abstract to external space. We have not yet demarcated this latter point.

Time dependence of (12′)-type may be the source of electron states in the continuum:

$$\begin{aligned}\big(0_{|i=0\ k(0)\rangle\otimes|1\omega\rangle}(t)\ldots 0_{|i=0\ k(0);1\omega\rangle\otimes|0\omega\rangle}(t)\ldots 0_{|i=n^*\ k(n^*);0\omega\rangle\otimes|0\omega\rangle}\cdots\\ 1_{|n\ k(n)\rangle\otimes||0\omega\rangle}(t)\ldots\big)\end{aligned} \tag{12′}$$

State (12′) could gate a free-electron state sustained by a proper I-frame and positive-ion-state partite. This requires a q-event sharing some similarity to "tunneling."

We return to (10), which has to be adjusted in order to include electron spin base states. The electron states correspond to 2-spinors. Each electronic quantum number includes spin S and a projection along an arbitrary direction M_S; the dimension $2S+1$ gives the range $-S\leq M_S\leq +S$. For $S=1/2$, orthogonal base states: $|S=1/2\ M_S=-1/2|\rangle$ and $|S=1/2\ M_S=+1/2|\rangle$ are well known; these base states are sustained by two elementary materials: two electrons. For the atomic K-shell, the electronic configuration reads: $1s^2$. For (12′), there will be one electron at state $|1s^2\rangle$ and a hole state: $|1s^1 1s^0\rangle$, where $1s^0$ indicates a hole at the K-shell. The state (12″) must be reshaped:

$$\begin{aligned}\big(1\alpha_{|i=0\ k(0)\rangle\otimes|1\omega\rangle}(t)\ldots 0_{|i=0\ k(0);1\omega\rangle\otimes|0\omega\rangle}\ldots 0_{|i=n^*\ k(n^*);0\omega\rangle\otimes|0\omega\rangle}\cdots\\ 1\beta_{|n\ k(n)\rangle\otimes|0\omega\rangle}(t)\ldots\big)\end{aligned} \tag{12″}$$

α spinor (1 0) and β spinor (0 1) so in this form they can be correlated. For an ionized state corresponding to release of one-electron state measurable at lab space, the corresponding q-event is implicit (not computable). With this caveat, the one-electron level is assigned an arbitrary spin 1/2 formally as:

$$\left(\left(C_\alpha C_\beta\right)_{|i=0\ k(0)\rangle\otimes|1\omega\rangle}\ldots 0_{|i=0\ k(0);1\omega\rangle\otimes|0\omega\rangle}\ldots 0_{|i=n^*\ k(n^*);0\omega\rangle\otimes|0\omega\rangle}\ldots 0\ldots\right)$$

This equation may signal an unpolarized spin state and an implicit hole at electronic level $i=0$. The precursor to the emitted electron state must be a p-state to compensate spin one taken away by the ionizing excitation. As low-frequency probes act on this state, this may lead to coherent-like states propagating in time:

$$\left(\left(C_\alpha\ C_\beta\right)(t)_{|i=0\ k(0)\rangle\otimes|1\omega\rangle}\ldots C_{|i=0\ k(0);1\omega\rangle\otimes|0\omega\rangle}(t)\ldots C_{|i=n^*\ k(n^*);0\omega\rangle\otimes|0\omega\rangle}(t)\ldots 0\ldots\right) \tag{13}$$

Once a laboratory event happens, this path becomes a dead end and replaced by a bi-partite situation (laboratory electron state release).

Now, intra-q-state electronic transitions become possible sustained by an ionic partite. For the spherical symmetric hole state, any target state associated to AM equal to 1 would present an allowed transition and consequently one envisages a corresponding photon emission. The source of this process (root state) displays symmetry $l=1$ and the corresponding target states requires an energy release with symmetry touching either a state $l=0$ (s-state) or $l=2$ (d-state). If energy is sufficient to put the state in the continuum, a secondary electron will be detected after a q-event. This is known as Auger electron and the full process is known as an Auger process.

Note that if the energy recovered by annihilating a hole state does not suffice to ionize the site again, it would be possible to propagate energy internally or to other sites if one handles solid-state cases.

In solid-state cases, the energy released in filling a hole state may be used to set up an electron state in the conduction band. We may thence have electric charge flux in the elementary material sustaining the q-states.

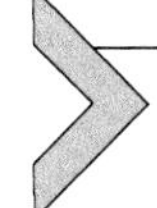

7. CHEMISTRY FROM A PHOTONIC QUANTUM PHYSICAL PERSPECTIVE

The concept of molecular structure and chemical structure underlies most of descriptive chemical phenomenology. In quantum chemistry, the corresponding algorithms (eg, based in Born–Oppenheimer framework) lead to electronic wavefunctions depending parametrically on nuclear positions and typified by nodal plane patterns (NPPs). The NPPs yield a partitioning of the nuclear space, which is useful as a communication tool. The point is that both semiclassic and photonic frameworks address the same quantum system and, consequently, it would be possible to develop "ties" (bridge, relations) in order to develop a more comprehensive understanding of chemical processes, particularly in this age of tabletop lasers.[21] The approaches have to be complementary, not excluding, which is the intention of our work.

The basis set for the photonic scheme has the form given in Eq. (4). Thus, in principle, a huge number of energy states can come into a chemical horizon; many of them almost degenerate though with zero-value amplitudes at a given time.

Chemistry is about forming/breaking bonds or in present language changing the number of partite states. From a one- to a two-partite situation, the abstract quantum description must relate domains accepting different numbers of I-frames. These latter are essential to describe laboratory events, eg, in the scattering of two-partite elements. Care must be exercised because domains might not be commensurate. The bi- and multipartite system permits their laboratory space localization, which is an advantage. This is not a Hilbert–Fock space character, in spite of the fact that semiclassic schemes mix up all of them.

A q-event may thus connect both spaces under particular circumstances, eg, laboratory probing. Consider a general basis:

$$\begin{aligned}&(|1g2g\rangle\otimes|1_{\omega}\rangle\ldots|1g2g;1_{\omega}\rangle\otimes|0_{\omega}\rangle\ldots|1g2e;0_{\omega}\rangle\otimes|0_{\omega}\rangle\ldots|1e2g;0_{\omega}\rangle\otimes|0_{\omega}\rangle\ldots)\\&\to|\text{basis}\rangle\end{aligned} \tag{14}$$

Both subsystems stand at the "door" of an entanglement event that quantum states like (15) may fix their qualities. The second slot being an entangled base state so that the photon field energy now appears "smeared" into the material constituent state.

For quantum states such as $\left(C_{|1g2g\rangle\otimes|1\omega\rangle}\cdots C_{|1g2g;1\omega\rangle\otimes|0\omega\rangle}\cdots 0_{|1g2e;0\omega\rangle\otimes|0\omega\rangle}\cdots 0_{|1e2g;0\omega\rangle\otimes|0\omega\rangle}\cdots\right)$, amplitudes cover radiation–matter interaction as if it were a portal gate that eventually could be used to open the system and trap energy by closing off $C_{|1g2g\rangle\otimes|1\omega\rangle}=0$. Through entanglement or taking a time dependence driving in the opposite direction, one would obtain $1_{|1g2g\rangle\otimes|1\omega\rangle}$ and $0_{|1g2g;1\omega\rangle\otimes|0\omega\rangle}$, where amplitudes in one sense or the other refer to a one-photon state that eventually can be emitted if a q-event were to happen.

When amplitudes show a continuing propagation, the state below signals a family of excited states: $(0_{|1g2g>\otimes|1\omega\rangle}\ldots 0_{|1g2g;1\omega\rangle\otimes|0\omega}\rangle\ldots C_{|1g2e;0\omega\rangle\otimes|0\omega}\rangle\ldots C_{|1e2g;0\omega\rangle\otimes|0\omega}\rangle\ldots)$. For this q-state, there are no possibilities for the material system to emit a one-photon state. And chemistry can possibly henceforth proceed via a quantum dissociation event. To illustrate this, let us construct two normalized linear superpositions $|\pm\rangle$:

$$|\pm\rangle \rightarrow \Big(0_{|1g2g\rangle\otimes|1\omega}\rangle\ldots 0_{|1g2g;1\omega\rangle\otimes|0\omega}\rangle\ldots 1/\sqrt{2}_{|1g2e;0\omega\rangle\otimes|0\omega}\rangle\ldots \pm 1/\sqrt{2}_{|1e2g;0\omega\rangle\otimes|0\omega}\rangle\ldots\Big) \tag{15}$$

Coupling the states $|\pm\rangle$, one can get either (16) or (17):

$$|+\rangle+|-\rangle \rightarrow \Big(0_{|1g2g\rangle\otimes|1\omega}\rangle\ldots 0_{|1g2g;1\omega\rangle\otimes|0\omega}\rangle\ldots 1_{|1g2e;0\omega\rangle\otimes|0\omega}\rangle\ldots 0_{|1e2g;0\omega\rangle\otimes|0\omega}\rangle\ldots\Big) \tag{16}$$

$$|+\rangle-|-\rangle \rightarrow \Big(0_{|1g2g\rangle\otimes|1\omega}\rangle\ldots 0_{|1g2g;1\omega\rangle\otimes|0\omega}\rangle\ldots 0_{|1g2e;0\omega\rangle\otimes|0\omega}\rangle\ldots 1_{|1e2g;0\omega\rangle\otimes|0\omega}\rangle\ldots\Big) \tag{17}$$

Thus, reading from labels either one detects an exited state at one end, and independent of distance, there will be a ground state in the opposite direction. Eg, take (17) and identify a detector to label D-X1; coincidence with the label amplitude $1_{|1e2g;0\omega\rangle\otimes|0\omega}\rangle$ permits assignment of the q-event to amplitude at base state $_{|1e2g;0\omega\rangle\otimes|0\omega}\rangle$. As we construct theory and detector apparatus, a conclusion follows: if at the antipode of D-X1, one sets up another detector, it will not respond. The conclusion may appear puzzling. This is because, in one way or another, labels implicitly hide the R-parameter, and since energies are R-independent, the base state would share this character.

To actualize an R-dependence, the bi-partite state should be made effective. Note that the interaction responsible for the recombination leading to (16), for example, includes simultaneous information on both ends. The R-dependence will show up now if a q-event happens.

The q-state (17), if probed and detected, would also show separate base state response, namely, ground state at one end and excited states at the other that, by construction, are R-separated. If this is true, having detected excitation at one end, the conclusion will be that the q-state shows a ground state system. There would be no signal involved; this result reflects the nature of the entangled state. The transition between one- and bi-partite states results from q-interactions that do not fully belong to the entangled state under scrutiny; it lies on a bridge between spaces. Large R, superluminal signals have no rational place within the present model. Entangled states, once "activated," contain all necessary information.

7.1 Dissociation Competition

For a reactant system that from ground state shows two bond-dissociation possibilities with different dissociation limits, the lowest energy one would show up as product thence followed by products from the higher energy one. Take a one-photon excitation of a chromophore partite with energy above the highest channel. Now the energy gaps are inverted and one would expect to sense the product signals in reverse order.

For the photonic approach, the response from the given materiality to external probes matters. The structural element enters as a graph label only. The spectroscopic idea replaces that of structure. In this sense, photonic and semiclassic frameworks would complement each other as they target the same type of processes from opposite sides (worlds), as it were. The procedure translated to laboratory information tells us something simple: if the excitation is detected at one end, this means that the q-state at the other end comes up as information and can be used with certainty. We do not need to carry out any further measure. There is no use for a psychophysical hypothesis (action), and registering would occur in the absence of an observer (q-event). In other words, there would be no classic decoherence, and materiality does not occupy energy label states.

8. INFORMATION TRANSFERS

In apprehending the meaning of wavefunctions, even from early stages, the information concept has occupied a central place, though the initial representational content is not retained here. Responses toward external probes connect to amplitudes that appear actually to control the interaction. Amplitude variation of this kind can be activated either by a quantum event or by a q-interaction in Hilbert space (interaction operators); the later ones do not comport real energy exchange, only the possibility domain.

Reading from labels is one basic aspect. It is through this that one can follow histories and permit telling stories. The state (6) concerns both q-fields, ie, photon and matter q-fields, that vary at the same time; no free photon states are available once entangled and consequently there is no room for justifying a semiclassic picture. Lab processes related to changes involving many-partite states require entanglement first between partite states to occur in a reacting q-space. Such states do not belong to partite states taken separately. These special states are added to the base state vector, and the number of I-frames goes down by one unit. As a result, these entangled states play a fundamental role in linking spaces that otherwise are incommensurate.

This latter piece of information is central to apprehend processes seen from abstract space perspectives: *spaces sustained as possibilities*. Physics without objects is the logical way to implement a q-theory for understanding chemical processes including photonics; one must become familiar with this new state of affairs before examining the type of description associated with chemical processes. Finkelstein noticed that q-systems ought to be seen as a plexus of q-processes (q-events for us) and not a plenum of q-objects.[22]

Yet elementary materiality must be present. This materiality would be seen as information carriers to the extent they sustain extended q-states. EM energy actually is a carrier of energy quanta as well as quantized angular momenta. The classical representation with the help of electric and magnetic fields is useful to construct coupling operators, but misses q-events.

Along the line of information transfer, constructing bridges to semiclassic models would add a supplementary dimension to modeling approaches. See, for instance, Berrada et al.'s work on entanglement generation involving the model obtained from Eq. (3) and a beam-splitting device.[23]

Inclusion of SRT information, configuration space supported labels read: $|\mathbf{x},ict\rangle$ and $|\mathbf{p},iE/c\rangle$ with connecting function[24]:

$$\langle \mathbf{x}, -ict|\mathbf{p}, iE/c\rangle = (2\pi\hbar)^{-2}\exp\left(i(\mathbf{p}.\mathbf{x} - Et)/\hbar\right) \tag{18}$$

The arguments have kinematic meanings in the realm of SRT, while in abstract space they are employed as labels: c speed of light and E with dimension of energy.

Conservation laws, eg, energy, angular, and linear momenta, enter the photonic framework in a natural manner as soon as processes producing variation of the I-frame numbers. This advantage permits the analyses of physical and chemical routes, as illustrated earlier. To the extent that biological materials present similarities with photonics, it is natural to use the present approach to think biological processes in quantum physical terms as well. Yet they will never represent objects.

Opening access to spin triplet indicates a multiphoton mechanism regulated by conservation laws. Excitation wandering, for example, in any solar cell will perform well (including red shifts). The key is quantum coherence propagating via particular materiality. Coherence/decoherence would fix pathways via vertices able to prompt photon state emission. This sort of photon recycling would increase efficiency.[25]

9. DISCUSSION

That a q-state emerges as image if appropriately recorded follows from the present photonic scheme. Tonomura's experiment for a double-slit setup substantiate it,[16] for initially collected q-events seem to stand for a random process. However, after gathering sufficient number of q-events, a supportive image develops corresponding more and more to what is named an interference pattern. This is a sort of "impressionist" rendering of the q-state. From the first "click," the q-state (amplitudes) is present as a possibility.

The above situation, if anything, shows that a q-state is not an object, but it hangs somehow on sensitive surfaces, revealing an image constructed from q-events: q-energy and AM exchanges—such is the reality of a q-state. No doubt, materiality sustaining the q-state must arrive to the detecting surface, but information transported is richer than what a classical particle impact would convey.

Between a q-system and a q-detector, there are energy *and* information exchanges: q-events. These q-events actually suppress the standard view of decoherence. Now, it is a physical process that mediates probing, not observers with their friends producing decoherence.

Activation events with one-photon state were explored using simple examples.[26] Photon/matter state entanglement plays a key role hinging to q-events, a sort of mediating channel. Reversing direction, the entangled state prompts a one-photon event possibility. Its lifetime would characterize a situation not included so far.

The opening of a spin-triplet state required a path starting from the nearest (parent) spin-singlet excited electronic state. A base state affected by a zero-valued amplitude does not respond to an external probe. Use of the *S–T* energy gap injection as photon states with resonant frequency turns out to be a key to opening the channel toward setting a nonzero value amplitude at the triplet state. A second resonant photon leads to a cascade accomplishing the triplet state activation. Measuring the process from the spin-singlet ground electronic state, two units of angular momentum are necessary. The role of two *S–T* gap energy quanta permitted AM conservation.

Thus, injection of *one* energy quantum (photon) does not lead to a dynamical process. The role of low-frequency radiation injected (shone) onto the system will help support quantum dynamical processes. The important result is that no time evolution can be expected without the presence of low-frequency EM energy quanta.

The possibility to produce multipartite states permits analyses of chemical processes. They can be seen related to entanglement forming/breaking sustainment. These entanglement changes are to be taken as quantum processes, not as mechanical ruptures/sewing of chemical bonds. Yet they are related via q-events. The q-events are related to real energy and AM transfer processes linking the q-system to probing devices.[2,3,14,26]

Note that q-events and q-entanglements do not elicit computational algorithms in themselves. It is a matter of principle. However, registering becomes a resource employed to indicate relative distances and/or orientations. This is another feature of the present quantum approach.

The population idea fades away as the representational character of the theory is absent in the photonic scheme. Neither photons nor molecules are seen as objects. The energy exchanged between q-materials and photon fields corresponds to the quantity of energy assigned to the photon field. This reason explains why we do refer to q-events, because of the nonrepresentational character assigned to symbols. Q-events embody a quantum of energy and angular momentum. The q-event can be spatially localized and timed. This type of process is associated to changes in the number of I-frames, thereby relating the event to particular real space localizations. Taking Aspect's version of EPR (Einstein–Podolski–Rosen) experiment,[15] we use our language to examine this case. A one-partite system acts as source as it "decomposes" into a bi-partite state: each one with an I-frame. This change in the number of I-frames signals a q-event that in this case corresponds with a two-photon state emission. Conservation of linear momentum imposes photon states **k** vectors signal opposite directions. However, the 1-partite states have internal quantum numbers, eg, eliciting polarization: an entangled state. Note that the axis between **k**-vectors may be "tumbling" if the original I-frame or, more properly, the photon states display space angular momentum. Use of classical ideas of particle or waves is not granted.

Here we use our q-states (15)–(17) to display quantum changes preparing dissociation. The states $|\pm\rangle$ are entangled q-states, and a q-event is modeled by either state (16) or (17). Because they are related to one component, they will have amplitudes affected by $\exp(i\mathbf{kr})$ and $\exp(-i\mathbf{kr})$. So one can put detectors to sense excitation that now are identified by either $1_{|1g2e;0\omega\rangle\otimes|0\omega\rangle}$ or $1_{|1e2g;0\omega\rangle\otimes|0\omega\rangle}$. But reading the labels, one detects excitation at a point, while on the antipode probing, there will be no excitation to detect. It is apparent that here there is no signal sent. The quantum system has its own abstract space that has little to do with real space.

The introduction of I-frames also provides the opportunity to introduce the concept of NPPs. There are possibilities to take into account different types of symmetries without being stopped by the parametric-dependent electronic wave functions. Also, prospects of constructing bridges toward semiclassic schemes result from this.[4,27]

To wit, quantum mechanics does not need interpretation in the photonic framework as presented here. From this perspective, what was apparently missing in standard quantum mechanics was the means required for transferring information, and formulating connections between isolated partite subsystems. This could not be done without quantum concepts of q-entanglement and quantum events as expounded here.

ACKNOWLEDGMENTS

The author is indebted to Prof. E. Ludena for cogent and enlightening discussions, as well as to one referee for constructive criticisms that helped improve presentation.

REFERENCES

1. Löwdin, P. O. Quantum Theory of Many-Particle Systems. I. Physical Interpretations by Means of Density Matrices, Natural Spin-Orbitals, and Convergence Problems in the Method of Configurational Interaction. *Phys. Rev.* **1955**, *97*, 1474.
2. Tapia, O. Quantum-Matter Photonic Framework Perspective of Chemical Processes: Entanglement Shifts in HCN/CNH Isomerization. *Int. J. Quantum Chem.* **2015**, *115*, 1490.
3. Tapia, O. Quantum Photonic Base States: Concept and Molecular Modeling. Managing Chemical Process Descriptions Beyond Semi-Classic Schemes. *J. Mol. Mod.* **2014**, *20*, 2110.
4. Tapia, O. Electro-Nuclear Quantum Mechanics Beyond the Born-Oppenheimer Approximation. Towards a Quantum Electronic Theory of Chemical Reaction Mechanisms. In Hernández-Laguna, A., et al., Eds.; Quantum Systems in Chemistry and Physics, Vol. 2; Kluwer Academic Publishers: The Netherlands, 2000; pp 195–212.
5. Woolley, R. G. On Gauge Invarance and Molecular Dynamics; In Maruani, J., et al., Eds.; New Trends in Quantum Systems in Chemistry and Physics, Vol. 2; Kluwer Academic Publishers: The Netherlands, 2000; pp 3–21.
6. Dutra, S. M. *Cavity Quantum Electrodynamics*; Wiley-Interscience: New York, 2005.
7. Pusey, M.; et al. On the Reality of the Quantum State. *Nat. Phys.* **2012**, *8*, 475–478.
8. Ringbauer, M.; et al. Measurements on the Reality of the Wave Function. *Nat. Phys.* **2015**, *11*, 249–254.
9. Tapia, O. Quantum States for Quantum Measurements. *Adv. Quantum Chem.* **2011**, *61*, 49–106.

10. D'Espagnat, B. Are There Realistically Interpretable Local Theories?*J. Stat. Phys.* **1989**, *56*, 747–766.
11. Omnès, R. A New Interpretation of Quantum Mechanics and Its Consequences in Epistemology. *Found. Phys.* **1995**, *25*, 605–629.
12. Flick, J.; et al. Kohn-Sham Approach to Quantum Electrodynamical Density-Functional Theory: Exact Time-Dependent Effective Potentials in Real Space. *PNAS* **2015**, *112*, 15285–15290. http://dx.doi.org/10.1073/pnas.1518224112.
13. Newton, R. G. What Is a State in Quantum Mechanics?*Am. J. Phys.* **2004**, *72*, 348.
14. Tapia, O. *Quantum Entanglement and Decoherence: Beyond Particle Models. A Farewell to Quantum Mechanics' Weirdness*; Cornell University Library: Cornell, USA, 2014. Available from: <arXiv:1404.0552> [quant-ph].
15. Aspect, A. Bell's Theorem: The Naïve View of an Experimentalist. Available from: http://arxiv.org/ftp/quant-ph/papers/0402/0402001.pdf.
16. Tonomura, A.; et al. Demonstration of Single-electron Buildup of an Interference Pattern. *Am. J. Phys.* **1989**, *57*, 117–120.
17. Dirac, P. A. M.; Fock, V. A.; Podolsky, B. On Quantum Electrodynamics. In: *Selected Papers on Quantum Electrodynamics*; Schwinger, J. Ed.; Dover: New York, 1958; pp 29–40. Paper 3.
18. Selesnick, S. A. *Quanta, Logic and Spacetime*; World Scientific: Singapore, 2003.
19. Clarke, R. H.; Hofeldt, R. H. Optically Detected Zero Field Magnetic Resonance Studies of the Photoexcited Triplet States of Chlorophyll a and b. *J. Chem. Phys.* **1974**, *61*, 4582–4587.
20. Takeda, K.; Takegoshi, K.; Terao, T. Zero-Field Electron Spin Resonance and Theoretical Studies of Light Penetration into Single Crystal and Polycrystalline Material Doped with Molecules Photoexcitable to the Triplet State via Intersystem Crossing. *J. Chem. Phys.* **2002**, *117*, 4940–4946.
21. Chen, C.; et al. Tomographic Reconstruction of Circularly Polarized High-harmonic Fields: 3D Attosecond Metrology. *Sci. Adv.* **2016**, *2. e1501333.*
22. Finkelstein, D. Space-Time Code. III. *Phys. Rev.* **1972**, *D5*, 2922–2931.
23. Berrada, K.; et al. Beam Splitting and Entanglement Generation: Excited Coherent States. *Quantum Inf. Process.* **2013**, *12*, 69–82.
24. Brändas, E. Comment on Background Independence in Quantum Theory. *J. Chin. Chem. Soc.* **2016**, *63*, 11–19.
25. Yablonovitch, E. Lead Halides Join the Top Optoelectronic League. *Science* **2016**, *351*, 1401.
26. Tapia, O. *Quantum Physical States for Describing Photonic Assisted Chemical Change: I. Torsional Phenomenon at Femtosecond Time Scale.* Cornell University Library: Cornell, USA, 2013. Available from: <arXiv:1212.4990v2> [quant-ph].
27. Tapia, O.; Polo, V.; Andres, J. Generalized Diabatic Study of Ethylene "Isomerism". In: *Recent Advances in the Theory of Chemical and Physical Systems*; Julien, J.-P. et al., Eds.; Springer: New York, 2006; pp 177–196.

CHAPTER TWELVE

Quantum Chemistry with Thermodynamic Condition. A Journey into the Supercritical Region and Approaching the Critical Point

Marcelo Hidalgo Cardenuto, Kaline Coutinho, Sylvio Canuto[1]
Instituto de Física, Universidade de São Paulo, Cidade Universitária, São Paulo, SP, Brazil
[1]Corresponding author: e-mail address: canuto@if.usp.br

Contents

Abstract

Combining Statistical Mechanics and Quantum Chemistry it is possible to study solvent effects in spectroscopy and understand chemical reactivity in solution. However, once the thermodynamic condition can be incorporated, it is possible to advance in other important regions of the phase diagram. Hence supercritical fluids with temperature and pressure beyond the critical point can be studied. Supercritical fluids are of interest both for their remarkable physical chemical properties and the industrial interests. The critical point, however, is apparently not a thermodynamic condition amenable to quantum chemical calculations. This is because it is characterized by intense fluctuations and density inhomogeneity. The correlation length becomes infinite at the critical point. But for points close enough to the critical point the fluctuations disappear, and it is possible to get very close to this rather interesting point in the phase diagram. In this work we review some results for the spectroscopy of molecular systems in the supercritical region and the static dipole polarizability and the refractive index of Ar only 2 K above the critical point. The refractive index presents some peculiarities, but it is well behaved as we pass at the critical point. The numerical value obtained of 1.083 is in very good agreement with the experimental value of 1.086. We contend that the proximity of the critical point is amenable to theoretical quantum mechanical studies possibly accessing new physical phenomena.

Advances in Quantum Chemistry, Volume 74
ISSN 0065-3276
http://dx.doi.org/10.1016/bs.aiq.2016.06.006

1. INTRODUCTION

Since the advent of the Schrödinger equation the use of quantum mechanics has unraveled the properties of atomic and molecular systems. A new field has emerged, denoted as Quantum Chemistry. It successfully evolved from its infancy in the early decades of last century into a mature and robust discipline today.[1] This development has occurred very much as predicted by Dirac that recognized that the entire Chemistry could be studied by solving the Schrödinger equation, but its accurate solution would necessitate incredible mathematical developments. However, one aspect of utmost importance cannot be overlooked as chemical reactions and the entire field of biochemistry necessitates the consideration of the thermodynamic condition. In fact, several chemical reactions do not take place if not in proper temperature. In special one could consider all reactions that are needed for the existence of life. The Schrödinger equation does not include explicitly a thermodynamic condition and this has to be included ad hoc, as an external additional ingredient. Today the combination of Quantum Chemistry and Statistical Mechanics makes the hybrid method known generally by QM/MM (quantum mechanics and molecular mechanics).[2–4] The important pioneering work of Levitt and Warshel[5] and also the efforts made by Karplus[6] were recognized with the Nobel Prize in 2013. The method is essentially divided into two parts: (a) the molecular mechanics where the thermodynamic condition is specified and (b) the quantum mechanical where the electronic properties and reactivity are considered. There are variants of the QM/MM methods, and we have used a sequential methodology.[7–9] In this, classical Monte Carlo (MC) or molecular mechanics simulations are performed first and configurations are sampled for subsequent quantum mechanical calculations. The sampling can be made very efficiently such that statistical convergence can be assured with a relatively small number of QM calculations.[10] The use of classical simulations has proved to be efficiently done thanks to the developments of good force fields. However, this is not free from risks and in some cases the use of quantum molecular dynamics may be advisable.[11,12] The possibility of incorporating thermodynamic condition led to great developments in the theoretical description of molecular spectroscopy and reactivity in solution. Solvent effects can be incorporated realistically contributing to a more accurate description of the spectroscopic results but also adding some new aspects such as the calculation of inhomogeneous line broadening. It is clearly a natural step to proceed further and advance into other regions of the phase diagram. Recently there has been an increased theoretical interest in the supercritical region.[11,13–17] For thermodynamic conditions beyond the critical point, the supercritical region, the fluid exhibits novel characteristics with unusual properties very different from normal and regular liquids. For instance, small changes in the thermodynamic condition of the fluid in the supercritical region change pronouncedly the dielectric constant. Hence by tuning the thermodynamic condition it is possible to gauge the dielectric constant and hence control the environment condition aiming at, for instance, better reaction rates. Indeed supercritical fluids are acquiring great importance also for being a benign solvent. The critical point itself that defines the beginning of the supercritical region is found at the end of the coexistence line between the gas and the liquid phases. It is the convergence point of the increasing density of the gas and decreasing density of the liquid and therefore submitted to intense fluctuations and density inhomogeneity. At the critical point some properties present singular behavior.[18] The specific heat is known to diverge at the critical point. The theoretical treatment relies on the theory of critical phenomena and critical exponents[19–21] and renormalization theory.[21,22] The fact that the correlation length becomes infinite at the critical point precludes the use of a finite representative system and hence quantum chemical calculations. However, the difficulties associated to the divergences at the critical point disappear if the thermodynamic condition changes slightly. It is thus possible to approach the critical point and perform quantum chemical studies in the close vicinities. In a previous study[23] this was accomplished and for the first time it was possible to follow the lead of Robert S. Mulliken and see "what are the electrons really doing" in the vicinities of the critical point. This investigation is along these lines and we will report some spectroscopic results in a supercritical environment and also approach the critical point.

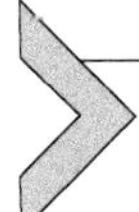

2. SHORT REVIEW OF SOME MOLECULAR STUDIES IN THE SUPERCRITICAL REGION

Studies of molecular spectroscopy in supercritical environment have been successfully reported by several groups. We briefly review some of the results obtained in our laboratory discussing some specific cases.[24–30]

The UV–vis absorption spectrum of benzophenone has been obtained experimentally[31,32] in normal (NW) and supercritical water (SCW). The intense $\pi-\pi^*$ transition is seen to suffer a blue shift of 1700 cm^{-1} upon change from NW ($P=1$ atm and $T=298$ K; for simulation in the NVT ensemble we use $\rho=0.997$ g/cm^3) to SCW (in the condition of $P=340.2$ atm and $T=673$ K). Using the sequential QM/MM this has been analyzed.[24] We find that the $\pi-\pi^*$ band is composed by four transitions with the most intense showing a calculated blue shift of 1425 cm^{-1} using the semiempirical ZINDO method. Using time-dependent density functional theory (TD-DFT) within the CAM B3LYP/6-311+G(2d,p) model it is found that the shift is 1715 cm^{-1} in very good agreement with experiment. This good result is obtained only after careful consideration of the electronic polarization of benzophenone in the solvent environment. This increases the dipole moment from the gas phase by 35% in SCW and 88% in NW.[24] A similar analysis[25] has been made of the solvatochromic shift of the $n-\pi^*$ transition of acetone in SCW, in the same supercritical condition described earlier. The experimentally inferred result[32] of 500–700 cm^{-1} can be compared with the calculated value of 670 cm^{-1} obtained using 170 explicit solvent molecules using ZINDO. Using TD-DFT within the B3LYP/6-31+G(p) it is found that the shift is of 825 cm^{-1} including only the electrostatic solute–solvent interaction. Additional studies including NMR parameters have also been made.[26] For instance, the nuclear isotropic shielding constants $\sigma(^{17}O)$ and $\sigma(^{13}C)$ of the carbonyl bond of acetone in water again at the same supercritical ($P=340.2$ atm and $T=673$ K) and NW conditions have been studied theoretically using MC simulation and quantum mechanics calculations based on the B3LYP/6-311++G(2d,2p) method. In SCW, there is a decrease in the magnitude of $\sigma(^{13}C)$ but a sizable increase in the magnitude of $\sigma(^{17}O)$ when compared with the results obtained in NW. Changing from NW to SCW, the calculations[26] give a ^{13}C chemical shift of 11.7 ± 0.6 ppm for acetone in good agreement with the experimentally inferred result of 9–11 ppm.[33]

One interesting case is *para*-nitroaniline (*p*-NA) in SCW that has been reported[34] to suffer a red shift of 17 nm, corresponding to 0.24 eV (or 1935 cm^{-1}), in the intense charge transfer transition. The SCW condition used experimentally was $T=655$ K and $\rho=0.12$ g/cm^3. It is well known that the absorption spectrum of *p*-NA is due to a charge transfer transition involving an electron acceptor and an electron donor group both amenable to hydrogen bond with water. One group is a hydrogen bond donor (NH_2), and the other is a hydrogen bond acceptor (NO_2) both affecting the solvatochromic shift. Using again configurations generated by classical simulation at the proper thermodynamic condition the calculated transition using CAM-B3LYP/aug-cc-pVDZ is obtained at 305 nm[27] compared to the experimental value of 309 nm,[34] differing by only 4 nm. Moreover, considering the result for the isolated (in vacuum) and the SCW condition a theoretical shift of 17 nm (0.24 eV) has been obtained in sharp agreement with experiment. When reporting solvent changes in isotropically nuclear shielding, the use of a reference system is unnecessary because it cancels out (using eg, tetramethylsylane, TMS): $\delta\sigma = [\sigma(\mathrm{A}) - \sigma(\mathrm{TMS})] - [\sigma(\mathrm{B}) - \sigma(\mathrm{TMS})] = \sigma(\mathrm{A}) - \sigma(\mathrm{B})$.

In all the cases considered above classical force fields have been used in the simulations to generate the solute–solvent configurations. This may require some care in cases that have either not been tested previously or in more drastic thermodynamic conditions. One special case is the supercritical CO_2, where some polar aspects may appear contrary to normal expectations.[11,35–37]

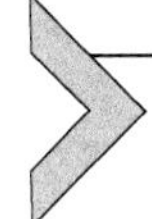

3. SUPERCRITICAL Ar AND THE VICINITIES OF THE CRITICAL POINT

We will discuss results for the static dipole polarizability and the refractive index of Ar in the liquid phase, the supercritical region and only 2 K above the critical point.[23] It is known[18–22] that the behavior at the critical point is universal, and hence the use of atomic Ar is very convenient to observe the behavior of the refractive index as we pass the vicinity of the critical point (experimental values: $T_c = 150.7$ K, $P_c = 48.2$ atm, and $\rho_c = 0.531$ g/cm^3).

We adopted the sequential QM/MM methodology[7–9] performing classical MC simulation for generating the fluid structure to be subsequently submitted to QM calculations. The MC simulation of atomic fluids is equivalent to a molecular dynamics because of the lack of internal degrees of freedom. We used the program DICE[38] for all MC atomic simulations. The atoms interact via the conventional Lennard-Jones potential, and the parameters used are the ones suggested by Maitland and Smith.[39] This potential has proved its accuracy in different circumstances. Within our perspective it is important to mention that this potential obtains a very good value of the critical temperature,[23] missing the experimental result by only 2 K. Hence, the theoretical critical temperature is found to be 148.7 K in comparison with the experimental result of 150.7 K. In addition, the calculated radial distribution is in due agreement with experimental results obtained from X-ray experiments.[40] The MC simulations were made in the *NVT* and the *NPT* ensembles, depending on the convenience. If a clear specification of the density was needed, we used the *NVT* ensemble but if the thermodynamic location in the phase diagram was important we used the *NPT* ensemble. As usual we adopt the image method and periodic boundary conditions.[41] The simulation box includes 2500 Ar atoms, and after analysis of the configurations a total of 150 statistically uncorrelated configurations composed of 14 Ar atoms were used in the QM calculations. The emphasis here is in the calculation of the static dipole polarizability. For this we used DFT employing the B3P86 functional[42,43] and the aug-cc-pVDZ basis set.[44] The QM calculations were made using the Gaussian-09 program.[45] After obtaining the dipole polarizability it is possible to obtain the dielectric constant from the Clausius–Mossotti equation and thereby the refractive index. Each value of the static dipole polarizability, α, is obtained from an average of 150 QM results, $\alpha = \langle \alpha_i \rangle$, $i = 1, 150$. For obtaining the refractive index n, we use the polarizability values α_i to obtain the dielectric constant ε_i and from the average $\varepsilon = \langle \varepsilon_i \rangle$ we obtain n^2.

We will first discuss the results for the dipole polarizability in the liquid phase, shown in Fig. 1. As it can be seen the value of the dipole polarizability does not change with an increased value of the pressure. This is a consequence of the relative stability of the density with increased pressure in the liquid regime. For instance, related to Fig. 1, the temperature is fixed at 92 K but increasing the pressure by a factor of nearly 20 only increases the density by less than 1%, negligible compared to the statistical error. The dipole polarizabilities in these thermodynamic conditions are not

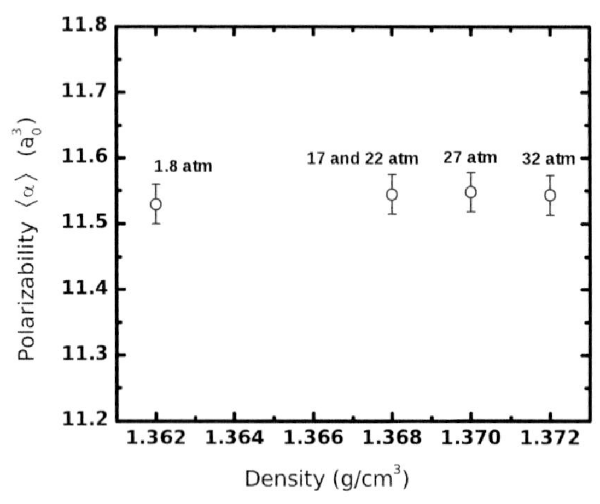

Fig. 1 The calculated dipole polarizability for different thermodynamic conditions in the liquid phase (temperature is fixed at 92 K).

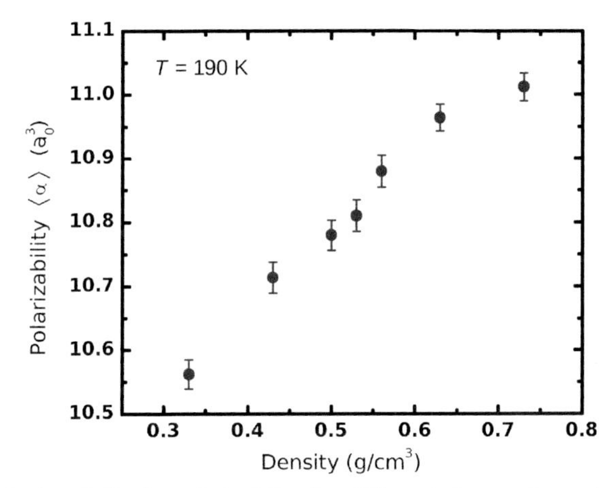

Fig. 2 The calculated dipole polarizability for different thermodynamic conditions in the supercritical phase (temperature is fixed at 190 K). Pressure values range between 85 atm ($\rho = 0.33$ g/cm^3) and 165 atm ($\rho = 0.73$ g/cm^3).

known directly, but from these values we obtain a refractive index of 1.232 in sharp agreement with the experiment value of 1.233.[46]

This situation is very different in the supercritical region, as illustrated in Fig. 2. The decreased value of the density in this regime can be noted. The pressure values change but are always above the critical pressure to ensure that this corresponds to the supercritical region (ie, $T > T_c$ and $P > P_c$). The behavior is a linear increase of the polarizability with the density. This linear increase indicates also a linear increase of the refractive index. In the interval adopted the refractive index changes between 1.050 and 1.116. Experimental values are apparently not known in this part of the supercritical region.

Now we analyze the results close to the critical point. As we have seen in the liquid regime, the dielectric constant is essentially unchanged for increasing pressure but increases linearly with the density in the supercritical region. The behavior close to the critical point is not known, although it has long been argued that for nonpolar fluids it should remain finite.[47] This has been studied using the theory of critical phenomena. More recently, a combined Statistical Mechanics and quantum mechanics study has been made thus revealing the electronic structure close to the critical point.[23] These results are illustrated in Fig. 3 showing the refractive index for temperature 2 K above the critical temperature. This is necessary to avoid the fluctuations characteristic of the critical point where the correlation length goes to

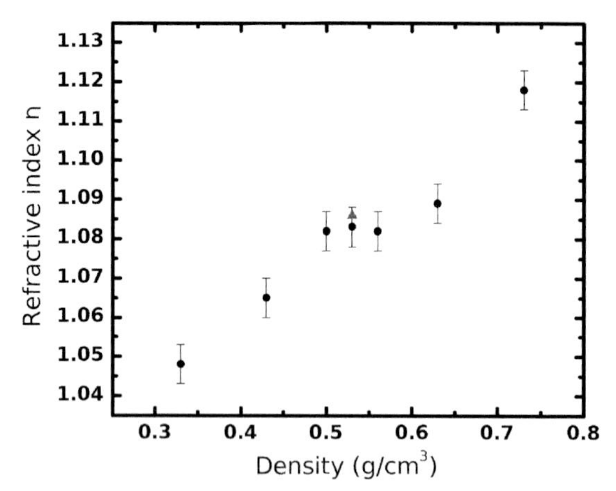

Fig. 3 The calculated refractive index for different thermodynamic conditions near the critical point (temperature is fixed 2 K above the critical temperature). The experimental value at the critical density is also shown (*triangle*).

infinity. In addition for temperatures above T_c the possibility of phase transition is avoided. Fig. 3 then shows that the refractive index increases for increasing densities but remains insensitive for small changes close to the critical density. The accuracy of these results can be certified. The theoretical value of 1.083 ± 0.005 is found in very good agreement with the experimental value of 1.086.[48]

Now we briefly discuss the behavior of the pressure and the first virial in the supercritical region and around the critical point. In the results reported in Figs. 2 and 3 each point shown is obtained by fixing the temperature and the density, hence in the NVT ensemble. In this ensemble the pressure is obtained as

$$\langle P \rangle = \frac{NkT + \langle W \rangle}{V}, \tag{1}$$

where NkT/V is the ideal gas contribution and $\langle W \rangle$ is the first virial,

$$\langle W \rangle = -\frac{1}{3}\left\langle r\frac{\partial U}{\partial r}\right\rangle. \tag{2}$$

The pressure departs from the ideal gas by the first virial term $\langle W \rangle$ that is proportional to the gradient of the interatomic potential U with the opposite sign. As expected, the fluctuations of $\langle W \rangle$ are very high giving average pressure with large fluctuations. Fig. 4 shows the variation of the pressure with

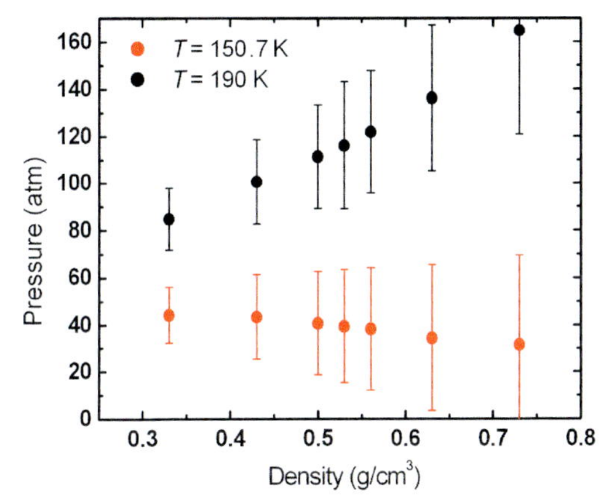

Fig. 4 Variation of the calculated pressure with respect to the density in the isotherms of the supercritical region ($T = 190$ K) and around the critical point ($T = T_c + 2$ K). The difference in the inclination is the result of the change in the contribution of the first virial.

the density in the two isotherms considered, the supercritical region ($T = 190$ K) and in the vicinities of the critical point ($T = 150.7$ K). In the supercritical region the pressure increases with the density. Therefore, this region has the same behavior as the ideal gas, where the pressure is proportional to the density, and also interacting systems with a positive first virial, $\langle W \rangle > 0$. But in the vicinities of the critical point the pressure slightly decreases with the density. Thus, in this region the interacting system has a negative first virial, $\langle W \rangle < 0$. This opposite behavior of the first virial in both regions is a consequence of the sign of the interatomic potential gradient (see Eq. 2). In the supercritical region the first virial is positive due to an average negative interatomic potential gradient that means having the interacting atoms on average closer than the equilibrium distance. However, in the vicinities of the critical point the opposite sign of the average interatomic potential gradient means that the interacting atoms on average are beyond the equilibrium distance.

4. SUMMARY AND CONCLUSIONS

This report attempts to emphasize that the explicit use of thermodynamic conditions in quantum mechanics is a powerful procedure to study atomic and molecular systems in different regions of the phase diagram. After successful developments for dealing with both spectroscopy and reactivity of molecular systems in liquid environment, QM/MM procedures are available to advance into yet unexplored conditions. The supercritical region is one good example where both basic science and technological appealing situations are to be considered a great motivation for advancing in this long and promising avenue. Along these lines progress has indeed been obtained. However, getting close to the critical point has been avoided and the knowledge of the phenomenology of critical phenomena has so far been exclusively obtained by using Statistical Mechanics. This can be changed and electronic structure of systems close to the critical point can be studied. As the correlation length is infinite at the critical point and the associated fluctuations and density inhomogeneity are difficult to deal, we must admit the enormous difficulty in positioning at the precise location of the critical point. However, these fluctuations rapidly disappear and the proximity of the critical point is amenable to theoretical Quantum Mechanical studies. This is an interesting perspective because it may not only confirm or supplement previous results but may also access new physical phenomena.

ACKNOWLEDGMENTS

Dedicated to the memory of Prof. Per-Olov Löwdin on the occasion of his 100th anniversary. S.C. recognizes that Prof. Löwdin's scientific authority and friendly guidance has been influential in his life and now feels the honor and privilege to have been a student of the Quantum Chemistry Group in Uppsala, where teachers and colleagues became friends forever. This work has been partially supported by CNPq, CAPES, FAPESP, INCT-FCx, and BioMol (Brazil).

REFERENCES

1. Dykstra, C.; Frenking, G.; Kim, K.; Scuseria, G., Eds. *Theory and Applications of Computational Chemistry: The First Forty Years*; Elsevier Science: Amsterdam, The Netherlands, 2005.
2. Senn, H. M.; Thiel, W. QM/MM Methods for Biomolecular Systems. *Angew. Chem. Int. Ed. Engl.* **2009**, *48*, 1198.
3. Amara, P.; Field, M. J. Combined Quantum Mechanical and Molecular Mechanical Potentials. *Encyclopedia of Computational Chemistry*; John Wiley: New York, 2002.
4. See the special issue: Canuto, S. Ed. Combining Quantum Mechanics and Molecular Mechanics. Some Recent Progresses in QM/MM Methods. In Advances in Quantum Chemistry, Vol. 59; 2010; pp 1–416.
5. Warshel, A.; Levitt, M. Theoretical Studies of Enzymic Reactions: Dielectric, Electrostatic and Steric Stabilization of the Carbonium Ion in the Reaction of Lysozyme. *J. Mol. Biol.* **1976**, *103*, 227.
6. Field, M. J.; Bash, P. A.; Karplus, M. A Combined Quantum Mechanical and Molecular Mechanical Potential for Molecular Dynamics Simulations. *J. Comput. Chem.* **1990**, *11*, 700.
7. Canuto, S. Ed. *Solvation Methods on Molecules and Biomolecules. Computational Methods and Applications*; Springer: Berlin, 2008.
8. Coutinho, K.; Canuto, S. Solvent Effects from a Sequential Monte Carlo—Quantum Mechanical Approach. *Adv. Quantum Chem.* **1997**, *28*, 89.

9. Coutinho, K.; Canuto, S.; Zerner, M. C. A Monte Carlo-Quantum Mechanics Study of the Solvatochromic Shifts of the Lowest Transition of Benzene. *J. Chem. Phys.* **2000**, *112*, 9874; Rocha, W. R.; Martins, V. M.; Coutinho, K.; Canuto, S. Solvent Effects on the Electronic Absorption Spectrum of Formamide Studied by Sequential Monte Carlo Quantum Mechanical Approach. *Theor. Chem. Acc.* **2002**, *108*, 31; Rivelino, R.; Cabral, B. J. C.; Coutinho, K.; Canuto, S. Electronic Polarization of Liquid Acetonitrile: A Sequential Monte Carlo Quantum Mechanical Investigation. *Chem. Phys. Lett.* **2005**, *407*, 13.
10. Fileti, E. E.; Coutinho, K.; Malaspina, T.; Canuto, S. Electronic Changes Due to Thermal Disorder of Hydrogen Bonds in Liquids: Pyridine in an Aqueous Environment. *Phys. Rev. E* **2003**, *67*, 61504.
11. Cabral, B. J. C.; Rivelino, R.; Coutinho, K.; Canuto, S. A First Principles Approach to the Electronic Properties of Liquid and Supercritical CO_2. *J. Chem. Phys.* **2015**, *142*, 024504.
12. Martiniano, H. F. M. C.; Cabral, B. J. C. Structure and Electronic Properties of a Strong Dipolar Liquid: Born-Oppenheimer Molecular Dynamics of Liquid Hydrogen Cyanide. *Chem. Phys. Lett.* **2013**, *556*, 119.
13. Arai, Y.; Sako, T.; Takebayashi, Y., Eds. *Supercritical Fluids. Molecular Interactions, Physical Properties and New Applications*; Springer: Heidelberg, 2013.
14. See the special issue: Noyori, R. Ed. Supercritical Fluids. In Chemical Reviews, Vol. 99 (2); 1999; pp 353–634.
15. Beckman, E. J. Supercritical and Near-Critical CO_2 in Green Chemical Synthesis and Processing. *J. Supercrit. Fluids* **2004**, *28*, 121.
16. Besnard, M.; Tassaing, T.; Daten, Y.; Andanson, J.-M.; Soetens, J.-C.; Cansell, F.; Loppinet-Serani, A.; Reveron, H.; Aymonier, C. Bringing Together Fundamental and Applied Science: The Supercritical Fluids Route. *J. Mol. Liq.* **2006**, *125*, 88.
17. Skarmoutsos, I.; Samios, J. Local Density Inhomogeneities and Dynamics in Supercritical Water: A Molecular Dynamics Simulation Approach. *J. Phys. Chem. B* **2006**, *110*, 21931.
18. Fisher, M. E. Specific Heat of a Gas Near the Critical Point. *Phys. Rev. A* **1964**, *136*, 1599.
19. Fisher, M. E. The Theory of Equilibrium Critical Phenomena. *Rep. Prog. Phys.* **1967**, *30*, 615.
20. Kadanoff, L. P. Theories of Matter: Infinities and Renormalization. In *The Oxford Handbook of Philosophy Physics*; Batterman, R. Ed.; Oxford University Press: Oxford, 2011.
21. Stanley, H. E. Scaling, Universality, and Renormalization: Three Pillars of Modern Critical Phenomena. *Rev. Mod. Phys.* **1999**, *71*, S358.
22. Goldenfeld, N. *Lectures on Phase Transitions and the Renormalization Group*; Addison-Wesley: Reading, MA, 1993.
23. Hidalgo, M., Coutinho, K.; Canuto, S. Behavior of the Dielectric Constant of Ar Near the Critical Point. *Phys. Rev. E* **2015**, *91*, 032115.
24. Fonseca, T. L.; Georg, H. C.; Coutinho, K.; Canuto, S. Polarization and Spectral Shift of Benzophenone in Supercritical Water. *J. Phys. Chem. A* **2009**, *113*, 5112.
25. Fonseca, T. L.; Coutinho, K.; Canuto, S. Probing Supercritical Water with the n-π^* Transition of Acetone: A Monte Carlo/Quantum Mechanics Study. *J. Chem. Phys.* **2007**, *126*, 034508.
26. Fonseca, T. L.; Coutinho, K.; Canuto, S. The Isotropic Nuclear Magnetic Shielding Constants of Acetone in Supercritical Water: A Sequential Monte Carlo/Quantum Mechanics Study Including Solute Polarization. *J. Chem. Phys.* **2008**, *129*, 034502.
27. Cardenuto, M. H.; Coutinho, K.; Cabral, B. J. C.; Canuto, S. Electronic Properties in Supercritical Fluids: The Absorption Spectrum of p-Nitroaniline in Supercritical Water. *Adv. Quantum Chem.* **2015**, *71*, 323.

28. Fonseca, T. L.; Coutinho, K.; Canuto, S. Hydrogen Bond Interaction Between Acetone and Supercritical Water. *Phys. Chem. Chem. Phys.* **2010**, *12*, 6660.
29. Hidalgo, M.; Rivelino, R.; Canuto, S. Origin of the Red Shift for the Lowest Singlet π-π* Charge-Transfer Absorption of p-Nitroaniline in Supercritical CO_2. *J. Chem. Theory Comput.* **2014**, *10*, 1554.
30. Cabral, B. J. C.; Rivelino, R.; Coutinho, K.; Canuto, S. Probing Lewis Acid Base Interactions with Born Oppenheimer Molecular Dynamics: The Electronic Absorption Spectrum of p-Nitroaniline in Supercritical CO_2. *J. Phys. Chem. B* **2015**, *119*, 8397.
31. Dilling, W. L. The Effect of Solvent on the Electronic Transitions of Benzophenone and Its o- and p-Hydroxy Derivatives. *J. Org. Chem.* **1966**, *31*, 1045.
32. Bayliss, N. S.; Wills-Johnson, G. Solvent Effects on the Intensities of the Weak Ultraviolet Spectra of Ketones and Nitroparaffins. *Spectrochim. Acta A* **1968**, *24*, 551; Bennett, G. E.; Johnston, K. P. UV-Visible Absorbance Spectroscopy of Organic Probes in Supercritical Water. *J. Phys. Chem.* **1994**, *98*, 441.
33. Takebayashi, Y.; Yoda, S.; Sugeta, T.; Otake, K.; Sako, T.; Nakahara, M. Acetone hydration in Supercritical Water: ^{13}C-NMR Spectroscopy and Monte Carlo Simulation. *J. Chem. Phys.* **2004**, *120*, 6100.
34. Oka, H.; Kajimoto, O. UV Absorption Solvatochromic Shift of 4-Nitroaniline in Supercritical Water. *Phys. Chem. Chem. Phys.* **2003**, *5*, 2535.
35. Muñoz Losa, A.; Martins-Costa, M. T.; Ingrosso, F.; Ruiz-López, F. Correlated Ab Initio Molecular Dynamics Simulation of the Acetone-Carbon Dioxide Complex: Implications for Solubility in Supercritical CO_2. *Mol. Simul.* **2014**, *40*, 154.
36. Raveendran, P.; Ikushima, Y.; Wallen, S. L. Polar Attributes of Supercritical Carbon Dioxide. *Acc. Chem. Res.* **2005**, *38*, 478.
37. Saharay, M.; Balasubramanian, S. Ab Initio Molecular Dynamics Study of Supercritical Carbon Dioxide. *J. Chem. Phys.* **2004**, *120*, 9694.
38. Coutinho, K.; Canuto, S. *DICE: A Monte Carlo Program for Molecular Liquid Simulation, v: 2.9*; University of São Paulo: Brazil, 2010.
39. Maitland, G. C.; Smith, E. B. The Intermolecular Pair Potential of Argon. *Mol. Phys.* **1971**, *22*, 861.
40. Eisenstein, A.; Gingrich, N. S. The Diffraction of X-rays by Argon in the Liquid, Vapor and Critical Regions. *Phys. Rev.* **1942**, *62*, 261; Mikolaj, P. G.; Pings, C. J. Structure of Liquids III. An X-ray Diffraction Study of Fluid Argon. *J. Chem. Phys.* **1967**, *46*, 1401.
41. Allen, M. P.; Tildesley, D. J. *Computer Simulation of Liquids*; Oxford University Press: Oxford, 1987.
42. Becke, A. D. Density-Functional Thermochemistry. III. The Role of Exact Exchange. *J. Chem. Phys.* **1993**, *98*, 5648.
43. Perdew, J. P. Density-Functional Approximation for the Correlation Energy of the Inhomogeneous Electron Gas. *Phys. Rev. B* **1986**, *33*, 8822.
44. Woon, D. E.; Dunning, T. H., Jr. Gaussian Basis Sets for Use in Correlated Molecular Calculations. III. The Atoms Aluminum Through Argon. *J. Chem. Phys.* **1993**, *98*, 1358.
45. Frisch, M. J.; Trucks, G. W.; Schlegel, H. B.; Scuseria, G. E.; Robb, M. A.; Cheeseman, J. R.; Scalmani, G.; Barone, V.; Mennucci, B.; Petersson, G. A.; Nakatsuji, H.; Caricato, M.; Li, X.; Hratchian, H. P.; Izmaylov, A. F.; Bloino, J.; Zheng, G.; Sonnenberg, J. L.; Hada, M.; Ehara, M.; Toyota, K.; Fukuda, R.; Hasegawa, J.; Ishida, M.; Nakajima, T.; Honda, Y.; Kitao, O.; Nakai, H.; Vreven, T.; Montgomery, J. A., Jr.; Peralta, J. E.; Ogliaro, F.; Bearpark, M.; Heyd, J. J.; Brothers, E.; Kudin, K. N.; Staroverov, V. N.; Kobayashi, R.; Normand, J.; Raghavachari, K.; Rendell, A.; Burant, J. C.; Iyengar, S. S.; Tomasi, J.; Cossi, M.; Rega, N.; Millam, J. M.; Klene, M.; Knox, J. E.; Cross, J. B.; Bakken, V.; Adamo, C.; Jaramillo, J.; Gomperts, R.; Stratmann, R. E.; Yazyev, O.; Austin, A. J.;

Cammi, R.; Pomelli, C.; Ochterski, J. W.; Martin, R. L.; Morokuma, K.; Zakrzewski, V. G.; Voth, G. A.; Salvador, P.; Dannenberg, J. J.; Dapprich, S.; Daniels, A. D.; Farkas, O.; Foresman, J. B.; Ortiz, J. V.; Cioslowski, J.; Fox, D. J. *Gaussian 09*. Gaussian, Inc.: Wallingford, CT, 2009.

46. Lide, D. R. Ed.; *Handbook of Chemistry and Physics*, 73rd ed.; CRC Press: Boca Raton, 1993; p 9.51.
47. Stell, G.; HØye, J. S. Dielectric Constant and Mean Polarizability in the Critical Region. *Phys. Rev. Lett.* **1974**, *33*, 1268.
48. Teague, R. K.; Pings, C. J. Refractive Index and the Lorentz–Lorenz Function for Gaseous and Liquid Argon, Including a Study of the Coexistence Curve Near the Critical State. *J. Chem. Phys.* **1968**, *48*, 4973.

CHAPTER THIRTEEN

Electron Propagator Theory: Foundations and Predictions

Héctor H. Corzo, J. Vince Ortiz[1]
Auburn University, Auburn, AL, United States
[1]Corresponding author: e-mail address: ortiz@auburn.edu

Contents

Abstract

Electron propagator theory is an efficient means to accurately calculating electron binding energies and associated Dyson orbitals that is systematically improvable and easily interpreted in terms of familiar concepts of valence theory. After a brief discussion of the physical meaning of the poles and residues of the electron propagator, the Dyson quasiparticle equation is derived. Practical approximations of the self-energy operator in common use are defined in terms of the elements of the Hermitian superoperator Hamiltonian matrix. Methods that retain select self-energy terms in all orders of the fluctuation potential include the two-particle-one-hole Tamm–Dancoff approximation, the renormalized third-order method, the third-order algebraic diagrammatic construction, and the renormalized, nondiagonal second-order approximation. Methods based on diagonal second-order and third-order elements of the self-energy matrix, such as the diagonal second-order, diagonal third-order, outer valence Green's function, partial third-order, and renormalized partial third-order approximations, provide efficient alternatives. Recent numerical tests on valence, vertical ionization energies of representative, small molecules, and a comparison of arithmetic and memory requirements provide guidance to users of electron propagator software. A survey of recent applications and extensions illustrates the versatility and interpretive power of electron propagator methodology.

Advances in Quantum Chemistry, Volume 74
ISSN 0065-3276
http://dx.doi.org/10.1016/bs.aiq.2016.05.001

1. INTRODUCTION

Two historical missions have been the constant companions of the field in which Per-Olov Löwdin exercised lasting influence, quantum chemistry. The first is the fulfillment of a reductionist project: to determine molecular properties solely from fundamental constants and equations of physics. The derivation of approximate methods, the design of efficient algorithms and their adaptation to modern computing platforms, and numerical testing of the accuracy of calculations thus enabled are means to producing tools for the prediction of molecular properties. Steady advances on all of these fronts have made quantum chemistry an indispensable part of a chemist's education and have enabled the emergence of a new specialty, computational chemistry. The capabilities of the latter field are especially valuable when competing experimental techniques are costly, unsafe, or slow. These advantages notwithstanding, quantum chemistry is not merely a branch of contemporary computational chemistry because of the former field's second historical mission. As the modern successor of valence theory, quantum chemistry continues to generate concepts of chemical bonding that inform the thinking of specialists who synthesize or characterize new or important forms of matter. These concepts enable the recognition and understanding of patterns of structure, energetics, reactivity, and physical properties and therefore stimulate the formulation and execution of renewed experimental activity. They also inform the education of the next generation of scientists.

There is an inherent tension between these two aspirations. Fulfillment of the first mission inevitably leads to wave function Ansätze, perturbative arguments, density functionals, and parametrization schemes of increasing complexity. Concepts that are products of the second mission, to have any purchase on the minds of experimentalists, must be general and verifiable. Accurate predictions are a necessary condition for influence. However, if such data are not clearly interpreted or if they fail to provide an insight into related systems, their significance may be limited. Qualitative theories based on relatively simple concepts have promoted the recognition of broad patterns of chemical phenomena, but their rigorous numerical realization often leads to inaccurate predictions that undermine their authority.

Theories that are systematically improvable and that generate qualitative concepts with a rigorous foundation therefore have inherent advantages. One such approach to quantum chemistry is based on electron propagator (or one-electron Green's function) theory.[1–8] An additional advantage arises from relevance to experiments that have been prominent in the development of quantum theory from its inception: measurements of electron binding energies. Finally, there are advantages of computational economy that are related to the conceptual elegance of the propagator approach to spectra. All of these features of electron propagator theory are related to the Dyson quasiparticle equation, wherein a nonlocal, energy-dependent operator has eigenvalues that are, in principle, exact electron binding energies and corresponding eigenfunctions known as Dyson orbitals that describe how electronic structure changes when an electron is removed or added.

In this review, some fundamental aspects of electron propagator theory's realization in the context of quantum chemistry are discussed. A derivation of the Dyson quasiparticle equation that employs some of Löwdin's most useful concepts and some prominent approximation schemes are presented. Finally, numerical tests that demonstrate the advantages of electron propagator methods for the computational chemist and some recent applications are discussed.

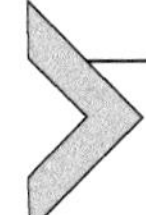

2. POLES AND RESIDUES OF THE ELECTRON PROPAGATOR

The exact electron propagator suffices to determine the electron binding energies, Dyson orbitals, one-electron properties, and total energy of an N-electron system.[1,5] Let $(\chi_r, \chi_s, \chi_t, \chi_u, \ldots)$ be a set of orthonormal spin-orbitals where the corresponding field operators obey the following relationships:

$$\left[a_r^\dagger, a_s^\dagger\right]_+ = \left[a_r, a_s\right]_+ = 0$$
$$\left[a_r^\dagger, a_s\right]_+ = \delta_{rs}.$$

In this notation, the second-quantized Hamiltonian is given by

$$H = \Sigma_{rs} h_{rs} a_r^\dagger a_s + \frac{1}{4} \Sigma_{rstu} \langle rs \| ut \rangle a_r^\dagger a_s^\dagger a_t a_u,$$

where **h** is the matrix of the one-electron operator and where antisymmetrized two-electron integrals in Dirac notation, such that

$$\langle rs \| ut \rangle = \langle rs | ut \rangle - \langle rs | tu \rangle$$
$$\langle rs | tu \rangle = \int \chi_r^*(1)\, \chi_s^*(2)\, g(1,2)\, \chi_t(1)\, \chi_u(2) d(1) d(2),$$

appear in the last term. Solutions of the Schrödinger equation read

$$H|N,0\rangle = E_0(N)|N,0\rangle$$

for the initial, N-electron, ground state and

$$H|N+1,m\rangle = E_m(N+1)|N+1,m\rangle$$
$$H|N-1,n\rangle = E_n(N-1)|N-1,n\rangle$$

for final states with $N \pm 1$ electrons. Elements of the electron propagator matrix, **G**(E), are expressed as

$$G_{rs}(E) = \Sigma_m V_{rm} V_{sm}^* (E - A_m)^{-1} + \Sigma_n U_{rn}^* U_{sn} (E - D_n)^{-1},$$

where A_m is the m-th electron attachment energy, $E_m(N+1) - E_0(N)$, D_n is the n-th electron detachment energy, $E_0(N) - E_n(N-1)$, and where the overlap amplitudes are given by

$$V_{rm} = \langle N+1, m|a_r^\dagger|N,0\rangle$$
$$U_{sn} = \langle N-1, n|a_s|N,0\rangle.$$

Dyson orbitals for electron attachment and detachment which read

$$\phi_m(x_1) = (N+1)^{1/2} \int dx_2\, dx_3\, dx_4 \ldots dx_{N+1}\, \psi_{N+1,m}(x_1, x_2, x_3, \ldots, x_{N+1}) \times \psi_{N,0}(x_2, x_3, x_4, \ldots, x_{N+1})$$

$$\phi_n(x_1) = (N)^{1/2} \int dx_2\, dx_3\, dx_4 \ldots dx_N\, \psi_{N,0}(x_1, x_2, x_3, \ldots, x_N) \times \psi_{N-1,n}(x_2, x_3, x_4, \ldots, x_N)$$

are related to the overlap amplitudes through

$$\phi_m(x_1) = \Sigma_r V_{rm}^* \chi_r(x_1)$$
$$\phi_n(x_1) = \Sigma_r U_{rn} \chi_r(x_1).$$

Whereas the poles of the electron propagator equal electron binding energies, the corresponding residues are related to Dyson orbitals.

3. DERIVATION OF THE DYSON QUASIPARTICLE EQUATION

A generalized notation for propagators with field operators or field operator products that change the number of electrons by one, designated by $\mu^\dagger$ and ν, may be introduced such that

$$\langle\langle \mu^\dagger; \nu \rangle\rangle = \Sigma_m V'_{\mu m} V'^*_{\nu m} (E - A_m)^{-1} + \Sigma_n U'^*_{\mu n} U'_{\nu n} (E - D_n)^{-1}$$

$$V'_{\mu m} = \langle N + 1, m | \mu^\dagger | N, 0 \rangle$$

$$U'_{\nu n} = \langle N - 1, n | \nu | N, 0 \rangle.$$

In this notation,

$$G_{rs}(E) = \langle\langle a_r^\dagger; a_s \rangle\rangle.$$

A generalized propagator defined in terms of $\mu^\dagger$ and ν may be related to a more complicated propagator through

$$E\langle\langle \mu^\dagger; \nu \rangle\rangle = \langle N, 0 | [\mu^\dagger, \nu]_+ | N, 0 \rangle + \langle\langle \mu^\dagger; [\nu, H] \rangle\rangle.$$

Let a binary product and an accompanying metric matrix for the operators be defined by

$$(\mu|\nu) = \langle N, 0 | [\mu^\dagger, \nu]_+ | N, 0 \rangle = \mathrm{Tr}(\rho [\mu^\dagger, \nu]_+),$$

where

$$\rho = |N, 0\rangle\langle N, 0|.$$

(In his lectures for aspiring quantum chemists, Löwdin often discussed the axioms, such as Hermitian symmetry, that a binary product must satisfy.)[9] Hamiltonian and identity superoperators[7,10] then may be defined by

$$\hat{H}\mu = \mu H - H\mu = [\mu, H]$$

$$\hat{I}\mu = \mu.$$

A chain of propagators may be generated, where for example

$$E\langle\langle a_r^\dagger; a_s\rangle\rangle = \langle N,0|[a_r^\dagger, a_s]_+|N,0\rangle + \langle\langle a_r^\dagger; [a_s, H]\rangle\rangle$$

$$= (a_r|a_s) + \langle\langle a_r^\dagger; \hat{H}a_s\rangle\rangle$$

$$E\langle\langle a_r^\dagger; \hat{H}a_s\rangle\rangle = \langle N,0|[a_r^\dagger, \hat{H}a_s]_+|N,0\rangle + \langle\langle a_r^\dagger; [[a_s, H], H]\rangle\rangle$$

$$= (a_r|\hat{H}a_s) + \left\langle\left\langle a_r^\dagger; \hat{H}^2 a_s\right\rangle\right\rangle$$

$$E\langle\langle a_r^\dagger; \hat{H}^\xi a_s\rangle\rangle = \langle N,0|[a_r^\dagger, \hat{H}^\xi a_s]_+|N,0\rangle + \langle\langle a_r^\dagger; [\hat{H}^\xi a_s, H]\rangle\rangle$$

$$= (a_r|\hat{H}^\xi a_s) + \left\langle\left\langle a_r^\dagger; \hat{H}^{\xi+1} a_s\right\rangle\right\rangle.$$

The electron propagator therefore may be written as a series, where

$$\langle\langle a_r^\dagger; a_s\rangle\rangle = E^{-1}(a_r|a_s) + E^{-2}(a_r|\hat{H}a_s) + E^{-3}\left(a_r|\hat{H}^2 a_s\right) + E^{-4}\left(a_r|\hat{H}^3 a_s\right) + \cdots,$$

or in terms of the superoperator resolvent, $(E\hat{I} - \hat{H})^{-1}$, such that

$$\langle\langle a_r^\dagger; a_s\rangle\rangle = \left(a_r|(E\hat{I} - \hat{H})^{-1} a_s\right) = G_{rs}(E).$$

A projection technique discussed by Löwdin[11] may be used to replace the superoperator resolvent by an inverse matrix expressed in a complete manifold of operators, $\mathbf{w}$, such that

$$(E\hat{I} - \hat{H})^{-1} = |\mathbf{w})\left(\mathbf{w}|(E\hat{I} - \hat{H})\mathbf{w}\right)^{-1}(\mathbf{w}|.$$

This substitution yields the following form of the electron propagator matrix:

$$\mathbf{G}(E) = (\mathbf{a}|\mathbf{w})\left(\mathbf{w}|(E\hat{I} - \hat{H})\mathbf{w}\right)^{-1}(\mathbf{w}|\mathbf{a})$$

where $\mathbf{a}$ is a column vector of annihilation operators. The operator manifold, $\mathbf{w}$, may be partitioned into a primary space of simple field operators, $\mathbf{a}$, and an orthogonal, secondary space with higher field operator products, $\mathbf{f}$, where

$$|\mathbf{w}) = |\mathbf{a};\mathbf{f}),$$

$$(\mathbf{a}|\mathbf{a}) = \mathbf{1}_a,\ (\mathbf{a}|\mathbf{f}) = \mathbf{0}_{a\times f},\ (\mathbf{f}|\mathbf{a}) = \mathbf{0}_{f\times a},\ (\mathbf{f}|\mathbf{f}) = \mathbf{1}_f.$$

After partitioning,

$$\mathbf{G}(E) = [\mathbf{1}_a \mathbf{0}_{a\times f}] \left(\mathbf{a};\mathbf{f}|(E\hat{I} - \hat{H})\mathbf{a};\mathbf{f}\right)^{-1} [\mathbf{1}_a \mathbf{0}_{a\times f}]^{\dagger}.$$

Only the upper left block of the inverse matrix is needed because of the orthogonalization of the primary and secondary operator spaces. Now let $\hat{\mathbf{H}} = \left(\mathbf{a};\mathbf{f}|\hat{H}\mathbf{a};\mathbf{f}\right)$. After solving the Hermitian eigenvalue equation,

$$\hat{\mathbf{H}}\boldsymbol{\Omega} = \boldsymbol{\Omega}\boldsymbol{\omega},$$

the electron propagator matrix reads

$$\mathbf{G}(E) = \left[\mathbf{1}_a \mathbf{0}_{a\times f}\right] \left[\boldsymbol{\Omega}(E\mathbf{1} - \boldsymbol{\omega})^{-1}\boldsymbol{\Omega}^{\dagger}\right] \left[\mathbf{1}_a \mathbf{0}_{a\times f}\right]^{\dagger}.$$

For electron detachment and attachment energies, $\omega_n = D_n$ and $\omega_m = A_m$, respectively. The elements of $\boldsymbol{\Omega}$ from the primary operator space provide the residues, where $\Omega_{sm} = V_{sm}$ and $\Omega_{sn} = U_{sn}$. Solving for the upper left block of $\left(\mathbf{a};\mathbf{f}|(E\hat{I} - \hat{H})\mathbf{a};\mathbf{f}\right)^{-1}$ yields

$$\mathbf{G}(E) = \left[\left(\mathbf{a}|(E\hat{I} - \hat{H})\mathbf{a}\right) - \left(\mathbf{a}|(E\hat{I} - \hat{H})\mathbf{f}\right)\left(\mathbf{f}|(E\hat{I} - \hat{H})\mathbf{f}\right)^{-1}\left(\mathbf{f}|(E\hat{I} - \hat{H})\mathbf{a}\right)\right]^{-1}.$$

For orthogonalized operator spaces,

$$\mathbf{G}(E) = \left[\left(\mathbf{a}|(E\hat{I} - \hat{H})\mathbf{a}\right) - \left(\mathbf{a}|\hat{H}\mathbf{f}\right)\left(\mathbf{f}|(E\hat{I} - \hat{H})\mathbf{f}\right)^{-1}\left(\mathbf{f}|\hat{H}\mathbf{a}\right)\right]^{-1}.$$

Because the first term in the inverse matrix involves a generalized Fock operator, F, where

$$\left(a_s|\hat{H}a_r\right) = h_{rs} + \Sigma_{tu}\langle rt||su\rangle\langle N,0|a_t^{\dagger}a_u|N,0\rangle = F_{rs},$$

the inverse of the electron propagator matrix may be written as follows:

$$\mathbf{G}^{-1}(E) = E\mathbf{1}_a - \mathbf{F}^{t} - \left(\mathbf{a}|\hat{H}\mathbf{f}\right)\left(\mathbf{f}|(E\hat{I} - \hat{H})\mathbf{f}\right)^{-1}\left(\mathbf{f}|\hat{H}\mathbf{a}\right).$$

A reference electron propagator, G_0, whose poles and residues are defined by eigenvalues and eigenfunctions of a Hermitian one-electron operator, H_0, may be expressed as

$$\mathbf{G}_0(E) = (E\mathbf{1}_a - \mathbf{H}_0)^{-1}.$$

The reference and exact electron propagators therefore may be related to each other by

$$\mathbf{G}^{-1}(\mathrm{E}) = \mathbf{G}_0^{-1}(\mathrm{E}) - (\mathbf{F}^{\mathrm{t}} - \mathbf{H}_0) - (\mathbf{a}|\hat{\mathrm{H}}\mathbf{f})(\mathbf{f}|(\mathrm{E}\hat{\mathrm{I}} - \hat{\mathrm{H}})\mathbf{f})^{-1}(\mathbf{f}|\hat{\mathrm{H}}\mathbf{a}).$$

The second and third terms are, respectively, the energy-independent and energy-dependent parts of the self-energy matrix, $\boldsymbol{\Sigma}(\mathrm{E})$:

$$\mathbf{F}^{\mathrm{t}} - \mathbf{H}_0 = \boldsymbol{\Sigma}(\infty)$$

$$(\mathbf{a}|\hat{\mathrm{H}}\mathbf{f})(\mathbf{f}|(\mathrm{E}\hat{\mathrm{I}} - \hat{\mathrm{H}})\mathbf{f})^{-1}(\mathbf{f}|\hat{\mathrm{H}}\mathbf{a}) = \boldsymbol{\sigma}(\mathrm{E})$$

$$\boldsymbol{\Sigma}(\infty) + \boldsymbol{\sigma}(\mathrm{E}) = \boldsymbol{\Sigma}(\mathrm{E}).$$

The energy-independent term is given an argument of ∞ to emphasize that the limit of $\boldsymbol{\Sigma}(\mathrm{E})$ as $|\mathrm{E}|$ increases without bound is $\boldsymbol{\Sigma}(\infty)$. The inverse form of the Dyson equation reads

$$\mathbf{G}^{-1}(\mathrm{E}) = \mathbf{G}_0^{-1}(\mathrm{E}) - \boldsymbol{\Sigma}(\mathrm{E}).$$

The regular form of the Dyson equation is

$$\mathbf{G}(\mathrm{E}) = \mathbf{G}_0(\mathrm{E}) + \mathbf{G}_0(\mathrm{E})\boldsymbol{\Sigma}(\mathrm{E})\mathbf{G}(\mathrm{E}).$$

Poles of the electron propagator are values of E where

$$\det \mathbf{G}^{-1}(\mathrm{E}) = 0.$$

These values correspond to solutions of

$$[\mathbf{F}^{\mathrm{t}} + \boldsymbol{\sigma}(\mathrm{E})]\mathbf{C} = \mathbf{C}\mathrm{E}.$$

The latter equation may be solved self-consistently with respect to E. The energy-dependent, nonlocal $\sigma(\mathrm{E})$ operator and the one-electron density matrix that determines the Fock operator may be systematically improved until exact electron binding energies result. The energy dependence of the $\sigma(\mathrm{E})$ operator implies that the number of poles is larger than the dimension of the corresponding matrix. The associated eigenfunctions determined by $\mathbf{C}$ are proportional to the Dyson orbitals. The norm of the Dyson orbital corresponding to the electron detachment energy $\mathrm{D_n}$, known as its pole strength, is given by

$$\mathrm{P_n} = \langle \phi_{\mathrm{n}} | \phi_{\mathrm{n}} \rangle = \left[1 - \mathbf{C}_{\mathrm{n}}^{\dagger} \boldsymbol{\sigma}'(\mathrm{E}) \mathbf{C}_{\mathrm{n}}\right]^{-1},$$

where

$$\sigma'_{rs}(E) = d\sigma_{rs}(E)/dE$$

is evaluated at the nth pole, $E_n = D_n$, and $\mathbf{C}_n^{\dagger}\mathbf{C}_n = 1$. Valid pole strengths vary between zero and unity. A generalized form of the Dyson quasiparticle equation at self-consistency reads

$$[F + \sigma(D_n)]\phi_n = D_n\phi_n$$

for electron detachments. For electron attachments, where pole strengths are obtained from

$$P_m = \langle\phi_m|\phi_m\rangle = \left[1 - \mathbf{C}_m^{\dagger}\boldsymbol{\sigma}'(E)\mathbf{C}_m\right]^{-1},$$

the Dyson quasiparticle equation at self-consistency is

$$[F + \sigma(A_m)]\phi_m = A_m\phi_m.$$

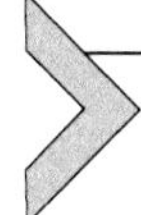

4. APPROXIMATIONS IN THE DYSON QUASIPARTICLE EQUATION

Approximations to the exact **F** and $\boldsymbol{\sigma}$(E) matrices may be defined in terms of the manifold of operators retained in the secondary space, **f**, and the reference-state density matrices (ie, ρ density operators) used to evaluate matrix elements of the Hamiltonian superoperator, $\hat{H}$. The union of all 2hp, 2ph, 3h2p, 3p2h, 4h3p, 4p3h, ... operators (ie, Manne's $\mathbf{f}_3$, $\mathbf{f}_5$, $\mathbf{f}_7$, ..., $\mathbf{f}_{2N+1}$ operator manifolds) suffices for a complete **f** space provided that the defining reference determinant that distinguishes particles (ie, p or virtual orbitals with indices a,b,c,...) from holes (ie, h or occupied orbitals with indices i,j,k,...) is not orthogonal to the exact reference state, $|N,0\rangle$.[12] The nonredundant members of the **a**, $\mathbf{f}_3$, and $\mathbf{f}_5$ operator manifolds are shown in Table 1. For a complete operator manifold, no improvements over this single determinant are needed in $|N,0\rangle$ to produce exact eigenvalues of $\hat{H}$ and poles of **G**(E). This principle is similar to Löwdin's conclusion that the poles of the resolvent's expectation value, $\langle 0|(E1 - H)^{-1}|0\rangle$, are the same for exact and approximate $|0\rangle$ provided that the approximate reference state is not orthogonal to its exact counterpart.[11,13,14] Therefore, it is possible to obtain exact poles of **G**(E) using an approximate reference state that consists of a single determinant. However, to obtain exact residues of **G**(E), the reference state also must be exact.

Table 1 Operator Manifolds

	General Operator	nh(n − 1)p Operator	Indices[a]	np(n − 1)h Operator	Indices[a]
a	a_p	a_i	i occ	a_a	a vir
$\mathbf{f}_3$	$a_p^\dagger a_q a_r$	$a_a^\dagger a_i a_j$	i<j occ a vir	$a_i^\dagger a_a a_b$	a<b occ i vir
$\mathbf{f}_5$	$a_p^\dagger a_q^\dagger a_r a_s a_t$	$a_a^\dagger a_b^\dagger a_i a_j a_k$	i<j<k occ a<b vir	$a_i^\dagger a_j^\dagger a_a a_b a_c$	a<b<c vir i<j occ

[a]i,j,k,l occupied; a,b,c,d virtual.

Improvements over a single-determinant approximation for the reference state may reduce the need for higher $\mathbf{f}_w$ operators and also restore the Hermiticity of $\hat{\mathbf{H}}$, a property which can be lost when approximate choices for $|N,0\rangle$ are made. A study of the effects of perturbative improvements over Hartree–Fock reference states credits Linderberg with the observation that non-Hermitian terms in $\hat{\mathbf{H}}$ may be expressed as

$$\begin{aligned}(Y|\hat{H}X) - (X|\hat{H}Y)^* &= \mathrm{Tr}\left(\rho\left[X, Y^\dagger\right]_+ H - \rho H\left[X, Y^\dagger\right]_+\right)\\ &= \mathrm{Tr}\left([H, \rho]\left[X, Y^\dagger\right]_+\right),\end{aligned}$$

where ρ is a general density operator.[15] For example, one may choose ρ to be a pure-state density operator: $\rho = |N,0\rangle\langle N,0|$. The usual assumptions of perturbation theory, where

$$\begin{aligned}H &= H_0 + \lambda V,\\ [H_0, \rho_0] &= 0\end{aligned}$$

and

$$\rho = \rho_0 + \lambda\rho_1 + \lambda^2\rho_2 + \cdots,$$

imply that

$$[H_0, \rho_k] + [V, \rho_{k-1}] = 0.$$

If ρ is correct through order n, then the non-Hermitian terms are of order n+1:

$$(Y|\hat{H}X) - (X|\hat{H}Y)^* = \mathrm{Tr}\left([V, \rho_n]\left[X, Y^\dagger\right]_+\right).$$

This conclusion also is valid for the polarization propagator, where a commutator replaces the anticommutator in the definition of the superoperator metric.

For all choices of ρ, non-Hermitian terms vanish when X and Y pertain to the primary (**a**) operator space, for $\left[a_p^\dagger, a_q\right]_+ = \delta_{pq}$. These terms also equal zero when X is an nh(n – 1)p operator and Y is an n′p(n′ – 1)h operator (where n and n′ = 1, 2, 3, ...), as the anticommutator term, $[X,Y^\dagger]_+$, vanishes in such cases.

First-order non-Hermitian terms may appear when ρ is based on a single Slater determinant. When X is an nh(n – 1)p operator and Y is an (n + 1)hnp operator, the anticommutator yields a combination of single-replacement operators and therefore the non-Hermitian terms vanish through first order only when the reference determinant satisfies Brillouin's condition. For example, the expression

$$\left(a_a^\dagger a_j a_k \middle| \hat{H} a_i\right) - \left(a_i \middle| \hat{H} a_a^\dagger a_j a_k\right) = \left(1 - P_{jk}\right)\delta_{ik}F_{aj}$$

is equal to zero when Hartree–Fock orbitals are assumed. (When X is an np(n – 1)h operator and Y is an (n + 1)pnh operator, the same conclusion is reached.) However, when X is an nh(n – 1)p operator and Y is an (n + 2)h(n + 1)p operator, $\left[X, Y^\dagger\right]_+$ becomes a combination of double-replacement operators. Nonvanishing, first-order terms in the $\left(a\middle|\hat{H}\mathbf{f}_5\right)$ matrix therefore appear. For example,

$$\left(a_a^\dagger a_b^\dagger a_j a_k a_l \middle| \hat{H} a_i\right) - \left(a_i \middle| \hat{H}\, a_a^\dagger a_b^\dagger a_j a_k a_l\right) = \left(1 - P_{lk} - P_{lj}\right)\delta_{il}\langle jk \| ab\rangle.$$

These terms are cancelled when first-order corrections to ρ are included. The resulting zero matrix is identical to the adjoint of the $\left(\mathbf{f}_5\middle|\hat{I}\mathbf{a}\right)$ matrix. Blocks of $\hat{\mathbf{H}}$ that are more remote from the diagonal, where nh(n – 1)p and (n + m)h(n + m – 1)p operators or np(n – 1)h and (n + m)p(n + m – 1)h operators with $m > 2$ are coupled, have matrix elements that vanish through first order and therefore there are no corresponding non-Hermitian terms. Diagonal blocks pertaining to nh(n – 1)p and np(n – 1)h operators also have no non-Hermitian terms in first order.

An alternative strategy that originates in equation-of-motion theory[3] is to introduce a density operator, derive $\hat{\mathbf{H}}$ and solve for eigenvalues of the

Hermitized matrix, $1/2\left(\hat{\mathbf{H}}+\hat{\mathbf{H}}^{\dagger}\right)$. Disconnected terms that may appear in $\hat{\mathbf{H}}$ should be removed before the Hermitization step.

The Manne operator manifold is orthonormal when ρ is obtained from a single determinant. However, when other density operators are employed, the superoperator overlap matrix may no longer be diagonal, ie, $(X|Y) \neq \delta_{XY}$. To retain the usual form of $\sigma(E)$ where E-dependence occurs only in the $\left(\mathbf{f}|(E\hat{I}-\hat{H})\mathbf{f}\right)^{-1}$ matrix, orthogonalized secondary operator spaces must be introduced. A convenient technique is to employ a symmetric orthogonalization, where, for example, the operator $a_a^{\dagger}a_i a_j = f_{aij}$ is replaced by

$$f'_{aij} = a_a^{\dagger}a_i a_j - \frac{1}{2}\Sigma_b\left(a_b|a_a^{\dagger}a_i a_j\right)a_b - \frac{1}{2}\Sigma_{bkl}\left(a_b^{\dagger}a_k a_l|a_a^{\dagger}a_i a_j\right)a_b^{\dagger}a_k a_l + \cdots.$$

Several widely used approximations retain all terms in $\mathbf{F}^{t}+\boldsymbol{\sigma}(E)$ through a certain order of the fluctuation potential. When Hartree–Fock orbitals are assumed, one-electron density matrix elements, $\langle N,0|a_r^{\dagger}a_s|N,0\rangle$ or P_{rs}, have vanishing contributions in first order. Nonzero terms in $\mathbf{P}$ appear in second and higher orders. Therefore, correlation contributions to $\mathbf{F}$ begin in third order,[4] where for real orbitals

$$\Sigma_{pq}^{(3)}(\infty) = \Sigma_{rs}\langle pr\,||\,qs\rangle P_{rs}^{(2)}$$

$$P_{ij}^{(2)} = -\frac{1}{2}\Sigma_{abk}t_{jkab}^{(1)}t_{ikab}^{(1)}$$

$$P_{ab}^{(2)} = \frac{1}{2}\Sigma_{ijc}t_{ijac}^{(1)}t_{ijbc}^{(1)}$$

$$P_{ia}^{(2)} = (\varepsilon_i-\varepsilon_a)^{-1}\left[\Sigma_{jbc}t_{ijbc}^{(1)}\langle bc\,||\,aj\rangle + \Sigma_{jkb}t_{jkab}^{(1)}\langle ib\,||\,jk\rangle\right]$$

$$P_{ai}^{(2)} = P_{ia}^{(2)}$$

$$t_{ijab}^{(1)} = \langle ij\,||\,ab\rangle\left(\varepsilon_i+\varepsilon_j-\varepsilon_a-\varepsilon_b\right)^{-1}.$$

Whereas the $P_{ij}^{(2)}$ and $P_{ab}^{(2)}$ terms arise from products of first-order, double-replacement amplitudes, the $P_{ia}^{(2)}$ term involves their second-order, single-replacement counterparts.

For most bases other than the canonical, Hartree–Fock orbitals, F, contains exchange as well as correlation terms. For example, in a canonical, Kohn–Sham basis,

$$\Sigma_{pq}(\infty) = -\Sigma_i \langle pi|iq\rangle - [v_{xc}]_{pq} + \Sigma_{rs}\langle pr \| qs\rangle(P_{rs} - \delta_{rs} n_r),$$

where matrix elements of the exchange-correlation potential (v_{xc}) occur in the second-term and spin-orbital occupation numbers, n_r, appear in the last term. The term in parentheses may be set to zero by assuming that the density matrix of the reference determinant is close to exact. In this case, the role of $\Sigma_{pq}(\infty)$ is to recover the usual exchange operator in the Dyson quasiparticle equation.

To define approximate forms of $\boldsymbol{\sigma}(E)$, one may specify **f** operator manifolds and the density matrices that are used to evaluate matrix elements of $\hat{H}$. For example, the two-particle-one-hole Tamm–Dancoff approximation (2ph-TDA)[16] for the self-energy is defined by an operator manifold that includes **a** and $\mathbf{f}_3$ in combination with a Hartree–Fock (ie, zero order) reference state:

$$\boldsymbol{\sigma}^{2ph-TDA}(E) = (\mathbf{a}|\hat{H}\mathbf{f}_3)_0(\mathbf{f}_3|(E\hat{I} - \hat{H})\mathbf{f}_3)_0^{-1}(\mathbf{f}_3|\hat{H}\mathbf{a})_0.$$

For the Hartree–Fock reference state, couplings between 2hp and 2ph operators equal zero and enable a separation of $\boldsymbol{\sigma}(E)$ into two terms:

$$\begin{aligned}\boldsymbol{\sigma}^{2ph-TDA}(E) = &(\mathbf{a}|\hat{H}\mathbf{f}_{2hp})_0(\mathbf{f}_{2hp}|(E\hat{I} - \hat{H})\mathbf{f}_{2hp})_0^{-1}(\mathbf{f}_{2hp}|\hat{H}\mathbf{a})_0 \\ &+ (\mathbf{a}|\hat{H}\mathbf{f}_{2ph})_0(\mathbf{f}_{2ph}|(E\hat{I} - \hat{H})\mathbf{f}_{2ph})_0^{-1}(\mathbf{f}_{2ph}|\hat{H}\mathbf{a})_0.\end{aligned}$$

When the usual Møller–Plesset choice for $\hat{H}_0$ is made, zeroth order contributions to the primary–secondary couplings vanish and resulting first-order expressions read

$$\begin{aligned}(a_a^\dagger a_i a_j|\hat{H}a_p)_0 &= \langle pa \| ji\rangle \\ (a_i^\dagger a_a a_b|\hat{H}a_p)_0 &= \langle pi \| pa\rangle.\end{aligned}$$

Through first order in V, the 2hp–2hp and 2ph–2ph couplings are given by

$$\begin{aligned}(a_a^\dagger a_i a_j|\hat{H}a_b^\dagger a_k a_l)_0 = &\delta_{ab}\delta_{ik}\delta_{jl}(\varepsilon_i + \varepsilon_j - \varepsilon_a) - \delta_{ab}\langle kl \| ij\rangle \\ &+ (1 - P_{ij})(1 - P_{kl})\delta_{ik}\langle al \| bj\rangle,\end{aligned}$$

for i<j, k<l, and

$$\left(a_i^\dagger a_a a_b|\hat{H}a_j^\dagger a_c a_d\right)_0 = \delta_{ij}\delta_{ac}\delta_{bd}(\varepsilon_a + \varepsilon_b - \varepsilon_i)$$
$$+\delta_{ij}\langle cd \| ab\rangle - (1 - P_{ab})(1 - P_{cd})\delta_{ac}\langle id \| jb\rangle,$$

for a < b, c < d. The latter terms generate ring, ladder, and mixed ring-ladder diagrams in all orders of V. Because the Hartree–Fock reference state generates no energy-independent terms,

$$\boldsymbol{\sigma}^{2ph-TDA}(E) = \boldsymbol{\Sigma}^{2ph-TDA}(E).$$

The chief arithmetic bottleneck in 2ph-TDA calculations arises from matrix multiplications that involve 2ph–2ph couplings. These contractions scale as ov^4, where o is the number of occupied orbitals and v is the number of virtual orbitals.

By ignoring the first-order terms in $\left(a_a^\dagger a_i a_j|\hat{H}a_b^\dagger a_k a_l\right)_0$ and $\left(a_i^\dagger a_a a_b|\hat{H}a_j^\dagger a_c a_d\right)_0$, the self-energy expression becomes

$$\boldsymbol{\Sigma}^{(2)}(E) = \left(\mathbf{a}|\hat{H}\mathbf{f}_{2hp}\right)_0\left(\mathbf{f}_{2hp}|E\hat{I} - \hat{H}_0)\mathbf{f}_{2hp}\right)_0^{-1}\left(\mathbf{f}_{2hp}|\hat{H}\mathbf{a}\right)_0$$
$$+\left(\mathbf{a}|\hat{H}\mathbf{f}_{2hp}\right)_0\left(\mathbf{f}_{2hp}|(E\hat{I} - \hat{H}_0)\mathbf{f}_{2hp}\right)_0^{-1}\left(\mathbf{f}_{2hp}|\hat{H}\mathbf{a}\right)_0$$

and the matrix elements acquire their familiar, second-order form:

$$\Sigma_{pq}^{(2)}(E) = \Sigma_{a\,i<j}\langle qa \| ji\rangle\left(E + \varepsilon_a - \varepsilon_i - \varepsilon_j\right)^{-1}\langle ji \| pa\rangle$$
$$+\Sigma_{i\,a<b}\langle qi \| ba\rangle(E + \varepsilon_i - \varepsilon_a - \varepsilon_b)^{-1}\langle ba \| pi\rangle$$

For electron detachment energies in which the assumptions of Koopmans's identity are qualitatively valid, neglect of off-diagonal elements of $\boldsymbol{\Sigma}(E)$ usually introduces deviations of 0.01–0.02 eV. The resulting diagonal, or quasiparticle, approximation leads to an especially simple form of the Dyson quasiparticle equation in which

$$E = \varepsilon_i + \Sigma_{ii}(E)$$

When the diagonal, second-order (D2) self-energy approximation is used with flexible basis sets, valence, vertical ionization energies (VIEs) of closed-shell molecules are predicted to be too small, with mean absolute deviations from reliable data of approximately 0.4 eV. Larger errors obtain for small basis sets. Predicted VIEs increase as more functions are added. Whereas negatives of canonical Hartree–Fock orbital energies usually give VIEs that are too large, chiefly because of their neglect of final-state relaxation of orbitals, D2 results that are extrapolated with respect to basis saturation provide an informal lower bound. For every Koopmans result that is refined with D2 corrections, only a partial integral transformation to the Hartree–Fock basis is required. The latter transformation has fourth-power arithmetic scaling. Because evaluation of $\Sigma_{ii}(E)$ matrix elements has cubic scaling, it constitutes no bottleneck.

A noniterative formula in which the pole is approximated by

$$E \approx \varepsilon_i + \Sigma_{ii}^{(2)}(\varepsilon_i)$$

is identical to the second-order result of Rayleigh–Schrödinger perturbation theory in which the N-electron, Møller–Plesset fluctuation potential also is used to generate total energies for states with $N-1$ electrons. (The same conclusion may be reached in third, but not higher orders.)[17] However, iterations with respect to E commonly yield nonnegligible shifts of approximately 0.05 eV. Pole searches may be accelerated by evaluating the derivatives of $\Sigma_{ii}(E)$ with respect to E and using Newton's method to estimate the next guess. Convergence to within 0.01 eV usually follows after the third iteration.

A perturbative analysis of the second-order self-energy[7,18] discloses that final-state relaxation effects for electron detachment (or attachment) energies are attributable to terms in the 2hp (or 2ph) summation where i or j (a or b) equals p. The remaining terms in these summations account for final-state correlation effects. Second-order, pair correlation energies in the N-electron reference state that are destroyed (created) in final states with $N-1$ $(N+1)$ electrons are given by the 2ph (2hp) summation.

To efficiently estimate VIEs for a set of chemically related molecules, spin-scaled D2 approximations have been introduced.[19,20] Several scaled versions of the D2 self-energy with the general formula,

$$\Sigma_{pp}^{(2)}(E) = \frac{1}{2}\Sigma_{aij}\langle pa|ij\rangle\left(E+\varepsilon_a-\varepsilon_i-\varepsilon_j\right)^{-1}\left[C_{C-2hp}\langle ij|pa\rangle + C_{E-2hp}\langle ji|pa\rangle\right]$$
$$+\frac{1}{2}\Sigma_{iab}\langle pi|ab\rangle\left(E+\varepsilon_i-\varepsilon_a-\varepsilon_b\right)^{-1}\left[C_{C-2ph}\langle ab|pi\rangle + C_{E-2ph}\langle ba|pi\rangle\right],$$

have been examined for the purpose of enabling calculations on large molecules. Coulomb (C subscript) and exchange (E subscript) contributions and the 2hp and 2ph terms have been given four separate weights. Because D2 often succeeds in identifying Koopmans defects, where the order of final states predicted with canonical orbital energies is incorrect, it is a better basis for parametrizations than the Hartree–Fock equations. When a linear fit of Hartree–Fock orbital energies to reliable data necessarily fails because of Koopmans defects, parametrized versions of the D2 quasiparticle equation may yield useful tools for interpolation.

The D2 method, without the introduction of scaling parameters, also has been used in the context of semiempirical Hamiltonians.[21] Successful assignments of photoelectron spectra in which the Koopmans ordering of final states is incorrect have been realized for several classes of organic and inorganic compounds.

For valence electron binding energies where a frozen-orbital determinant is a reasonable description of the final state, the D2 approximation provides valuable, semiquantitative corrections to the results of Koopmans's identity. However, for core ionization energies, where final-state orbital relaxation is strong, the orbital energy provided by the transition operator method (TOM) is a superior zero-order approximation.[22–24] Matrix elements of the transition operator read

$$F_{pq}^{TOM} = h_{pq} + \Sigma_r \langle pr \parallel qr \rangle n_r.$$

Occupation numbers of 0 and 1 are assigned to each spin-orbital, save for a single spin-orbital (ie, the transition orbital) assigned to an occupation number of 1/2. (This choice of occupation numbers is a special case of grand-canonical Hartree–Fock theory.)[25,26] The eigenvalue corresponding to the transition spin-orbital, ε_p^{TOM}, incorporates orbital relaxation effects. For the grand-canonical Hartree–Fock density operator, a generalized form for the second-order self-energy obtains, where

$$\Sigma_{pq}^{(2)}(E) = \frac{1}{2}\Sigma_{rst} N_{rst} \langle qr \parallel st \rangle (E + \varepsilon_r - \varepsilon_s - \varepsilon_t)^{-1} \langle st \parallel pr \rangle$$

$$N_{rst} = n_r(1 - n_s)(1 - n_t) + (1 - n_r) n_s n_t.$$

The second-order, transition-operator (TOEP2) method[24] also employs the diagonal self-energy approximation. Poles satisfy the equation

$$E = \varepsilon_p^{TOM} + \Sigma_{pp}^{(2)}(E).$$

This method provides a useful, semiquantitative account of core and valence electron binding energies, with mean absolute errors of approximately 0.35 eV for valence IEs.

Despite its retention of nondiagonal elements of the self-energy operator and its inclusion of ring, ladder, and mixed ring-ladder terms beyond second-order, the 2ph-TDA yields larger average errors for valence VIEs than the less computationally demanding D2 method. The chief advantage of 2ph-TDA is its ability to produce a first-order account of correlation states in which 2hp or 2ph configurations dominate over the h and p configurations assumed in Koopmans's identity. In the inner-valence region of a photoelectron spectrum, numerous poles with low strengths obliterate the Koopmans picture of one-hole final states.[27] Canonical Hartree–Fock orbital energies in the inner-valence region therefore have little physical meaning with respect to specific transitions. For typical molecules, the He II photoelectron spectrum (ie, up to approximately 40 eV) may be qualitatively assigned with 2ph-TDA calculations.

To generate all third-order terms in $\mathbf{F} + \boldsymbol{\sigma}(E)$ or $\boldsymbol{\Sigma}(E)$ with Hartree–Fock orbitals, the following approximation suffices:

$$\boldsymbol{\Sigma}^{3+}(E) = \boldsymbol{\Sigma}^{(3)}(\infty) + \left(\mathbf{a}|\hat{H}\mathbf{f}_3\right)_1 \left(\mathbf{f}_3|\left(E\hat{I} - \hat{H}\right)\mathbf{f}_3\right)_0^{-1} \left(\mathbf{f}_3|\hat{H}\mathbf{a}\right)_1.$$

Third-order, energy-independent terms are added to an expression for $\boldsymbol{\sigma}(E)$ in which the primary–secondary couplings are correct through second-order (because of first-order terms in the density operator) and the inverse, first-order matrix of secondary–secondary couplings generates terms in all orders. Whereas all third-order terms in $\boldsymbol{\Sigma}(E)$ are present, there are many higher-order terms as well. Therefore, this approximation may be denominated 3+. Expressions for the primary–secondary couplings with real orbitals read

$$\left(a_a^\dagger a_i a_j|\hat{I}Ia_p\right)_1 = \left(a_a^\dagger a_i a_j|\hat{H}a_p\right)_0 + \frac{1}{2}\Sigma_{bc} t_{jibc}\langle pa \| bc\rangle + \left(1 - P_{ij}\right)\Sigma_{bk} t_{kjba}\langle pk \| bi\rangle$$
$$- \hat{\mathbf{H}}^{(1)}_{aij,p} + \hat{\mathbf{H}}^{(2)}_{aij,p}$$

$$\left(a_i^\dagger a_a a_b|\hat{H}a_p\right)_1 = \left(a_i^\dagger a_a a_b|\hat{H}a_p\right)_0 + \frac{1}{2}\Sigma_{jk} t_{jkba}\langle pi \| jk\rangle + (1 - P_{ab})\Sigma_{jc} t_{ijbc}\langle pc \| ja\rangle$$
$$= \hat{\mathbf{H}}^{(1)}_{iab,p} + \hat{\mathbf{H}}^{(2)}_{iab,p}$$

For each value of p, there are fifth-power contractions. Determination of all primary–secondary couplings therefore is a sixth-power process and constitutes the chief noniterative bottleneck in 3+ calculations. (In practice, iterative processes with fifth-power scaling may require more arithmetic operations.) This additional effort is rewarded with superior accuracy in calculations on valence ionization energies and electron affinities.[28]

One may extend the 3+ self-energy by adding more energy-independent terms.[29] One such approach is based on a relationship between the electron propagator matrix and the one-electron density matrix that reads

$$\mathbf{P} = (2\pi i)^{-1} \int_C \mathbf{G}(E)dE,$$

where C denotes a contour in the complex plane that includes all electron detachment poles and where electron propagator matrix elements with a complex argument are given by

$$G_{rs}(E) = \lim_{\eta \to 0} \left[\Sigma_m V_{rm} V^*_{sm} (E - A_m + i\eta)^{-1} + \Sigma_n U^*_{rn} U_{sn} (E - D_n - i\eta)^{-1} \right].$$

By truncating the expansion of the Dyson equation,

$$\begin{aligned}\mathbf{G}(E) = \mathbf{G}_0(E) + \mathbf{G}_0(E)\mathbf{\Sigma}(E)\mathbf{G}_0(E) + \mathbf{G}_0(E)\mathbf{\Sigma}(E)\mathbf{G}_0(E)\mathbf{\Sigma}(E)\mathbf{G}_0(E) + \mathbf{G}_0(E) \\ \times \mathbf{\Sigma}(E)\mathbf{G}_0(E)\mathbf{\Sigma}(E)\mathbf{G}_0(E)\mathbf{\Sigma}(E)\mathbf{G}_0(E) + \cdots,\end{aligned}$$

after the second term, the approximation

$$\mathbf{P} \approx (2\pi i)^{-1} \int_C [\mathbf{G}_0(E) + \mathbf{G}_0(E)\mathbf{\Sigma}(E)\mathbf{G}_0(E)]dE$$

is obtained. An approximate density matrix defined in this way yields a new $\mathbf{\Sigma}(\infty)$ and therefore a new $\mathbf{\Sigma}(E)$. By setting

$$\mathbf{\Sigma}(E) = \mathbf{\Sigma}(\infty) + \boldsymbol{\sigma}^{3+}(E),$$

where

$$\boldsymbol{\sigma}^{3+}(E) = \left(\mathbf{a}|\hat{H}\mathbf{f}_3\right)_1 \left(\mathbf{f}_3|\left(E\hat{I} - \hat{H}\right)\mathbf{f}_3\right)_0^{-1} \left(\mathbf{f}_3|\hat{H}\mathbf{a}\right)_1,$$

one may obtain $\mathbf{P}$ and $\mathbf{\Sigma}(\infty)$ self-consistently. This extension of the 3+ self-energy is the most common version of the third-order algebraic diagrammatic construction, or ADC(3).[2] It suffices to recover all fourth order terms and many higher-order terms in $\mathbf{\Sigma}(\infty)$. In the ADC(3) method, sixth-power contractions that scale as o^2v^4 are performed iteratively in the determination of $\mathbf{\Sigma}(\infty)$. Whereas the correlation contribution to $\mathbf{P}$ has a vanishing trace for the 3+ self-energy and for the exact case, its ADC(3) counterpart does not have this property. This deviation stems from the retention of only some

$\mathbf{\Sigma}(\infty)$ terms in fifth and higher orders of the fluctuation potential and may become problematic for large molecules.

The design of the 3+ and ADC(3) methods assumes the need to include all third-order terms in $\mathbf{\Sigma}(E)$ and embraces the inclusion of higher-order terms. An alternative approach is based on examination of improvements to the second-order self-energy and retention of terms that suffice to provide reliable predictions of electron binding energies.[30] For the calculation of electron detachment energies, an asymmetric superoperator metric is adopted, where

$$(X|Y)_D = \left\langle HF\left|\left[X^\dagger, Y\right]_+\left(1 + T_2^{(1)}\right)\right|HF\right\rangle,$$

and where the reference Hartree–Fock determinantal wave function, $|HF\rangle$, and the first-order Møller–Plesset wave function define the double excitation amplitudes of $T_2^{(1)}$. The operator manifold comprises the $\mathbf{a}$ and $\mathbf{f}_3$ spaces. Second-order terms appear only when Y is a 2ph operator or X is a 2hp operator. For electron detachment energies, all of these second-order terms except those occurring in the 2hp-h block of $\hat{\mathbf{H}}$ may be neglected. First-order terms in the 2ph–2ph block of $\hat{\mathbf{H}}$ also may be omitted. After Hermitizing $\hat{\mathbf{H}}$, the resulting self-energy matrix elements are expressed as

$$\begin{aligned}
\mathbf{\Sigma}_{ij}^{NR2-D}(E) &= \left[\hat{\mathbf{H}}_{i,2hp}^{(1)} + \frac{1}{2}\hat{\mathbf{H}}_{i,2hp}^{(2)}\right]\left(\mathbf{f}_{2hp}\middle|\left(E\hat{I} - \hat{H}\right)\mathbf{f}_{2hp}\right)_0^{-1}\left[\hat{\mathbf{H}}_{2hp,j}^{(1)} + \frac{1}{2}\hat{\mathbf{H}}_{2hp,j}^{(2)}\right] \\
&\quad + \hat{\mathbf{H}}_{i,2ph}^{(1)}\left(\mathbf{f}_{2ph}\middle|\left(E\hat{I} - \hat{H}_0\right)\mathbf{f}_{2ph}\right)_0^{-1}\hat{\mathbf{H}}_{2ph,j}^{(1)} \\
\mathbf{\Sigma}_{ia}^{NR2-D}(E) &= \left[\hat{\mathbf{H}}_{i,2hp}^{(1)} + \frac{1}{2}\hat{\mathbf{H}}_{i,2hp}^{(2)}\right]\left(\mathbf{f}_{2hp}\middle|\left(E\hat{I} - \hat{H}\right)\mathbf{f}_{2hp}\right)_0^{-1}\hat{\mathbf{H}}_{2hp,a}^{(1)} \\
&\quad + \hat{\mathbf{H}}_{i,2ph}^{(1)}\left(\mathbf{f}_{2ph}\middle|\left(E\hat{I} - \hat{H}_0\right)\mathbf{f}_{2ph}\right)_0^{-1}\hat{\mathbf{H}}_{2ph,a}^{(1)} \\
\mathbf{\Sigma}_{ai}^{NR2-D}(E) &= \left[\mathbf{\Sigma}_{ia}^{NR2-D}(E)\right]^* \\
\mathbf{\Sigma}_{ab}^{NR2-D}(E) &= \hat{\mathbf{H}}_{a,2hp}^{(1)}\left(\mathbf{f}_{2hp}\middle|\left(E\hat{I} - \hat{H}\right)\mathbf{f}_{2hp}\right)_0^{-1}\hat{\mathbf{H}}_{2hp,b}^{(1)} \\
&\quad + \hat{\mathbf{H}}_{a,2ph}^{(1)}\left(\mathbf{f}_{2ph}\middle|\left(E\hat{I} - \hat{H}_0\right)\mathbf{f}_{2ph}\right)_0^{-1}\hat{\mathbf{H}}_{2ph,b}^{(1)}
\end{aligned}$$

The designation NR2 was chosen because this self-energy approximation is nondiagonal, renormalized, and complete through second order. Ring and ladder renormalizations are generated by the first-order terms that occur in

$\left(\mathbf{f}_{2hp}|\left(E\hat{I}-\hat{H}\right)\mathbf{f}_{2hp}\right)_0^{-1}$. There are no energy-independent terms. Because 2ph rings and ladders and second-order p–2hp terms are neglected, the need for electron repulsion integrals with four virtual indices is eliminated in NR2 calculations of electron detachment energies. The most arithmetically intensive contraction has o^3v^3 scaling and is the only step that requires electron repulsion integrals with three virtual indices.

For the calculation of electron attachment energies, the metric is chosen according to

$$(X|Y)_A=\left\langle HF\left|\left(1+T_2^{(1)}\right)^{\dagger}\left[X^{\dagger},Y\right]_+\right|HF\right\rangle$$

and the roles of particles and holes are reversed in the selection of self-energy terms. For electron attachment energies,

$$\begin{aligned}
\mathbf{\Sigma}_{ab}^{NR2-A}(E)&=\left[\hat{\mathbf{H}}_{a,2ph}^{(1)}+\frac{1}{2}\hat{\mathbf{H}}_{a,2ph}^{(2)}\right]\left(\mathbf{f}_{2ph}|\left(E\hat{I}-\hat{H}\right)\mathbf{f}_{2ph}\right)_0^{-1}\left[\hat{\mathbf{H}}_{2ph,b}^{(1)}+\frac{1}{2}\hat{\mathbf{H}}_{2ph,b}^{(2)}\right]\\
&\quad+\hat{\mathbf{H}}_{a,2hp}^{(1)}\left(\mathbf{f}_{2hp}|\left(E\hat{I}-\hat{H}_0\right)\mathbf{f}_{2hp}\right)_0^{-1}\hat{\mathbf{H}}_{2hp,b}^{(1)}\\
\mathbf{\Sigma}_{ai}^{NR2-A}(E)&=\left[\hat{\mathbf{H}}_{a,2ph}^{(1)}+\frac{1}{2}\hat{\mathbf{H}}_{a,2ph}^{(2)}\right]\left(\mathbf{f}_{2ph}|\left(E\hat{I}-\hat{H}\right)\mathbf{f}_{2ph}\right)_0^{-1}\hat{\mathbf{H}}_{2ph,i}^{(1)}\\
&\quad+\hat{\mathbf{H}}_{a,2hp}^{(1)}\left(\mathbf{f}_{2hp}|\left(E\hat{I}-\hat{H}_0\right)\mathbf{f}_{2hp}\right)_0^{-1}\hat{\mathbf{H}}_{2hp,i}^{(1)}\\
\mathbf{\Sigma}_{ia}^{NR2-A}(E)&=\left[\mathbf{\Sigma}_{ai}^{NR2-A}(E)\right]^*\\
\mathbf{\Sigma}_{ij}^{NR2-A}(E)&=\hat{\mathbf{H}}_{i,2ph}^{(1)}\left(\mathbf{f}_{2ph}|\left(E\hat{I}-\hat{H}\right)\mathbf{f}_{2ph}\right)_0^{-1}\hat{\mathbf{H}}_{2ph,j}^{(1)}\\
&\quad+\hat{\mathbf{H}}_{i,2hp}^{(1)}\left(\mathbf{f}_{2hp}|\left(E\hat{I}-\hat{H}_0\right)\mathbf{f}_{2hp}\right)_0^{-1}\hat{\mathbf{H}}_{2hp,j}^{(1)}.
\end{aligned}$$

In the evaluation of $\hat{\mathbf{H}}_{a,2ph}^{(2)}$, there is a contraction with o^2v^4 arithmetic scaling that involves electron repulsion integrals with three virtual indices. Inclusion of the first-order 2ph–2ph elements of $\hat{\mathbf{H}}$ entails a need for electron repulsion integrals with four virtual indices. NR2 calculations of electron attachment energies may be expected to require more arithmetic operations and memory than their counterparts for electron detachment energies.

Arithmetic bottlenecks encountered in 3+ and NR2 calculations may be reduced by introducing two additional approximations. To evaluate products of $\left(\mathbf{f}_3|\left(E\hat{I}-\hat{H}\right)\mathbf{f}_3\right)_0^{-1}$ with other matrices, repeated multiplications are required. As a result, self-energy terms in all orders are retained. However, if the inverse matrix is approximated according to

$$\left(\mathbf{f}_3|\left(\mathrm{E}\hat{\mathrm{I}}-\hat{\mathrm{H}}\right)\mathbf{f}_3\right)_0^{-1} \approx \left(\mathbf{f}_3|\left(\mathrm{E}\hat{\mathrm{I}}-\hat{\mathrm{H}}_0\right)\mathbf{f}_3\right)_0^{-1} + \left(\mathbf{f}_3|\left(\mathrm{E}\hat{\mathrm{I}}-\hat{\mathrm{H}}_0\right)\mathbf{f}_3\right)_0^{-1}$$
$$\left(\mathbf{f}_3|\left(\hat{\mathrm{H}}-\hat{\mathrm{H}}_0\right)\mathbf{f}_3\right)_0 \left(\mathbf{f}_3|\left(\mathrm{E}\hat{\mathrm{I}}-\hat{\mathrm{H}}_0\right)\mathbf{f}_3\right)_0^{-1},$$

only terms up to fourth order remain and all third-order terms are conserved. By applying these arguments to the 3+ self-energy, the third-order self-energy's structure is shown:

$$\begin{aligned}\boldsymbol{\Sigma}^{(3)}(\mathrm{E}) = &\boldsymbol{\Sigma}^{(2)}(\mathrm{E}) + \boldsymbol{\Sigma}^{(3)}(\infty) + \hat{\mathbf{H}}_{13}^{(2)}\left(\mathbf{f}_3|\left(\mathrm{E}\hat{\mathrm{I}}-\hat{\mathrm{H}}_0\right)\mathbf{f}_3\right)_0^{-1}\hat{\mathbf{H}}_{31}^{(1)}\\ &+\hat{\mathbf{H}}_{13}^{(1)}\left(\mathbf{f}_3|\left(\mathrm{E}\hat{\mathrm{I}}-\hat{\mathrm{H}}_0\right)\mathbf{f}_3\right)_0^{-1}\hat{\mathbf{H}}_{31}^{(2)}\\ &+\hat{\mathbf{H}}_{13}^{(1)}\left(\mathbf{f}_3|\left(\mathrm{E}\hat{\mathrm{I}}-\hat{\mathrm{H}}_0\right)\mathbf{f}_3\right)_0^{-1}\left(\mathbf{f}_3|\left(\hat{\mathrm{H}}-\hat{\mathrm{H}}_0\right)\mathbf{f}_3\right)_0\left(\mathbf{f}_3|\left(\mathrm{E}\hat{\mathrm{I}}-\hat{\mathrm{H}}_0\right)\mathbf{f}_3\right)_0^{-1}\hat{\mathbf{H}}_{31}^{(1)}\end{aligned}$$

The last term is responsible for the ring and ladder diagrams that appear in third order. Applying similar truncations to the NR2 self-energy formulae defines nondiagonal, partial third-order (NP3) approximations. For example, whereas the occupied–occupied block of the NP3 self-energy matrix for electron detachment energies reads

$$\begin{aligned}\boldsymbol{\Sigma}_{\mathrm{ij}}^{\mathrm{NP3-D}}(\mathrm{E}) = &\boldsymbol{\Sigma}_{\mathrm{ij}}^{(2)}(\mathrm{E}) + \frac{1}{2}\hat{\mathbf{H}}_{\mathrm{i,2hp}}^{(1)}\left(\mathbf{f}_{\mathrm{2hp}}|\left(\mathrm{E}\hat{\mathrm{I}}-\hat{\mathrm{H}}_0\right)\mathbf{f}_{\mathrm{2hp}}\right)_0^{-1}\hat{\mathbf{H}}_{\mathrm{2hp,j}}^{(2)}\\ &+\frac{1}{2}\hat{\mathbf{H}}_{\mathrm{i,2hp}}^{(2)}\left(\mathbf{f}_{\mathrm{2hp}}|\left(\mathrm{E}\hat{\mathrm{I}}-\hat{\mathrm{H}}_0\right)\mathbf{f}_{\mathrm{2hp}}\right)_0^{-1}\hat{\mathbf{H}}_{\mathrm{2hp,j}}^{(1)}\\ &+\hat{\mathbf{H}}_{\mathrm{i,2hp}}^{(1)}\left(\mathbf{f}_{\mathrm{2hp}}|\left(\mathrm{E}\hat{\mathrm{I}}-\hat{\mathrm{H}}_0\right)\mathbf{f}_{\mathrm{2hp}}\right)_0^{-1}\left(\mathbf{f}_{\mathrm{2hp}}|\left(\hat{\mathrm{H}}-\hat{\mathrm{H}}_0\right)\mathbf{f}_{\mathrm{2hp}}\right)_0\\ &\left(\mathbf{f}_{\mathrm{2hp}}|\left(\mathrm{E}\hat{\mathrm{I}}-\hat{\mathrm{H}}_0\right)\mathbf{f}_{\mathrm{2hp}}\right)_0^{1}\hat{\mathbf{H}}_{\mathrm{2hp,j}}^{(1)},\end{aligned}$$

the virtual–virtual block of its counterpart for electron attachments reads

$$\begin{aligned}\boldsymbol{\Sigma}_{\mathrm{ab}}^{\mathrm{NP3-A}}(\mathrm{E}) = &\boldsymbol{\Sigma}_{\mathrm{ab}}^{(2)}(\mathrm{E}) + \frac{1}{2}\hat{\mathbf{H}}_{\mathrm{a,2ph}}^{(1)}\left(\mathbf{f}_{\mathrm{2ph}}|\left(\mathrm{E}\hat{\mathrm{I}}-\hat{\mathrm{H}}_0\right)\mathbf{f}_{\mathrm{2ph}}\right)_0^{-1}\hat{\mathbf{H}}_{\mathrm{2ph,b}}^{(2)}\\ &+\frac{1}{2}\hat{\mathbf{H}}_{\mathrm{a,2ph}}^{(2)}\left(\mathbf{f}_{\mathrm{2ph}}|\left(\mathrm{E}\hat{\mathrm{I}}-\hat{\mathrm{H}}_0\right)\mathbf{f}_{\mathrm{2ph}}\right)_0^{-1}\hat{\mathbf{H}}_{\mathrm{2ph,b}}^{(1)}\\ &+\hat{\mathbf{H}}_{\mathrm{a,2ph}}^{(1)}\left(\mathbf{f}_{\mathrm{2ph}}|\left(\mathrm{E}\hat{\mathrm{I}}-\hat{\mathrm{H}}_0\right)\mathbf{f}_{\mathrm{2ph}}\right)_0^{-1}\left(\mathbf{f}_{\mathrm{2ph}}|\left(\hat{\mathrm{H}}-\hat{\mathrm{H}}_0\right)\mathbf{f}_{\mathrm{2ph}}\right)_0\\ &\left(\mathbf{f}_{\mathrm{2ph}}|\left(\mathrm{E}\hat{\mathrm{I}}-\hat{\mathrm{H}}_0\right)\mathbf{f}_{\mathrm{2ph}}\right)_0^{-1}\hat{\mathbf{H}}_{\mathrm{2ph,b}}^{(1)}\end{aligned}$$

The second approximation is neglect of the off-diagonal elements of $\mathbf{\Sigma}(E)$ in the canonical, Hartree–Fock orbital basis. For a given electron binding energy corresponding to spin-orbital r, only the elements of $\hat{\mathbf{H}}^{(2)}_{r,2hp}$ or $\hat{\mathbf{H}}^{(2)}_{r,2ph}$ must be evaluated and the corresponding sixth-power contractions are reduced to fifth power. Arithmetic operations are similarly reduced when $\left(\mathbf{f}_3|(E\hat{I}-\hat{H})\mathbf{f}_3\right)_0^{-1}\hat{\mathbf{H}}_{31}$ products are formed. The resulting diagonal third-order (D3) and P3[31,32] self-energies therefore have fifth-power arithmetic scaling. For real spin-orbitals, the D3 self-energy matrix elements may be written as

$$\begin{aligned}\Sigma^{(3)}_{rr}(E) =& \frac{1}{2}\Sigma_{aij}\langle ra\,||\,ij\rangle\left(E+\varepsilon_a-\varepsilon_i-\varepsilon_j\right)^{-1}\left[\langle ra\,||\,ij\rangle + 2\hat{H}^{(2)}_{raij} + U_{raij}(E)\right]\\ &+\frac{1}{2}\Sigma_{iab}\langle ri\,||\,ab\rangle(E+\varepsilon_i-\varepsilon_a-\varepsilon_b)^{-1}\left[\langle ri\,||\,ab\rangle + 2\hat{H}^{(2)}_{riab} + U_{riab}(E)\right]\\ &+\Sigma^{(3)}_{rr}(\infty),\end{aligned}$$

where

$$\begin{aligned}U_{raij}(E) =& -\frac{1}{2}\Sigma_{kl}\langle ra\,||\,kl\rangle(E+\varepsilon_a-\varepsilon_k-\varepsilon_l)^{-1}\langle kl\,||\,ij\rangle\\ &-\left(1-P_{ij}\right)\Sigma_{bk}\langle rb\,||\,jk\rangle\left(E+\varepsilon_b-\varepsilon_j-\varepsilon_k\right)^{-1}\langle ak\,||\,bi\rangle\\ U_{riab}(E) =& \frac{1}{2}\Sigma_{cd}\langle ri\,||\,cd\rangle(E+\varepsilon_i-\varepsilon_c-\varepsilon_d)^{-1}\langle cd\,||\,ab\rangle\\ &+(1-P_{ab})\Sigma_{jc}\langle rj\,||\,bc\rangle\left(E+\varepsilon_j-\varepsilon_b-\varepsilon_c\right)^{-1}\langle ic\,||\,ja\rangle.\end{aligned}$$

The usual computational bottleneck occurs in the $U_{riab}(E)$ expression, where a contraction with ov^4 scaling must be repeated for various values of E. For P3 electron detachment energies,

$$\begin{aligned}\Sigma^{P3-D}_{kk}(E) =& \frac{1}{2}\Sigma_{aij}\langle ka\,||\,ij\rangle\left(E+\varepsilon_a-\varepsilon_i-\varepsilon_j\right)^{-1}\left[\langle ka\,||\,ij\rangle + \hat{H}^{(2)}_{kaij} + U_{kaij}(E)\right]\\ &+\frac{1}{2}\Sigma_{iab}|\langle ki\,||\,ab\rangle|^2(E+\varepsilon_i-\varepsilon_a-\varepsilon_b)^{-1},\end{aligned}$$

and for P3 electron attachment energies,

$$\begin{aligned}\Sigma^{P3-A}_{cc}(E) =& \frac{1}{2}\Sigma_{iab}\langle ci\,||\,ab\rangle(E+\varepsilon_i-\varepsilon_a-\varepsilon_b)^{-1}\left[\langle ci\,||\,ab\rangle + \hat{H}^{(2)}_{ciab} + U_{ciab}(E)\right]\\ &+\frac{1}{2}\Sigma_{aij}|\langle ca\,||\,ij\rangle|^2\left(E+\varepsilon_a-\varepsilon_i-\varepsilon_j\right)^{-1}.\end{aligned}$$

In the former case, the evaluation of $\hat{\mathbf{H}}^{(2)}_{kaij}$ intermediates requires a contraction that scales as o^2v^3; iterations with respect to E require o^3v^2 contractions. For electron attachment energies, P3's noniterative, and iterative bottlenecks have o^2v^3 and ov^4 scaling factors.

D2 tends to overestimate corrections to canonical Hartree–Fock orbital energies and therefore to produce underestimates of electron detachment energies. However, D3 displays the opposite trend, especially as basis sets approach completeness. Estimates of higher-order terms usually are necessary to obtain results of predictive quality. For this purpose, the outer valence Green's function (OVGF) methods[2,33] contain two multiplicative factors for third-order terms in which ratios of third-order and second-order terms are formed. These factors read

$$X_r = -2\Big[\hat{\mathbf{H}}^{(1)}_{r,2hp}\left(\mathbf{f}_{2hp}|(E\hat{I}-\hat{H}_0)\mathbf{f}_{2hp}\right)_0^{-1}\hat{\mathbf{H}}^{(2)}_{2hp,r}$$

$$+\hat{\mathbf{H}}^{(1)}_{r,2ph}\left(\mathbf{f}_{2ph}|(E\hat{I}-\hat{H}_0)\mathbf{f}_{2ph}\right)_0^{-1}\hat{\mathbf{H}}^{(2)}_{2ph,r}\Big]\left[\Sigma^{(2)}_{rr}(E)\right]^{-1}$$

$$X_r^{2hp} = -2\left[\hat{\mathbf{H}}^{(1)}_{r,2hp}\left(\mathbf{f}_{2hp}|(E\hat{I}-\hat{H}_0)\mathbf{f}_{2hp}\right)_0^{-1}\hat{\mathbf{H}}^{(2)}_{2hp,r}\right]$$

$$\times\left[\hat{\mathbf{H}}^{(1)}_{r,2hp}\left(\mathbf{f}_{2hp}|(E\hat{I}-\hat{H}_0)\mathbf{f}_{2hp}\right)_0^{-1}\hat{\mathbf{H}}^{(1)}_{2ph,r}\right]^{-1}$$

$$X_r^{2ph} = -2\left[\hat{\mathbf{H}}^{(1)}_{r,2ph}\left(\mathbf{f}_{2ph}|(E\hat{I}-\hat{H}_0)\mathbf{f}_{2ph}\right)_0^{-1}\hat{\mathbf{H}}^{(2)}_{2ph,r}\right]$$

$$\times\left[\hat{\mathbf{H}}^{(1)}_{r,2ph}\left(\mathbf{f}_{2ph}|(E\hat{I}-\hat{H}_0)\mathbf{f}_{2ph}\right)_0^{-1}\hat{\mathbf{H}}^{(1)}_{2ph,r}\right]^{-1}.$$

In the A version, energy-independent and energy-dependent terms in third order are scaled as follows:

$$\Sigma^{OVGF\text{-}A}_{rr}(E) = \Sigma^{(2)}_{rr}(E) + (1+X_r)^{-1}\left[\Sigma^{(3)}_{rr}(E) - \Sigma^{(2)}_{rr}(E)\right].$$

Two scaling factors are applied to the energy-dependent, third-order terms in the B version, so that

$$\Sigma^{OVGF\text{-}B}_{rr}(E) = \Sigma^{(2)}_{rr}(E) + \Sigma^{(3)}_{rr}(\infty) + \left(1+X_r^{2hp}\right)^{-1}\Sigma^{3-2hp}_{rr}(E)$$

$$+\left(1+X_r^{2ph}\right)^{-1}\Sigma^{3-2ph}_{rr}(E).$$

In the C version, a more complicated formula is introduced for cases where second-order terms are small:

$$\Sigma_{rr}^{OVGF\text{-}C}(E) = \Sigma_{rr}^{(2)}(E) + \left(1 + X_r^C\right)^{-1}\left[\Sigma_{rr}^{(3)}(E) - \Sigma_{rr}^{(2)}(E)\right]$$

$$X_r^C = \left[X_r^{2hp}\Sigma_{rr}^{3-2hp}(E) + X_r^{2ph}\Sigma_{rr}^{3-2ph}(E)\right]$$

$$\left[\Sigma_{rr}^{3-2hp}(E) + \Sigma_{rr}^{3-2ph}(E)\right]^{-1}.$$

The scaling factors and self-energy matrix elements generally are evaluated at the D3 pole energy. Results of the A, B, and C versions are usually within 0.1 eV of each other. A recommended value which is reported typically as the recommended OVGF result emerges from the selection criteria of von Niessen[33]:

1. If $E^{OVGF\text{-}A} \leq 15\,eV$, the $E^{OVGF\text{-}B}$ value is chosen when $\sum_{rr}^{(2)}(E) \geq 0.6\,eV$.
2. If $E^{OVGF\text{-}A} \leq 15\,eV$, the $E^{OVGF\text{-}C}$ value is chosen when $\sum_{rr}^{(2)}(E) < 0.6$ eV.
3. If $E^{OVGF\text{-}A} \leq 15\,eV$ and $|X_r| \leq 0.85$, the $E^{OVGF\text{-}A}$ value is chosen unless $\sum_{rr}^{(2)}(E) < 0.6$ eV and $\left|X_r{}^C\right| \leq 0.85$. In the latter case, the $E^{OVGF\text{-}C}$ value is chosen.
4. If $E^{OVGF\text{-}A} > 15\,eV$, $|X_r| > 0.85$ and $\sum_{rr}^{(2)}(E) < 0.6\,eV$, the $E^{OVGF\text{-}C}$ value is chosen.
5. If $E^{OVGF\text{-}A} > 15\,eV$, $|X_r| > 0.85$ and $\sum_{rr}^{(2)}(E) \leq 0.6$ eV, the $E^{OVGF\text{-}B}$ value is chosen unless $\left|X_r^{2hp}\right| > 0.85$, $\left|X_r^{2ph}\right| > 0.85$, $E^{OVGF\text{-}C} < 15$ eV, or $\left|X_r^C\right| > 0.85$. In the latter cases, the $E^{OVGF\text{-}C}$ value is chosen.

In the P3+ method,[34] the self-energy reads

$$\Sigma_{kk}^{P3+D}(E) = \left(1 + Y_k^{2hp}\right)^{-1}\frac{1}{2}\Sigma_{aij}\langle ka\,||\,ij\rangle\left(E + \varepsilon_a - \varepsilon_i - \varepsilon_j\right)^{-1}$$

$$\left[\langle ka||ij\rangle + \hat{H}_{kaij}^{(2)} + U_{kaij}(E)\right]$$

$$+\frac{1}{2}\Sigma_{iab}\left|\langle ki\,||\,ab\rangle\right|^2\left(E + \varepsilon_i - \varepsilon_a - \varepsilon_b\right)^{-1},$$

where

$$Y_k^{2hp} = \left[-\frac{1}{2}\Sigma_{aij}\langle ka\,||\,ij\rangle\left(E + \varepsilon_a - \varepsilon_i - \varepsilon_j\right)^{-1}\hat{H}_{kaij}^{(2)}\right]\left[\Sigma_{kk}^{(2-2hp)}(E)\right]^{-1}.$$

For P3+ electron attachment energies,

$$\Sigma_{cc}^{P3+A}(E) = \left(1 + Y_c^{2ph}\right)^{-1} \frac{1}{2} \Sigma_{iab} \langle ci||ab\rangle (E + \varepsilon_i - \varepsilon_a - \varepsilon_b)^{-1} \left[\langle ci \,||\, ab\rangle + \hat{H}_{ciab}^{(2)} + U_{ciab}(E)\right] + \frac{1}{2} \Sigma_{aij} |\langle ca \,||\, ij\rangle|^2 \left(E + \varepsilon_a - \varepsilon_i - \varepsilon_j\right)^{-1}.$$

where

$$Y_c^{2ph} = \left[-\frac{1}{2} \Sigma_{iab} \langle ci \,||\, ab\rangle (E + \varepsilon_i - \varepsilon_a - \varepsilon_b)^{-1} \hat{H}_{ciab}^{(2)}\right] \left[\Sigma_{cc}^{(2-2ph)}(E)\right]^{-1}.$$

P3+ self-energy terms and Y factors are evaluated at the P3 pole energy. The OVGF and P3+ methods entail only trivial calculations beyond D3 and P3, respectively.

5. TEST CALCULATIONS

The predictive capabilities of the presently considered self-energy approximations for valence, VIEs have been examined recently.[35] In this study, coupled-cluster singles and doubles plus perturbative triples, ie, CCSD(T),[36] calculations have been performed with correlation-consistent double, triple, and quadruple ζ basis sets[37–39] on 21 molecules and on 52 cationic states. Basis-set extrapolations of these total energies provide standards of comparison. In Table 2, results obtained with the correlation-consistent quadruple ζ basis at the same molecular geometries are compared to these standards. In addition to the statistical measures of error (ie, mean signed error, mean absolute error, and root-mean-square error), the most taxing arithmetic bottlenecks and storage requirements are listed.

Diagonal self-energy approximations are in widest use. The most efficient of these methods, D2, consistently underestimates VIEs, but these errors decrease as the basis-set approaches completeness. D3 results tend to overestimate VIEs when large basis sets are used. OVGF produces more reliable data than D3 with the same effort. P3 is competitive with OVGF, for, with fewer arithmetic operations and smaller memory requirements, it is only slightly less accurate, despite having no selection criteria with numerical parameters. The P3+ method reduces the tendency of P3 to overestimate VIEs with almost no additional effort and is an efficient alternative to OVGF that involves no selection procedure.

Table 2 Errors of Calculated Vertical Ionization Energies[a] vs Extrapolated CCSD(T) Results

Electron Propagator Method	Mean Signed Error[b]	Mean Absolute Error[b]	Root-Mean-Square Error[b]	Iterative Arithmetic Bottleneck[c]	Noniterative Arithmetic Bottleneck[c]	Largest Intermediate Matrix[c]
Koopmans	−0.75	0.83	1.01			
D2	0.50	0.52	0.61	ov^2	NA[d]	ov^2
D3	−0.28	0.39	0.55	ov^4	o^2v^3	v^4
OVGF	−0.01	0.11	0.13	ov^4	o^2v^3	v^4
P3	−0.10	0.18	0.24	o^3v^2	o^2v^3	ov^3
P3+	0.01	0.13	0.16	o^3v^2	o^2v^3	ov^3
2ph-TDA	0.66	0.66	0.71	ov^4	NA[d]	v^4
3+	−0.08	0.17	0.23	ov^4	o^2v^4	v^4
ADC(3)	−0.12	0.16	0.23	o^2v^4	o^2v^4	v^4
NR2	0.11	0.16	0.19	o^2v^3	o^3v^3	ov^3

[a]52 Vertical ionization energies (eV) for 21 molecules calculated with the cc-pvqz basis.
[b]Positive signs correspond to underestimates of vertical ionization energies vs basis-set extrapolated CCSD(T) standards, ie, $\text{Error} = \text{VIE}_{\text{CCSD(T)}} - \text{VIE}_{\text{EP}}$.
[c]o = number of valence occupied orbitals; v = number of valence virtual orbitals.
[d]Not applicable: no energy-independent intermediates are necessary.

Nondiagonal self-energy methods also are compared in Table 2. For calculating valence VIEs, 2ph-TDA produces worse results than D2 with much higher effort. Unlike D2, 2ph-TDA is capable of giving a qualitatively meaningful description of shake-up (ie, chiefly 2hp) final states in photoelectron spectra and of core-excited (ie, chiefly 2ph) electron attachments. The 3+ method is a considerable improvement over 2ph-TDA for valence VIEs that retains the ability of the latter approximation to account for correlation final states. The iterative, sixth-power contractions that distinguish ADC(3) from 3+ do not appear to procure any advantage for this test set. 3+ and ADC(3) tend to overestimate VIEs; error criteria will increase slightly as larger basis sets are employed. For the nondiagonal methods, NR2 has the smallest error measures, arithmetic scaling factors, and memory requirements. Improvements in basis sets will reduce errors in the majority of cases where NR2 underestimates VIEs.

6. RECENT APPLICATIONS AND EXTENSIONS

Electron propagator methods have been applied to the calculation of ionization energies of common amino acids in the gas phase and, with the benefit of polarizable continuum models, in aqueous solution.[40] Applications to the photoelectron spectra of fullerenes, macrocyclic molecules, and nucleotide fragments have been reviewed.[41] Electron binding energies of compounds that are effective scavengers of free radicals have been determined.[42,43] Electron propagator methods have been used to predict bound, excited states of anionic fullerenes.[44] They have facilitated assignments of the photoelectron spectra of tetrazoles.[45,46] The nature of diffuse electronic structure in substituted aza-uracil and thio-uracil anions has been elucidated.[47–49] A systematic study of electron-accepting molecules that may be useful in photovoltaic devices showed the predictive power of electron propagator methods.[28] Electron binding energies of confined atoms, crucial quantities for understanding their electronic structure, have been determined with electron propagator methods.[50] Calculations on electron affinities of cations provided essential information for the determination of photoionization cross sections in the molecular quantum defect model.[51] The electronic structure of metallocenes and their Penning ionization spectra have been interpreted.[52] Electron propagator calculations provided an explanation for the remarkable changes in anion electronic structure that depend on coordination to noble-gas atoms.[53] They also demonstrated how Dyson orbitals may be localized or delocalized in halide-water complexes and in aqueous solution.[54–56] Anion photoelectron spectra of superalkalides were interpreted with electron propagator methods.[57,58]

A comparison of various approaches to calculating ionization energies of molecules reported superior performance for electron propagator methods.[59] Automated derivations of improved electron propagator methods have been incorporated into a general electronic structure package named for Löwdin.[60] An approach to self-energy expressions of any order and the convergence of perturbative expansions have been examined.[17]

7. CONCLUSIONS AND PROSPECTS

The Dyson quasiparticle equation provides a framework for accurate and efficient calculation of molecular electron binding energies. The simplest self-energy approximation, D2, often suffices to correct qualitative errors obtained with canonical Hartree–Fock orbital energies, eg, incorrect orderings of final states caused by neglect of final-state orbital relaxation or differential correlation effects. A restricted need for transformed electron repulsion integrals and low arithmetic demands indicate that D2 results should be obtained routinely after self-consistent-field iterations are complete. D2 provides a suitable foundation for parametrized interpolation schemes or semiempirical approaches that pertain to selected classes of molecules. Because the largest corrections to Koopmans results generally occur at the D2 level, this approximation can provide reliable diagnostics of basis-set effects and a means of estimating the results of higher-order calculations that are infeasible with large basis sets.

For predictions of valence, VIEs with mean, unsigned errors between 0.1 and 0.2 eV, the OVGF methods and their selection procedure constitute an efficient alternative to methods based on many-electron state functions or density functionals. This tool can be especially powerful when several final states of a given irreducible representation are needed. More computationally efficient alternatives of similar accuracy are provided by the partial third-order (P3) approximation and its renormalized extension (P3+). The quasiparticle, or diagonal self-energy, methods (ie, D2, OVGF, P3, and P3+) are most useful when the Koopmans description of an electron binding energy is qualitatively valid. In these cases, the Dyson orbital is a canonical Hartree–Fock orbital times the square root of the pole strength. The success of these methods implies that the chief flaw in the Koopmans description often pertains not to the quality of the occupied orbitals, but to the potential that determines their energies. In such cases, the addition of easily calculated, nonlocal, energy-dependent corrections to canonical Hartree–Fock orbital energies suffices for a prediction of electron binding energies that verifies the presence of gas-phase molecular species in experimental samples and enables assignments of photoelectron spectral peaks. For core ionization energies, where orbital relaxation effects are large, the TOM produces an orbital energy that may be improved with low-order, self-energy corrections, such as D2 generalized to grand-canonical Hartree–Fock reference ensembles.

More general (nonquasiparticle) approximations include all elements of the self-energy matrix. When correlation states in photoelectron spectra are under consideration, the collapse of the Koopmans picture can be diagnosed with 2ph-TDA calculations, although a quantitatively accurate description generally demands a self-energy with higher-order terms. Whereas the two-particle-one-hole Tamm–Dancoff approximation (2ph-TDA) fails to generate reliable results for valence, VIEs, other nondiagonal alternatives, such as renormalized third-order (3+), the third-order algebraic diagrammatic construction (ADC(3)) and the nondiagonal, renormalized, second-order (NR2) methods yield mean absolute errors between 0.1 and 0.2 eV. The NR2 method achieves competitive accuracy with smaller demands for arithmetic operations and memory. All of these nondiagonal methods are capable of describing 2hp correlation final states qualitatively. Their Dyson orbitals are linear combinations of canonical Hartree–Fock orbitals. Dyson orbitals for electron affinities are likely to require such flexibility. Recent studies of vertical electron affinities of electron-accepting molecules indicate that the NR2 method may be a promising approach.[28]

Higher accuracy for valence electron binding energies, quantitatively accurate calculations for correlation final states such as shakeups in photoelectron spectra or core-excited anions, descriptions of strong orbital relaxation effects for inner-shell ionization energies, and descriptions of more complex correlation effects that involve several open shells may be treated by introducing higher operator manifolds and more correlated reference states. Descriptions of low-spin, open-shell states that conserve exact spin quantum numbers may be accommodated with more flexible Ansätze for reference density matrices. Research along these lines is in progress.

REFERENCES

1. Linderberg, J.; Öhrn, Y. *Propagators in Quantum Chemistry*, 2nd ed.; Wiley-Interscience: Hoboken, NJ, 2004.
2. von Niessen, W.; Schirmer, J.; Cederbaum, L. S. Computational Methods for the One-Particle Green's Function. *Comput. Phys. Rep.* **1984**, *1*, 57–125.

3. Herman, M. F.; Freed, K. F.; Yeager, D. L. Analysis and Evaluation of Ionization Potentials, Electron Affinities, and Excitation Energies by the Equations of Motion—Green's Function Method. *Adv. Chem. Phys.* **1981**, *48*, 1–69.
4. Jørgensen, P.; Simons, J. *Second Quantization-Based Methods in Quantum Chemistry*; Academic Press: New York, 1981.
5. Ortiz, J. V. Electron Propagator Theory: An Approach to Prediction and Interpretation in Quantum Chemistry. *Wiley Interdiscip. Rev.: Comput. Mol. Sci.* **2013**, *3* (2), 123–142.
6. Ortiz, J. V. Toward an Exact One-Electron Picture of Chemical Bonding. *Adv. Quantum Chem.* **1999**, *35*, 33–52.
7. Pickup, B. T.; Goscincki, O. Direct Calculation of Ionization Energies I. Closed Shells. *Mol. Phys.* **1973**, *26*, 1013–1035.
8. Danovich, D. Green's Function Methods for Calculating Ionization Potentials, Electron Affinities, and Excitation Energies. *Wiley Interdiscip. Rev.: Comput. Mol. Sci.* **2011**, *1* (3), 377–387.
9. Löwdin, P.-O. *Linear Algebra for Quantum Theory*; John Wiley and Sons: New York, 1998.
10. Goscinski, O.; Lukman, B. Moment-Conserving Decoupling of Green Functions via Padé Approximants. *Chem. Phys. Lett.* **1970**, *7*, 573–576.
11. Löwdin, P. O. Studies in Perturbation Theory. X. Lower Bounds to Energy Eigenvalues in Perturbation-Theory Ground State. *Phys. Rev.* **1965**, *139* (2A), 357–372.
12. Manne, R. A Completeness Theorem for Operator Spaces. *Chem. Phys. Lett.* **1977**, *45* (3), 470–472.
13. Simons, J. The Electron Propagator and Superoperator Resolvent. *J. Chem. Phys.* **1976**, *64* (11), 4541–4543.
14. Löwdin, P.-O. Studies in Perturbation Theory XIII. Treatment of Constants of Motion in Resolvent Method, Partitioning Technique, and Perturbation Theory. *Int. J. Quantum Chem.* **1968**, *2* (6), 867–931.
15. Nehrkorn, C.; Purvis, G. D.; Öhrn, Y. Hermiticity of the Superoperator Hamiltonian in Propagator Theory. *J. Chem. Phys.* **1976**, *64* (4), 1752–1756.
16. Schirmer, J.; Cederbaum, L. S. The Two-Particle-Hole Tamm-Dancoff Approximation (2ph-TDA) Equations for Closed-Shell Atoms and Molecules. *J. Phys. B* **1978**, *11*, 1889–1900.
17. Hirata, S.; Hermes, M. R.; Simons, J.; Ortiz, J. V. General-Order Many-Body Green's Function Method. *J. Chem. Theory Comput.* **2015**, *11* (4), 1595–1606.
18. Born, G.; Kurtz, H. A.; Öhrn, Y. Elementary Finite Order Perturbation Theory for Vertical Ionization Energies. *J. Chem. Phys.* **1978**, *68*, 74–85.
19. Hu, C.-H.; Chong, D. P.; Casida, M. E. The Parametrized Second-Order Green Function Times Screened Interaction (pGW2) Approximation for Calculation of Outer Valence Ionization Potentials. *J. Electron Spectrosc. Relat. Phenom.* **1997**, *85* (1–2), 39–46.
20. Romero, J.; Charry, J. A.; Nakai, H.; Reyes, A. Improving Quasiparticle Second Order Electron Propagator Calculations with the Spin-Component-Scaled Technique. *Chem. Phys. Lett.* **2014**, *591*, 82–87.
21. Danovich, D. Green's Function Ionization Potentials in Semiempirical MO Theory. In *Encyclopedia of Computational Chemistry*; Schreiner, P. R. Ed.; Vol. 2; John Wiley: New York, 1998; pp 1190–1202.
22. Purvis, G. D.; Öhrn, Y. The Transition State, the Electron Propagator, and the Equation of Motion Method. *J. Chem. Phys.* **1976**, *65* (3), 917–922.
23. Ortiz, J. V.; Basu, R.; Öhrn, Y. Electron-Propagator Calculations with a Transition-Operator Reference. *Chem. Phys. Lett.* **1983**, *103*, 29–34.
24. Flores-Moreno, R.; Zakrzewski, V. G.; Ortiz, J. V. Assessment of Transition Operator Reference States in Electron Propagator Calculations. *J. Chem. Phys.* **2007**, *127*, 134106.

25. Abdulnur, S. F.; Linderberg, J.; Öhrn, Y.; Thulstrup, P. W. Atomic Central-Field Models for Open Shells with Application to Transition Metals. *Phys. Rev. A* **1972**, *6* (3), 889–898.
26. Jørgensen, P.; Öhrn, Y. Comparison of Two Statistical Approaches to Calculate Atomic and Molecular Orbitals. *Phys. Rev. A* **1973**, *8*, 112–119.
27. Cederbaum, L. S.; Domcke, W.; Schirmer, J.; Von Niessen, W. Correlation Effects in the Ionization of Molecules: Breakdown of the Molecular Orbital Picture. *Adv. Chem. Phys.* **1986**, *65*, 115–159.
28. Dolgounitcheva, O.; Díaz-Tinoco, M.; Zakrzewski, V. G.; Richard, R. M.; Marom, N.; Sherrill, C. D.; Ortiz, J. V. Accurate Ionization Potentials and Electron Affinities of Acceptor Molecules IV: Electron-Propagator Methods. *J. Chem. Theory Comput.* **2016**, *12* (2), 627–637.
29. Schirmer, J.; Angonoa, G. On Green-Function Calculations of the Static Self-Energy Part, the Ground State Energy, and Expectation Values. *J. Chem. Phys.* **1989**, *91*, 1754–1761.
30. Ortiz, J. V. A Nondiagonal, Renormalized Extension of Partial Third-Order Quasiparticle Theory: Comparisons for Closed-Shell Ionization Energies. *J. Chem. Phys.* **1998**, *108*, 1008–1014.
31. Ortiz, J. V. Partial Third-Order Quasiparticle Theory: Comparisons for Closed-Shell Ionization Energies and an Application to the Borazine Photoelectron Spectrum. *J. Chem. Phys.* **1996**, *104*, 7599–7605.
32. Ferreira, A. M.; Seabra, G.; Dolgounitcheva, O.; Zakrzewski, V. G.; Ortiz, J. V. Application and Testing of Diagonal, Partial Third-Order Electron Propagator Approximations. In *Understanding Chemical Reactivity*. Cioslowski, J. Ed.; Vol. 22: Kluwer Academic Publishers: Dordrecht, 2001; pp 131–160.
33. Zakrzewski, V. G.; Ortiz, J. V.; Nichols, J. A.; Heryadi, D.; Yeager, D. L.; Golab, J. T. Comparison of Perturbative and Multiconfigurational Electron Propagator Methods. *Int. J. Quantum Chem.* **1996**, *60*, 29–36.
34. Ortiz, J. V. An Efficient, Renormalized Self-Energy for Calculating the Electron Binding Energies of Closed-Shell Molecules and Anions. *Int. J. Quantum Chem.* **2005**, *105*, 803–808.
35. Corzo, H. H.; Galano, A.; Dolgounitcheva, O.; Zakrzewski, V. G.; Ortiz, J. V. NR2 and P3+: Accurate, Efficient Electron-Propagator Methods for Calculating Valence, Vertical Ionization Energies of Closed-Shell Molecules. *J. Phys. Chem. A* **2015**, *119*(33), 8813–8821.
36. Raghavachari, K.; Trucks, G. W.; Pople, J. A.; Head-Gordon, M. A Fifth-Order Perturbation Comparison of Electron Correlation Theories. *Chem. Phys. Lett.* **1989**, *157*(6), 479–483.
37. Dunning, T. H. Gaussian Basis Sets for Use in Correlated Molecular Calculations. I. The Atoms Boron Through Neon and Hydrogen. *J. Chem. Phys.* **1989**, *90*(2), 1007–1023.
38. Woon, D. E.; Dunning, T. H. Gaussian Basis Sets for Use in Correlated Molecular Calculations. III. The Atoms Aluminum Through Argon. *J. Chem. Phys.* **1993**, *98*(2), 1358–1371.
39. Davidson, E. R. Comment on "Comment on Dunning's Correlation-Consistent Basis Sets". *Chem. Phys. Lett.* **1996**, *260*(3–4), 514–518.
40. Close, D. M. Calculated Vertical Ionization Energies of the Common α-Amino Acids in the Gas Phase and in Solution. *J. Phys. Chem. A* **2011**, *115*(13), 2900–2912.
41. Zakrzewski, V. G.; Dolgounitcheva, O.; Zakjevskii, A. V.; Ortiz, J. V. Ab Initio Electron Propagator Calculations on Electron Detachment Energies of Fullerenes, Macrocyclic Molecules, and Nucleotide Fragments. *Adv. Quantum Chem.* **2011**, *62*, 105–136.

42. Pérez-González, A.; Galano, A. Ionization Energies, Proton Affinities, and pKa Values of a Large Series of Edaravone Derivatives: Implication for Their Free Radical Scavenging Activity. *J. Phys. Chem. B* **2011**, *115*(34), 10375–10384.
43. Pérez-González, A.; Galano, A.; Ortiz, J. V. Vertical Ionization Energies of Free Radicals and Electron Detachment Energies of Their Anions: A Comparison of Direct and Indirect Methods Versus Experiment. *J. Phys. Chem. A* **2014**, *118*(31), 6125–6131.
44. Zakrzewski, V. G.; Dolgounitcheva, O.; Ortiz, J. V. Electron Propagator Calculations on the Ground and Excited States of C_{60}^{-}. *J. Phys. Chem. A* **2014**, *118*, 7424–7429.
45. Pinto, R. M.; Dias, A. A.; Costa, M. L. Electronic Structure and Thermal Decomposition of 5-Aminotetrazole Studied by UV Photoelectron Spectroscopy and Theoretical Calculations. *Chem. Phys.* **2011**, *381*(1–3), 49–58.
46. Pinto, R. M.; Dias, A. A.; Costa, M. L. Electronic Structure and Thermal Decomposition of 5-Methyltetrazole Studied by UV Photoelectron Spectroscopy and Theoretical Calculations. *Chem. Phys.* **2012**, *392*(1), 21–28.
47. Chen, J.; Buonaugurio, A.; Dolgounitcheva, O.; Zakrzewski, V. G.; Bowen, K. H.; Ortiz, J. V. Photoelectron Spectroscopy of the 6-Azauracil Anion. *J. Phys. Chem. A* **2013**, *117*(6), 1079–1082.
48. Corzo, H. H.; Dolgounitcheva, O.; Zakrzewski, V. G.; Ortiz, J. V. Valence-Bound and Diffuse-Bound Anions of 5-Azauracil. *J. Phys. Chem. A* **2014**, *118*(34), 6908–6913.
49. Dolgounitcheva, O.; Zakrzewski, V. G.; Ortiz, J. V. Electron Propagator and Coupled-Cluster Calculations on the Photoelectron Spectra of Thiouracil and Dithiouracil Anions. *J. Chem. Phys.* **2011**, *134*(7), 074305.
50. Erwin, G.-H.; Cecilia, D.-G.; Rubicelia, V.; Jorge, G. Implementation of the Electron Propagator to Second Order on GPUs to Estimate the Ionization Potentials of Confined Atoms. *J. Phys. B At. Mol. Opt. Phys.* **2014**, *47*(18), 185007.
51. Velasco, A. M.; Lavín, C.; Dolgounitcheva, O.; Ortiz, J. V. Excitation Energies, Photoionization Cross Sections, and Asymmetry Parameters of the Methyl and Silyl Radicals. *J. Chem. Phys.* **2014**, *141*(7), 074308.
52. Kishimoto, N.; Kimura, M.; Ohno, K. Two-Dimensional Penning Ionization Electron Spectroscopy of Open-Shell Metallocenes: Outer Valence Ionic States of Vanadocene and Nickelocene. *J. Phys. Chem. A* **2013**, *117*(14), 3025–3033.
53. Streit, L.; Dolgounitcheva, O.; Zakrzewski, V. G.; Ortiz, J. V. Valence and Diffuse-Bound Anions of Noble-Gas Complexes with Uracil. *J. Chem. Phys.* **2012**, *137*(19), 194310.
54. Dolgounitcheva, O.; Zakrzewski, V. G.; Streit, L.; Ortiz, J. V. Microsolvation Effects on the Electron Binding Energies of Halide Anions. *Mol. Phys.* **2013**, *112*(3–4), 332–339.
55. Dolgounitcheva, O.; Zakrzewski, V. G.; Ortiz, J. V. Assignment of Photoelectron Spectra of Halide–Water Clusters: Contrasting Patterns of Delocalization in Dyson Orbitals. *J. Chem. Phys.* **2013**, *138*(16), 164317.
56. Canuto, S.; Coutinho, K.; Cabral, B. J. C.; Zakrzewski, V. G.; Ortiz, J. V. Delocalized Water and Fluoride Contributions to Dyson Orbitals for Electron Detachment from the Hydrated Fluoride Anion. *J. Chem. Phys.* **2010**, *132*, 214507.
57. Zein, S.; Ortiz, J. V. Interpretation of the Photoelectron Spectra of Superalkali Species: Li_3O and Li_3O^{-}. *J. Chem. Phys.* **2011**, *135*(16), 164307.
58. Zein, S.; Ortiz, J. V. Interpretation of the Photoelectron Spectra of Superalkali Species: Na_3O and Na_3O^{-}. *J. Chem. Phys.* **2012**, *136*(22), 224305.
59. McKechnie, S.; Booth, G. H.; Cohen, A. J.; Cole, J. M. On the Accuracy of Density Functional Theory and Wave Function Methods for Calculating Vertical Ionization Energies. *J. Chem. Phys.* **2015**, *142*(19), 194114.
60. Tamayo-Mendoza, T.; Flores-Moreno, R. Symbolic Algebra Development for Higher-Order Electron Propagator Formulation and Implementation. *J. Chem. Theory Comput.* **2014**, *10*(6), 2363–2370.

CHAPTER FOURTEEN

Cognition of Learning and Memory: What Have Löwdin's Orthogonalizations Got to Do With That?

Vipin Srivastava[*,†,1], **Suchitra Sampath**[*]

[*]Centre for Neural and Cognitive Sciences, University of Hyderabad, Hyderabad, Telangana, India
[†]School of Physics, University of Hyderabad, Hyderabad, Telangana, India
[1]Corresponding author: e-mail address: vipinsri02@gmail.com

Contents

Abstract

We present some initial results to show that Löwdin's two orthogonalization schemes, namely Symmetric and Canonical, can help us to understand certain important aspects of the brain's competence to learn and memorize. We propose that these orthogonalizations may constitute the physiological actions that the brain may perform to deal with certain types of memories.

1. INTRODUCTION

Converting a given set of linearly independent vectors into a set of mutually orthogonal vectors is an old problem, which has been studied at length in mathematics, physics, and chemistry. The orthonormal basis sets find wide ranging applications in all three disciplines. That the idea and schemes of orthogonalization could have bearing on understanding cognition

Advances in Quantum Chemistry, Volume 74
ISSN 0065-3276
http://dx.doi.org/10.1016/bs.aiq.2016.08.001

of learning and memory may appear far-fetched at first sight but, as we have shown, this is not all that implausible or unrealistic, as it may appear.

In a series of papers[1–3] we have shown that Gram–Schmidt orthogonalization may have a direct relevance to the brain's capability to learn, memorize, and discriminate between information such as appearances, smells, sounds, or anything perceived by the brain through sense organs. Our contention is that the brain possesses the competence to orthogonalize an incoming information with respect to those present in its memory in order to discriminate the new information from those already known to it.[4] In other words, when we come across a new piece of information and compare it with those that we already know about, then this cognitive action, we argue, tantamounts to the brain performing the mathematical operation of orthogonalization. That is, orthogonalization could be a physiological action that the brain is capable of performing. This proposal is a part of our larger scheme of thinking that the brain, to a good extent, may operate in a mathematical manner.

We have shown that the brain's capability to isolate similarities and differences while comparing two sets of information is essentially equivalent to the brain performing Gram–Schmidt orthogonalization.[1,2] Working within the framework of a Hopfield-like neuronal network[5] we have found that when a new information comes to be recorded, if the brain invokes Gram–Schmidt orthogonalization then it can compare the new with the stored information and also isolate its similarities and differences with the latter. And, if the brain chooses to store in the synapses the orthogonalized information then it means that it stores, not the full and complete information but, only the identified "similarities" and "differences" following the Hebbian hypothesis of synaptic plasticity.[6] The most important and curious aspect of this scheme of learning and storage of information is that even though it is only the similarities and differences of the new information with the stored memories that are lodged in the synapses, yet if we present the raw information (that initially came to be recorded) for retrieval by association then it is indeed retrieved accurately and efficiently. What is more, we will show later on that this associative recall is in fact content addressable in nature in that the lodged memory has a basin of attraction around it.[7]

Thus, we contend that, while the Gram–Schmidt scheme of orthogonalization may have been invented more than a century ago, nature may have bestowed on the brain with this extraordinary capability to orthogonalize in the Gram–Schmidt way at a rather early stage of evolution.

The Gram–Schmidt scheme has a sequential character in which the vectors are arranged in a mutually orthogonal basis structure in a multi-dimensional space in the particular sequence in which they are presented—a new vector is orthogonalized with respect to the previously orthogonalized vectors without affecting their orthogonalization and is simply added to the existing orthogonal basis set without changing the orientations of those already orthogonalized. In the context of learning and memory this would imply that Gram–Schmidt orthogonalization would be suitable for sequential learning processes like learning of languages, which typically are stored as the so-called semantic memories. Our success with use of Gram–Schmidt orthogonalization to understand cognition of learning and memory prompts us to explore if we can study along similar lines non-sequential processes involved in, say, episodic memories, which includes among other things, storage of snapshots of moments, or environments.

At the outset it appears that Löwdin's two orthogonalization schemes[8–11] should be useful in the above context because of their "democratic" character since all the given linearly independent vectors are employed at once to obtain orthonormal basis sets, regardless of the order in which they are presented. They can be useful to study situations where a lot of information relevant to a particular context, eg, information about surroundings where an episode has happened, is assimilated by the brain almost simultaneously. And later an encounter with a similar set of surroundings reminds the brain of the episode. That is, Löwdin's orthogonalizations should be able to help us understand the cognitive processes involved in storage and retrieval of memories where the context suffices as a clue.

Some time back one of us (V.S.) showed that Löwdin's two orthogonalization schemes possess curious and powerful characteristic properties.[12] Studying the squares of projections of the given vectors on the orthonormalized basis vectors, it was found that the sum of such squares is exactly the same for each individual basis vector in Löwdin's "Symmetric" orthogonalization, while for his "Canonical" orthogonalization the sum has the maximum value along one of the basis vectors followed by a hierarchy of values along other basis vectors (incidentally this maximum value is the maximum for any orthonormal basis set that one can, in principle, generate).

Before contemplating on the implications that the above special features of the two orthogonalization schemes might have to understand learning and memory, we would first speculate on what the basis vectors might mean in cognitive terms.

We represent the information that comes to be recorded in the brain by an N-dimensional vector, where N is the number of neurons in the model brain. The input vector comprises a string of randomly arranged $+/-1$'s and represents a feature bundle in the form of a checklist of "yes" (+1) and "no" (−1) for a series of features that the given information might require to be characterized.[1–3] When such vectors arrive they are first orthogonalized[1] by the brain and then their orthogonalized versions are stored following Hebbian hypothesis.[6] The vector components do not remain $+/-1$'s after orthogonalization, instead they become fractions, which can be interpreted as rates at which the neurons fire when the corresponding information is being processed and assimilated by the brain. We propose that the orthogonalized (or basis) vectors can represent prototypes of the various disparate pieces of information or the vectors that have been stored in the model brain. Each prototype is unique because the bases are mutually perpendicular to each other. While a particular prototype may emphasize (by assigning positive values to) certain features in varying degrees it may deemphasize (or deny the existence of) other features in varying degrees by attributing negative values to them. These pluses and minuses in the various prototypes adjust such that they become mutually distinctive or orthogonal.

We can now surmise what Symmetric and Canonical orthogonalizations could indicate in case they were to represent some aspects of cognition. The Symmetric orthogonalization could represent the situation where we take notice of our surroundings in a gross manner paying scanty but equal attention to everything. On the other hand the brain could invoke Canonical orthogonalization to promptly categorize information coming from surroundings and also prepare a hierarchy of groups—the largest, the next largest, and so on.

Having made these broad intuitive observations we will now examine what kind of studies we can possibly perform based on the premise that orthogonalizations can indeed be carried out by the brain. In the following we hint at a couple of possibilities.

For nearly two centuries it is being debated whether the brain consists of modules where each one performs specific tasks, or does it possess a general mechanism to tackle a range of problems?[13] Tsao et al.[14] apparently find a strong support in favor of the module hypothesis and show that there is a cortical region which deals only with face recognition. Thousands of faces that are distinguished from each other, all share the same set of basic features or characteristics. So a natural question would be: how do neurons in this cortical region code for the unique shape of each individual face?[13]

We expect that Löwdin's orthogonalizations can possibly form underlying mechanisms that deal with storage, recognition, and classification of faces.

Dyslexia is another special state of the brain where we suspect Löwdin's orthogonalizations can give us insight. For instance we can understand how dyslexics, who, in spite of having learning deficiencies, possess the special capability to identify patterns in otherwise random aggregations of things (eg, people, objects, numbers, and codes), which nondyslexics would find very hard to detect.

While we are planning simulations to explore the above problems, here we will show how well is content addressability achieved with orthogonalization. This means the following. If a set of patterns of +/−1's, represented by $\{\vec{\xi}^{\mu}\}$, is orthogonalized and inscribed in the network in Hebbian manner, and if an arbitrary pattern (comprising +/−1's) that is *not* the same as one of the input (ie, raw or preorthogonalized) patterns, say $\vec{\xi}^{\nu}$, but is similar to it, is presented to the brain for association (or retrieval), then the question we ask is will it converge to the raw input pattern $\vec{\xi}^{\nu}$? The answer is, it may! We reemphasize that we store the orthogonalized vectors, ie, $\vec{\eta}^{\mu}$'s (which consist of fractional numbers), but present $\vec{\xi}^{\nu}$ or a vector close to it for retrieval. The presented vectors, which are similar to $\vec{\xi}^{\mu}$'s, may converge to $\vec{\xi}^{\mu}$'s since our model brain acts like an attractor neural network where a basin of attraction exists around each inscribed memory $\vec{\xi}^{\mu}$ such that any vector within the basin would slide down to the global minimum $\vec{\xi}^{\mu}$ that represents a memory.

This is a significant result and also appears to mimic episodic memory, which is somewhat like this. We often remember the happening of an episode along with its surroundings. And, if we encounter a similar surroundings somewhere then we recall (or reconstruct) the original surroundings and along with it the episode. The property of content addressability possessed by our model brain thus seems to be similar to episodic memory as our simulations will substantiate. Before we move forward we should recall why we had to invoke orthogonalization in the first place,[1] and how the property of content addressability is special in our model as compared to that in the Hopfield's.[5]

Recall that the Hopfield model suffers from a serious constraint in which soon after a few memories are stored (only 14% of the number of neurons) a catastrophic breakdown happens and nothing that is stored is recalled.

The genesis of the problem lies in the fact that as memories accumulate in the synapses noise builds up and at one point the noise becomes so much that the signal submerges in it. To overcome this problem, which makes Hopfield model inappropriate for the real brain, it was proposed[1] that instead of storing information in its raw form, ie, in the original form in which it comes to be recorded, if we orthogonalize the information with respect to the stored information and store the orthogonalized version then the memory capacity increases to 100% of the number of neurons. And more importantly we are able to retrieve the original or the raw information. It is in this sense that content addressability is special in our model—what we store is not what we receive for retrieval, it is in some sense its tweaked version, yet when we present the original or the raw version for retrieval, we are able to retrieve it with 100% accuracy; and even if we present something similar to the original or the raw information, it also gets associated with the original after a small number of iterations of the association process, which in a way indicates that a small amount of effort is made by the brain before it is able to recall the original.

Before we move forward with the details of Löwdin's orthogonalizations and their possible connections with cognition, we like to point out an interesting (but instructive) coincidence—the considerations that led us to propose that orthogonalization could resolve a problem in cognition of learning and memory are analogous to those that led Löwdin to invent his orthogonalization schemes nearly three quarters of a century ago.[8] Just as orthogonalization helped us resolve the problem of "memory catastrophe," Löwdin's orthogonalization resolved the "nonorthogonality catastrophe."[15] And just as the memory catastrophe is caused by the mounting overlap among patterns as they come to be stored in the brain and their number increases, the nonorthogonality catastrophe happened due to increasing number of overlap integrals between orbitals of neighboring ions even though individually they may have been small.[16]

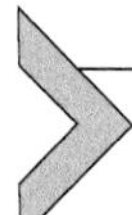

2. RECAPITULATION OF ORTHOGONALIZATION SCHEMES

If V represents a set of linearly independent vectors $\vec{v_1}, \vec{v_2}, \vec{v_3}, \ldots, \vec{v_N}$ in an N-dimensional space, then a linear transformation $\boldsymbol{A}$ can take the basis V to an orthonormal basis $\mathcal{Z}$

$$\mathcal{Z} = V\boldsymbol{A}, \tag{1}$$

$$\text{such that} < \mathcal{Z}|\mathcal{Z} > = \boldsymbol{I}. \tag{2}$$

Following Ref. 12, note that in

$$
\begin{aligned}
<\mathcal{Z}|\mathcal{Z}> &= <VA|VA>, \\
&= A^{\dagger} <\mathcal{Z}|\mathcal{Z}> A, \\
&= A^{\dagger}MA;
\end{aligned}
$$

the substitution,

$$A = M^{-\frac{1}{2}}B, \tag{3}$$

gives the general solution of the orthogonalization problem. Here M is the Hermitian metric matrix of the given basis V, and B is a unitary matrix. Two specific choices of B lead to Löwdin's two orthogonalization schemes: $B = I$ gives the Symmetric Orthogonalization, $\mathcal{Z} = \Phi = VM^{-\frac{1}{2}}$, while $B = U$, where U diagonalizes M,

$$U^{\dagger}MU = d, \tag{4}$$

gives the Canonical Orthogonalization, $\mathcal{Z} = \Lambda = VUd^{-\frac{1}{2}}$.

Some fascinating properties possessed by the two orthogonalization schemes are revealed[12] by studying the following matrix due to Schweinler and Wigner[17]

$$
\begin{pmatrix}
|(\vec{v_1}, \vec{z_1})|^2 & |(\vec{v_1}, \vec{z_2})|^2 & \ldots & |(\vec{v_1}, \vec{z_N})|^2 \\
|(\vec{v_2}, \vec{z_1})|^2 & |(\vec{v_2}, \vec{z_2})|^2 & \ldots & |(\vec{v_2}, \vec{z_N})|^2 \\
\vdots & \vdots & \ddots & \vdots \\
|(\vec{v_N}, \vec{z_1})|^2 & |(\vec{v_N}, \vec{z_2})|^2 & \ldots & |(\vec{v_N}, \vec{z_N})|^2
\end{pmatrix}.
$$

The elements are squares of projections of the given vectors $\{\vec{v_k}\}$ on the orthonormal basis vectors $\vec{z_\kappa}$. The elements in the rows simply add up to the squares of lengths of the given vectors. Interesting information is hidden in the additions of the elements in the columns. For a column they add up to a real positive number c_κ[12]:

$$\sum_k |(\vec{v_k}, \vec{z_\kappa})|^2 = (AMMA^{\dagger})_{\kappa\kappa} = (BMB^{\dagger})_{\kappa\kappa} = c_\kappa; \;\; k = 1, \ldots, N. \tag{5}$$

Note that,

$$\sum_{\kappa=1}^{N} c_\kappa = \sum_{k=1}^{N} |\vec{v_k}|^2, \text{ a constant for a given set } V, \tag{6a}$$

and

$$\sum_{\kappa} c_{\kappa}^{2} = m, \text{the SW parameter.}^{12,17} \tag{6b}$$

For normalized $\vec{v_k}$, we have shown[12] that,

$$m = m_{min} \text{ for the Symmetric basis } \boldsymbol{\mathcal{Z}} = \boldsymbol{\Phi}, \tag{7a}$$

and

$$m = m_{max} \text{ for the Canonical basis } \boldsymbol{\mathcal{Z}} = \boldsymbol{\Lambda}. \tag{7b}$$

That is, the c_{κ}'s have the maximally lopsided distribution for the Canonical basis, whereas for the Symmetric basis they have an average distribution—$c_1 = c_2 = \cdots = c_N = (c_1 + c_2 + \cdots + c_N)/N$. What these mean in simple terms are the following. In the Symmetric case, where $\boldsymbol{\mathcal{Z}} = \boldsymbol{\Phi} \equiv \{\phi_{\kappa}\}$,

$$\sum_{k} |(\vec{v_k}, \vec{\phi_{\kappa}})|^2 = \sum_{\kappa} |(\vec{v_k}, \vec{\phi_{\kappa}})|^2 = |\vec{v_k}|^2, \tag{8a}$$

that is for normalized $\vec{v_k}$'s the sum of projection squares of all the given vectors on each of $\vec{\phi_{\kappa}}$'s is the same. On the other hand for the Canonical case, where $\boldsymbol{\mathcal{Z}} = \boldsymbol{\Lambda} \equiv \{\vec{\lambda_{\kappa}}\}$,

$$\sum_{k} |(\vec{v_k}, \vec{\lambda_l})|^2 = \text{the maximum for, say } \kappa = l; \tag{8b}$$

$$\sum_{k} |(\vec{v_k}, \vec{\lambda_m})|^2 = \text{the next to maximum for, say } \kappa = m,$$

and so on and so forth. That is $\vec{\lambda_l}$ captures the maximum combined projections of all the $\vec{v_k}$'s. The sum of the projection squares of $\vec{v_k}$'s on $\vec{\lambda_{\kappa}}$'s thus has a hierarchy of values.

In simpler terms the orthogonal basis set $\{\vec{\phi_{\kappa}}\}$ is oriented in such a dramatic manner that all the given vectors $\{\vec{v_k}\}$ project on each of $\vec{\phi_{\kappa}}$ such that their squares add up to exactly the same value for each and every $\vec{\phi_{\kappa}}$—a remarkable property of $\{\vec{\phi_{\kappa}}\}$ that preserves symmetry properties, if any, in $\{\vec{v_k}\}$; on the other hand the orthogonal vectors $\{\vec{\lambda_{\kappa}}\}$ are oriented such that one of them, say $\vec{\lambda_l}$, samples maximally a bunch of $\vec{v_k}$'s, while another, say $\vec{\lambda_m}$, samples another bunch of $\vec{v_k}$'s to a smaller extent, and yet another, $\vec{\lambda_n}$, samples another bunch of $\vec{v_k}$'s but to an extent smaller than the previous two, and thus all these $\vec{\lambda_K}$'s can be assigned a descending order; this indicates that the Canonical orthogonalization can act as a very effective classification mechanism.

3. NUMERICAL DEMONSTRATION

Before coming to the possible cognitive implications of the two orthogonalization schemes due to Löwdin we will present a calculation on a small system to study how the properties of the two schemes, elicited in the previous section, show up.

We will consider 10-dimensional vectors $\{\vec{v_k}\}$ with components $+/-1$, which will be generated randomly. At the outset they will be normalized by dividing each element by $\sqrt{10}$. To start with we will consider two randomly generated vectors and orthogonalize them using the two schemes. Thereafter, we will add one randomly generated vector at a time and orthogonalize the whole lot of vectors together and look at the numbers. The data are given in Table 1 for Symmetric and Canonical orthogonalizations, respectively.

Both the schemes perform comparison of input vectors as is shown most clearly by the results for a pair of input vectors, ie, $p = 2$. In the Symmetric case the similar elements are designated by one number and the dissimilar elements by another number. Note that this number is larger for the dissimilarities in case they are fewer than the similarities and the other way round if the similarities are fewer than the dissimilarities. In the case of Canonical orthogonalization the identification of similarities and differences happens in a rather abrupt manner—between the similarities and differences, the one that is lesser in number is highlighted with larger magnitude and is always displayed in the second of the two vectors; moreover to further emphasize this fact the corresponding elements in the first pattern are zero; also those elements in the second vector that correspond to the majority, whether it is a similarity or a dissimilarity, are zero—that is, the majority feature is highlighted in the first vector and the minority in the second, as if the majority feature is paid attention to first and then the minority feature is noticed exclusively and with greater emphasis.

Table 1 (A) and (B) Five Randomly Generated 10-Dimensional Vectors Are Normalized and Orthogonalized Following Symmetric and Canonical Schemes

Unnormalized Input Vectors

$\vec{v_1}$: 1	−1	1	1	1	−1	1	1	−1	−1
$\vec{v_2}$: 1	−1	1	−1	−1	−1	−1	1	−1	−1
$\vec{v_3}$: −1	1	−1	−1	1	−1	−1	1	1	−1
$\vec{v_4}$: 1	−1	1	−1	−1	−1	1	−1	−1	−1
$\vec{v_5}$: −1	−1	1	1	1	1	1	1	1	−1
(A) Symmetric Orthogonalization									
of $\vec{v_1}$, $\vec{v_2}$ (normalized): ($p = 2$)									
0.2673	−0.2673	0.2673	0.4082	0.4082	−0.2673	0.4082	0.2673	−0.2673	−0.2673
0.2673	−0.2673	0.2673	−0.4082	−0.4082	−0.2673	−0.4082	0.2673	−0.2673	−0.2673
of $\vec{v_1}$, $\vec{v_2}$, $\vec{v_3}$ (normalized): ($p = 3$)									
0.2351	−0.2351	0.2351	0.3804	0.4531	−0.3078	0.3804	0.3078	−0.2351	−0.3078
0.2770	−0.2770	0.2770	−0.4000	−0.4223	−0.2547	−0.4000	0.2547	−0.2770	−0.2547
−0.2966	0.2966	−0.2966	−0.2743	0.3693	−0.3470	−0.2743	0.3470	0.2966	−0.3470
of $\vec{v_1}$, $\vec{v_2}$, $\vec{v_3}$, $\vec{v_4}$ (normalized): ($p = 4$)									
0.2259	−0.2259	0.2259	0.4207	0.4715	−0.2768	0.3392	0.3582	−0.2259	−0.2768
0.2578	−0.2578	0.2578	−0.2639	−0.3711	−0.1506	−0.5441	0.4307	−0.2578	−0.1506
−0.2835	0.2835	−0.2835	−0.3670	0.3344	−0.4178	−0.1763	0.2272	0.2835	−0.4178
0.1499	−0.1499	0.1499	−0.4220	−0.2313	−0.3405	0.4301	−0.5115	−0.1499	−0.3405

of $\vec{v_1}, \vec{v_2}, \vec{v_3}, \vec{v_4}, \vec{v_5}$ (normalized): $(p = 5)$

0.3470	−0.1135	0.1135	0.4084	0.4617	−0.4002	0.2779	0.2972	−0.3470	−0.1667
0.2019	−0.3096	0.3096	−0.2581	−0.3664	−0.0936	−0.5157	0.4590	−0.2019	−0.2013
−0.2813	0.2856	−0.2856	−0.3669	0.3345	−0.4202	−0.1773	0.2262	0.2813	−0.4158
0.0959	−0.2001	0.2001	−0.4161	−0.2265	−0.2855	0.4577	−0.4841	−0.0959	−0.3897
−0.3797	−0.3624	0.3624	0.1505	0.1462	0.3840	0.2547	0.2539	0.3797	−0.3580

Schweinler–Wigner Matrices for $p = 2, 3, 4, 5$

$p = 2$: $\begin{pmatrix} 0.9583 & 0.0417 \\ 0.0417 & 0.9583 \end{pmatrix} \begin{matrix} :1.0000 \\ :1.0000 \end{matrix}$

$p = 3$: $\begin{pmatrix} 0.9472 & 0.0422 & 0.0106 \\ 0.0422 & 0.9577 & 0.0001 \\ 0.0106 & 0.0001 & 0.9893 \end{pmatrix} \begin{matrix} :1.0000 \\ :1.0000 \\ :1.0000 \end{matrix}$

$p = 4$: $\begin{pmatrix} 0.9283 & 0.0341 & 0.0078 & 0.0298 \\ 0.0341 & 0.8655 & 0.0019 & 0.0985 \\ 0.0078 & 0.0019 & 0.9454 & 0.0448 \\ 0.0298 & 0.0985 & 0.0448 & 0.8269 \end{pmatrix} \begin{matrix} :1.0000 \\ :1.0000 \\ :0.9999 \\ :1.0000 \end{matrix}$

$p = 5$: $\begin{pmatrix} 0.8602 & 0.0406 & 0.0079 & 0.0358 & 0.0555 \\ 0.0406 & 0.8510 & 0.0019 & 0.0937 & 0.0128 \\ 0.0079 & 0.0019 & 0.9454 & 0.0447 & 0.0000 \\ 0.0358 & 0.0937 & 0.0447 & 0.8131 & 0.0127 \\ 0.0555 & 0.0128 & 0.0000 & 0.0127 & 0.9190 \end{pmatrix} \begin{matrix} :1.0000 \\ :1.0000 \\ :0.9999 \\ :1.0000 \\ :1.0000 \end{matrix}$

Continued

Table 1 (A) and (B) Five Randomly Generated 10-Dimensional Vectors Are Normalized and Orthogonalized Following Symmetric and Canonical Schemes—cont'd

(B) Canonical Orthogonalization

of $\vec{v_1}$, $\vec{v_2}$ (normalized): $(p = 2)$									
0.3780	−0.3780	0.3780	0	0	−0.3780	0	0.3780	−0.3780	−0.3780
0	0	0	−0.5774	−0.5774	0	−0.5774	0	0	0
of $\vec{v_1}$, $\vec{v_2}$, $\vec{v_3}$ (normalized): $(p = 3)$									
−0.4353	0.4353	−0.4353	−0.1027	0.0635	0.2690	−0.1027	−0.2690	0.4353	0.2690
−0.1414	0.1414	−0.1414	−0.4243	0.1414	−0.4243	−0.4243	0.4243	0.1414	−0.4243
−0.1027	0.1027	−0.1027	0.4353	0.7042	−0.1663	0.4353	0.1663	0.1027	−0.1663
of $\vec{v_1}$, $\vec{v_2}$, $\vec{v_3}$, $\vec{v_4}$ (normalized): $(p = 4)$									
0.4311	−0.4311	0.4311	−0.0717	−0.2142	−0.2886	0.1969	0.0200	−0.4311	−0.2886
−0.1002	0.1002	−0.1002	−0.3716	0.1691	−0.4405	−0.4143	0.4832	0.1002	−0.4405
0.0402	−0.0402	0.0402	0.5954	0.6310	−0.0758	0.2902	0.3810	−0.0402	−0.0758
−0.1512	0.1512	−0.1512	−0.2482	0.2297	−0.3268	0.5766	−0.4981	0.1512	−0.3268
of $\vec{v_1}$, $\vec{v_2}$, $\vec{v_3}$, $\vec{v_4}$, $\vec{v_5}$ (normalized): $(p = 5)$									
0.0843	−0.2014	0.2014	0.1859	−0.2886	0.3903	−0.5709	0.4837	−0.0843	0.2731
0.3869	0.2584	−0.2584	0.4342	0.3750	−0.3277	−0.0485	0.1650	−0.3869	0.3177
−0.1153	0.0764	−0.0764	−0.3088	0.2319	−0.4253	−0.3718	0.5272	0.1153	−0.4643
0.1778	0.2645	−0.2645	−0.4882	−0.4508	−0.2153	−0.3843	−0.3311	−0.1778	0.2271
−0.4380	0.4225	−0.4225	0.0838	0.2259	0.2960	−0.1865	−0.0101	0.4380	0.2805

Schweinler–Wigner Matrices for $p = 2, 3, 4, 5$

$$p = 2: \begin{pmatrix} 0.7000 & 0.7000 \\ 0.3000 & 0.3000 \end{pmatrix} \begin{matrix} :1.4000 \\ :0.6000 \end{matrix}$$

$$p = 3: \begin{pmatrix} 0.7236 & 0.5789 & 0.1447 \\ 0.0000 & 0.2000 & 0.8000 \\ 0.2764 & 0.2211 & 0.0553 \end{pmatrix} \begin{matrix} :1.4472 \\ :1.0000 \\ :0.5528 \end{matrix}$$

$$p = 4: \begin{pmatrix} 0.4984 & 0.5810 & 0.2150 & 0.7640 \\ 0.0120 & 0.2497 & 0.7398 & 0.0046 \\ 0.4884 & 0.0678 & 0.0014 & 0.1010 \\ 0.0012 & 0.1015 & 0.0438 & 0.1304 \end{pmatrix} \begin{matrix} :2.0585 \\ :1.0061 \\ :0.6585 \\ :0.2769 \end{matrix}$$

$$p = 5: \begin{pmatrix} 0.0079 & 0.1135 & 0.0428 & 0.1090 & 0.0026 \\ 0.1423 & 0.0108 & 0.0009 & 0.0571 & 0.1021 \\ 0.0342 & 0.2196 & 0.7359 & 0.0100 & 0.0038 \\ 0.3384 & 0.0651 & 0.0064 & 0.0491 & 0.8889 \\ 0.4772 & 0.5910 & 0.2141 & 0.7749 & 0.0025 \end{pmatrix} \begin{matrix} :0.2759 \\ :0.3131 \\ :1.0035 \\ :1.3479 \\ :2.0596 \end{matrix}$$

First, two vectors are orthogonalized, then new vectors are added, one at a time. The SW matrices are presented sequentially to examine how they change as new vectors are added. For the convenience of presentation, these numbers are presented as transpose of the SW matrix in the text.

When a third vector is added and all the three vectors are orthogonalized the numbers representing the elements of the vectors are all changed when compared with the previous case of $p = 2$. The numbers still reflect the patterns of similarities and dissimilarities in the input vectors in both the orthogonalization schemes.

Rationalization of the numbers and their reconciliation with the patterns of similarities and differences in the input vectors becomes increasingly difficult as p increases beyond 3. However, c_κ's are useful parameters to understand in a gross manner as to how the orthogonalized basis set reorganizes itself (by reorientation) as new vectors are added.

For Symmetric orthogonalization the property (8a) is satisfied precisely. Accordingly the SW matrices are perfectly Symmetric. They are also diagonally dominated indicating that $\vec{\phi}_1$ is almost aligned with $\vec{\nu_1}$, $\vec{\phi}_2$ with $\vec{\nu_2}$, and so on. Of course this alignment weakens as p increases but only marginally.

In Canonical orthogonalization the projections of $\vec{\nu_k}$'s are found to maximize along one of the $\vec{\lambda_k}$'s. For $p = 2 - 4$, it happens along $\vec{\lambda_1}$ but after $\vec{\nu_5}$ is included, the focus shifts to $\vec{\lambda_5}$, which captures bulk of the projections.

4. A MODEL FOR NEURONAL NETWORK

We consider a fully connected network of N neurons in which each neuron is synaptically connected to all the other $N - 1$ neurons. This may not represent the real network in which a neuron is actually connected to a relatively small fraction of all the other neurons, but the full connectivity assumption provides a mathematical convenience and does not come in the way of studying the network properties. This network is represented by the Hamiltonian

$$H = -\frac{1}{2}\sum_{\substack{j=1 \\ (j \neq i)}} J_{ij} S_i S_j; \quad S_i,\, S_j = +1 \text{ or } -1, \tag{9}$$

where S_i and S_j represent the states of neurons on sites i and j, $+1$ stands for a firing neuron and -1 for a quiescent neuron and J_{ij} represents the interaction between them. A pattern of "firing" and "not-firing" neurons will represent "an information." This set of $+/-1$'s will be represented by $\vec{\xi^\mu}$, which is the same as a vector $\vec{\nu_k}$ dealt with earlier in the chapter. Thus, ξ_i^μ is the value of the neuron S_i in the μ-th pattern or the information and itself can take value either +1 or −1. The interaction between i and j neurons in a particular pattern will be defined as,

$$J_{ij} = \frac{1}{N}\sum_{\mu=1}^{p} \xi_i^{\mu}\xi_j^{\mu}, \quad \text{with } J_{ij} = J_{ji} \text{ and } J_{ii} = 0, \tag{10a}$$

alternatively,

$$J_{ij} = \frac{1}{N}\sum_{\mu=1}^{p} (\xi_i^{\mu}\xi_j^{\mu} - \delta_{ij}\xi_i^{\mu}\xi_j^{\mu}). \tag{10b}$$

This is a mathematical representation of the Hebb's rule of synaptic plasticity.[6]

According to Hebb[6] when an information comes to be recorded or stored in the brain it alters the strength of the connection or the synapse between two neurons. The change in the strength of a synaptic connection depends on the activities of the neurons on either end of it. And these changes in synaptic strengths keep accumulating as more and more information are stored. Eq. (10) represents the strength J_{ij} of a particular synapse after p patterns or information have been stored. The change ΔJ_{ij} of a synaptic strength while storing a new pattern is thus the manifestation of learning of the new information.

In this way as an information represented by a vector $\vec{\xi^{\mu}}$ is learned or stored in a neuronal network it alters the values of all the synaptic connections in the network. The next vector changes the values of the synapses further and gets stored. Thus, information after information (represented by N-dimensional vectors of the type described earlier) are stored in the same set of neurons through cumulative changes of the synaptic strengths. The distributed manner of memory storage is unique to the brain and is substantiated experimentally.

To check if the learned p patterns are saved in the memory, we need to try and retrieve or recall them. We resort to associative memory. That is, we present one of the p patterns and examine if it is associated with itself in the memory. When the test pattern, say $\vec{\xi^{\nu}}$, is presented the neurons activated by it interact with each other through J_{ij}'s of Eq. (10) modified cumulatively by storage of p patterns. Each neuron will receive action potentials from all the other neurons weighted with the corresponding J_{ij}'s—the postsynaptic neuron i from presynaptic neurons j's—and a postsynaptic potential, h_i^ν will be built up as,

$$h_i^\nu = \frac{1}{2}\sum_{\substack{j=1 \\ j\neq i}}^{N} J_{ij}\xi_j^\nu, \tag{11}$$

This happens for each i. If the signs of h_i^ν's match with those of ξ_i^ν's for each individual i then we deduce that $\vec{\xi^\nu}$ is retrieved. This means that

$$h_i^\nu \xi_i^\nu = 1 - \frac{p}{N} + \frac{1}{N}\sum_{\substack{\mu=1 \\ (\mu\neq\nu)}}^{p} \xi_i^\mu \xi_i^\nu (\vec{\xi^\mu} \cdot \vec{\xi^\nu}), \tag{12}$$

should be > 0 for each i.

Simple considerations show that the above condition will begin to slacken as p increases and might become difficult to hold for a majority of i's beyond a certain value of p. Indeed, it is possible to ascertain analytically that the condition breaks down for $\frac{p}{N} > 0.14$.[18] This marks the so-called memory catastrophe where none of the stored patterns can be recalled.

Even though the above model, developed by Hopfield,[5] has the shortcoming of memory catastrophe, which we will address in the following, it possesses a remarkable property of "content addressability." In other words the Hopfield network is an attractor neural network (ANN).[7] The learned p patterns form p minima in an energy landscape which has a downward slope over a definite area around each minimum. A pattern, that is not actually learned but is similar to one of the learned patterns, if happens to fall in the slopy region around the learned pattern, will tend to slide down and coincide with the learned pattern at the bottom after a few iterations of the retrieval process of Eqs. (11) and (12). This region surrounding a minimum is appropriately called "basin of attraction."

Thus the energy landscape is embedded in a configuration space. Besides the "global" minima, which are caused by the learned or imprinted patterns, there can be many local minima punched here and there, which can also give the false impression of being imprinted in the retrieval dynamics. These small and shallow minima are due to the so-called spurious states[7] and can be overcome by introducing "temperature" in the system, ie, by inserting some random flips (by changing the signs of elements) in the vector after it appears to settle down in a minimum. This dynamical behavior has a significant cognitive connotation, which we will elaborate later. Before that we will return to the memory catastrophe.

The memory catastrophe is an unrealistic feature of a model for the brain, and therefore it must somehow be removed to make the model cognitively appealing. The memory blackout happens in the Hopfield model because of the fact that the incoming vectors always have nonzero overlap between themselves, and because of this, noise builds up in Eq. (12) as p increases and at a critical p the signal submerges in the noise and retrieval or recall from memory is prohibited. A way out was suggested by Srivastava and Edwards[1] that if the incoming vector is orthogonalized with respect to the vectors in the memory store, then the noise will be eliminated completely. Initially Gram–Schmidt orthogonalization was employed[12] and the results were promising. Here we are using Löwdin's orthogonalizations because of their "democratic" nature and hope to get new insight into memories that involve collective and simultaneous intake of information. Before looking at results we will point out two important facts about retrieval of information with orthogonalization incorporated in the Hopfield model.

First, suppose $\{\vec{\eta^\mu}\}$ represent the orthogonal basis set obtained from $\{\vec{\xi^\mu}\}$. We propose to store $\{\vec{\eta^\mu}\}$ rather than $\{\vec{\xi^\mu}\}$ according to the Hebbian hypothesis, and the J_{ij}'s are represented as,

$$J_{ij} = \sum_{\mu=1}^{p} \left(\eta_i^\mu \eta_j^\mu - \delta_{ij}\eta_i^\mu \eta_j^\mu\right). \tag{13}$$

This is the same as Eq. (10) with $\vec{\xi^\mu}$ replaced by $\vec{\eta^\mu}$. The first significant fact we point out is that even though the J_{ij}'s of (13) go into (11) the unorthogonalized $\{\vec{\xi^\mu}\}$ are still retrieved properly and efficiently. The second important fact is that the basins of attraction still exist, ie, content addressability is intact even though what are stored are $\vec{\eta^\mu}$'s, and not $\vec{\xi^\mu}$'s.

5. ADAPTATION TO COGNITIVE MEMORY

In Table 1 we have listed a set of five randomly generated vectors $\{\vec{v_k}\}$ and their orthogonal counterparts $\{\vec{\phi_\kappa}\}$ and $\{\vec{\lambda_\kappa}\}$ obtained using Symmetric and Canonical orthogonalization schemes, and also the corresponding SW matrix. The $\{\vec{v_k}\}$ are $\{\vec{\xi^\mu}\}$ that represent information that come to be recorded in the model brain. Depending on the situation the model brain can either pay a uniform and unbiased attention to each of the incoming information, or it may pay graded attention to them—the greatest to some, a little less to another lot, and so on. This means that depending on the situation the brain either performs Symmetric of Canonical orthogonalizations. Then $\{\vec{\phi_\kappa}\}$ and $\{\vec{\lambda_\kappa}\}$ would be identified, respectively, with two different sets of values of $\{\vec{\eta^\mu}\}$ and the numbers in the SW matrices will reflect the uniform or the graded attention paid to the various incoming information. Thus, $\{\vec{\phi_\kappa}\}$ or $\{\vec{\lambda_\kappa}\}$ are stored in J_{ij} following Eq. (13) as $\{\vec{\eta^\mu}\}$.

If we present $\{\vec{\xi^\mu}\}$ for retrieval as in Eqs. (11) and (12) they are retrieved perfectly in either case, whether $\{\vec{\phi_\kappa}\}$ are stored as $\{\vec{\eta^\mu}\}$ or $\{\vec{\lambda_\kappa}\}$ are stored as $\{\vec{\eta^\mu}\}$. The question relevant to content addressability or episodic memory is: what will happen to retrieval if we present, not $\{\vec{\xi^\mu}\}$ but patterns that are a little different from them? The answer is: they too are retrieved provided the presented patterns are within the basins of attraction of the $\vec{\xi^\mu}$'s.

So we see that our model brain can recall or connect with the original scenario (represented by a set of vectors, $\{\vec{\xi^\mu}\}$) when faced with similar scenarios represented by vectors that are different from $\vec{\xi^\mu}$'s but are similar to them to some extent. One can appreciate the fact that if the number of $\vec{\xi^\mu}$'s is large then the process of recollection of the original $\vec{\xi^\mu}$'s from a presented similar set would become more sensitive to the similarities that the individual presented vectors would have with the original set—the larger the number of $\vec{\xi^\mu}$'s the greater will be the sensitivity—by sensitivity we imply the closeness between the presented and the original set of vectors. This fact is reflected in the observation in Fig. 2 that the basins of attraction shrink in their expanse as p increases (Fig. 1 shows schematic of basins of attraction in the configuration space). The technical details of calculation of basin of attraction are not relevant here and will be given in a future paper. Right now we will discuss the essential results relevant to the present context.

Note that in Fig. 2 as p increases sufficiently the number of patterns having zero basin of attraction increases significantly. In fact, beyond $p = 87$ no pattern has nonzero basin of attraction. But all the patterns up to $p = N - 1$ are retrieved perfectly nevertheless. In cognitive terms we can interpret it

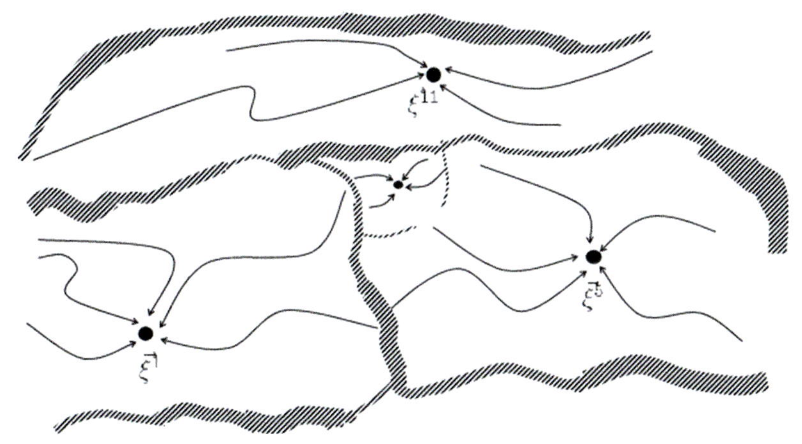

Fig. 1 Schematic diagram showing the basins of attractions for three arbitrary memories, $\vec{\xi}^1$, $\vec{\xi}^5$, and $\vec{\xi}^{11}$ inscribed in the configuration space—these form minima in an energy landscape created by Eq. (9). A configuration or pattern of $\pm$ 1's forming a vector similar to an inscribed pattern and falling in the latter's basin would converge to the inscribed vector or pattern. There can be other shallow minima within a basin of attraction of an inscribed patterns, but their basins of attraction are usually very small, and patterns falling within these basins eventually converge to the inscribed pattern by increasing the temperature. An example of such a minimum is also shown. Note that the minima are separated by hilly regions which are nonuniform in their height and width as shown by the hatched boundaries of the basins.

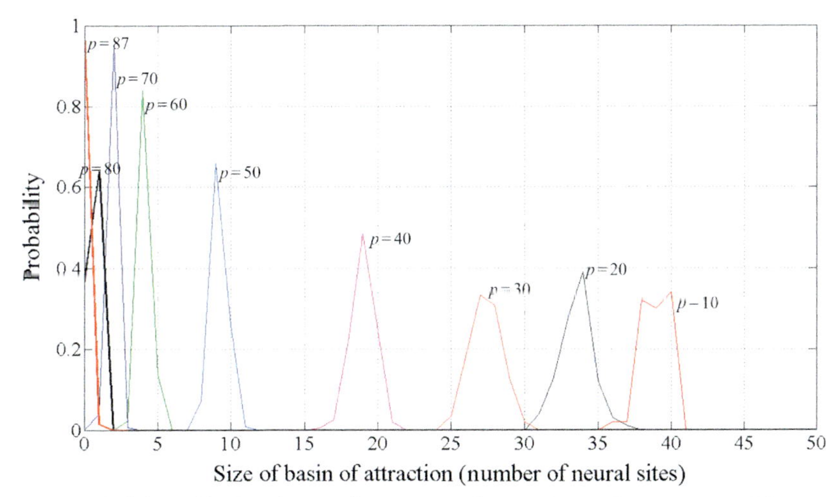

Fig. 2 Probability of finding basin of attraction of a certain size for $p = 10 - 87$ (in steps of 10). The calculation is for $N = 100$ and for each value of p it is repeated for five sets of randomly generated patterns. For $p = 10$, say, we will have 50 numbers to be plotted for the basin size for 50 different patterns. Each of these numbers is obtained by averaging over 10 samples where a sample represents a random sequence in which signs on the components of $\vec{\xi}^{\mu}$'s are flipped to identify the limit of its stability.

like this: as the number of patterns we assimilate simultaneously increases our capacity to recall them all when we encounter a set of similar patterns, would reduce steadily and there will be a situation beyond which we will be able to recall the original set only if exactly the same set is presented to us. We would like to assert that the states with zero basin of attraction are still stable in that they can be recalled associatively. Perhaps we can make a distinction between attractor and nonattractor states. The attractor states are stable and have nonzero basins of attraction, whereas the nonattractor states are stable even though they do not have a basin of attraction.

6. IN SUM

We find that Löwdin's orthogonalizations have a novel application in computational neuroscience. If our contention elaborated in this paper can be substantiated experimentally then it will be a pathbreaking finding that the orthogonalizations may actually be the physiological actions that the brain might be performing when we are storing the memories of an episode together with its surroundings.

We have attempted to give meanings in the cognitive context, to the orthonormal basis set that the two orthogonalizations due to Löwdin generate from a given set of linearly independent vectors and have tried to interpret the numbers representing the components of the orthonormal basis vectors, which change considerably as the set of input vectors grows in size. The Schweinler–Wigner matrix provides a useful tool to assign meanings, in gross cognitive sense, to the operations of orthogonalizations.

What we have presented are some initial results. There appears to be a great scope for understanding complex phenomena in the realm of cognitive neuroscience, and also psycholinguistics, using Löwdin's orthogonalizations.

ACKNOWLEDGMENT

V.S. thanks the Royal Society, London for supporting part of this research at the Department of Physiology, Development and Neuroscience, University of Cambridge.

REFERENCES

1. Srivastava, V.; Edwards, S. F. A Model of How the Brain Discriminates and Categorises. *Physica A* **2000**, *276*, 352–358.
2. Srivastava, V.; Parker, D. J.; Edwards, S. F. The Nervous System Might 'Orthogonalize' to Discriminate. *J. Theor. Biol.* **2008**, *253*(3), 514–517.
3. Srivastava, V.; Sampath, S.; Parker, D. J. Overcoming Catastrophic Interference in Connectionist Networks Using Gram-Schmidt Orthogonalization. *PLoS ONE* **2014**, *9*, e105619.

4. Srivastava, V.; Edwards, S. F. The Brain and Orthonormal Bases. In: *Physics and Biology: A Synergy*; Lakshmi, P. A., Srivastava, V., (Eds); Allied: Mumbai, 2009; pp. 199–205.
5. Hopfield, J. J. Neural Networks and Physical Systems with Emergent Collective Computational Abilities. *Proc. Natl. Acad. Sci.* **1982**, *79*, 2554–2558.
6. Hebb, D. O. *The Organization of Behavior: A Neuropsychological Approach*. John Wiley: New York, 1949.
7. Amit, D. J. *Modeling Brain Function: The World of Attractor Neural Networks*. Cambridge University Press: New York, NY, 1989.
8. Löwdin, P.-O. A Box Model of Alkali Halide Crystal. *Ark. Mat. Astr. Fys. A* **1947**, *35*, 30.
9. Löwdin, P.-O. Quantum Theory of Many-Particle Systems. I. Physical Interpretations by Means of Density Matrices, Natural Spin-Orbitals, and Convergence Problems in the Method of Configurational Interaction. *Phys. Rev.* **1955**, *97* (6), 1474.
10. Löwdin, P.-O. Quantum Theory of Cohesive Properties of Solids. *Adv. Phys.* **1956**, *5*(17), 1–171.
11. Löwdin, P.-O. On the Nonorthogonality Problem. *Adv. Quantum Chem.* **1970**, *5*, 185.
12. Srivastava, V. A Unified View of the Orthogonalization Methods. *J. Phys. A Math. Gen.* **2000**, *33* (35), 6219.
13. Kanwisher, N. What's in a Face? *Science(Washington)* **2006**, *311* (5761), 617–618.
14. Tsao, D. Y.; Freiwald, W. A.; Tootell, R. B.; Livingstone, M. S. A Cortical Region Consisting Entirely of Face-Selective Cells. *Science* **2006**, *311* (5761), 670–674.
15. Inglis, D. R. Non-Orthogonal Wave Functions and Ferromagnetism. *Phys. Rev.* **1934**, *46* (2), 135.
16. Slater, J. C. Cohesion in Monovalent Metals. *Phys. Rev.* **1930**, *35* (5), 509.
17. Schweinler, H. C.; Wigner, E. P. Orthogonalization Methods. *J. Math. Phys.* **1970**, *11*(5), 1693–1694.
18. Amit, D. J.; Gutfreund, H.; Sompolinsky, H. Storing Infinite Numbers of Patterns in a Spin-Glass Model of Neural Networks. *Phys. Rev. Lett.* **1985**, *55* (14), 1530.

CHAPTER FIFTEEN

Ab Initio Complex Potential Energy Surfaces From Standard Quantum Chemistry Packages

Arie Landau, Debarati Bhattacharya, Idan Haritan, Anael Ben-Asher, Nimrod Moiseyev[1]

Schulich Faculty of Chemistry, Technion—Israel Institute of Technology, Haifa, Israel

[1]Corresponding author: e-mail address: nimrod@tx.technion.ac.il

Contents

Abstract

Situations in which a molecule in a given configuration is electronically bound while in another configuration is autoionized are widespread in nature. In these situations, the change in molecular configuration due to nuclear dynamics is the reason the molecule emits free electrons to the surrounding, ie, autoionizes.

Such a situation may even happen to molecules in their ground electronic state, for example, it can happen to H_2^-: at some bond lengths, the molecule is autoionized, at some bond lengths its ground state is bound, and at sufficiently large internuclear distances a stable hydrogen atom and a stable negative charged hydrogen, H^-, in their ground electronic states, are obtained. In addition, such situations can be seen in electronic scattering from molecules and in cold molecular collisions. For example, in a collision between electronically excited helium atom and a hydrogen molecule in its ground state, metastable complex He^*–H_2 is formed. As time passes this complex decays to helium in its ground state, H_2^+, and a free electron.

Advances in Quantum Chemistry, Volume 74
ISSN 0065-3276
http://dx.doi.org/10.1016/bs.aiq.2016.10.001

In all these cases the molecular dynamics play a key role as the molecules are autoionized. This poses a problem, since the Born–Oppenheimer (BO) approximation is applicable only when the decay process due to ionization is ignored. Therefore, in order to study molecular dynamics and take autoionization into consideration, one should calculate the potential energy surfaces (PES) by imposing outgoing boundary conditions (OBCs) on the electronic wavefunctions. Doing so, the electronic molecular spectrum will be discrete (no continuum), where the PES will be either real (bound electronic states) or complex (metastable molecules that ionize). These complex potential energy surfaces (CPES) are what enables one to take into consideration the electronic autoionization in the molecular dynamics.

Nevertheless, calculating CPES by standard quantum chemistry packages (SQCPs) is not a trivial task, since they were designed to calculate bound electronic excited states. Bound states lie on the real plane, unlike metastable states (resonances); therefore, explicit calculation of resonances requires modification of SQCPs. Several different possibilities for calculating CPES by modifying SQCPs are discussed in this review. Yet, the holy grail is to be able to use SQCPs, which are highly efficient codes, for calculating resonances without changing the codes. The main focus of this review will be on new methods, we have developed, that enable calculating CPESs from SQCPs, ie, without any modifications of standard codes. Such methods allow the calculations of polyatomic CPESs, as indicated by our preliminary results.

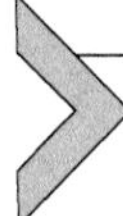

1. MOTIVATION AND DIFFICULTIES IN CALCULATING CPES

Atomic and molecular autoionization resonances are associated with complex eigenvalues of the Hamiltonian. The real part of the complex eigenvalues provides the energy position of the autoionization states, whereas the imaginary part provides the autoionization decay rate (inversely proportional to lifetime) of the states. This decay rate, also known as width, heavily depends on the structure of the molecular system under study. In some geometrical configurations, the molecules may have very short lifetimes, while in other geometrical configurations, the same molecules may have extremely long lifetime (very small width). In some molecules even the short-lived autoionization states become bound states by varying the geometrical structure. A simple example for this behavior is the molecular hydrogen anion, H_2^-: the ground electronic state of this molecular anion at the equilibrium distance has a lifetime which is about 1 fs, whereas beyond a critical bond length the ground state is bound. Moreover, at the dissociation limit, hydrogen and a stable H^- are obtained, and no ionization takes place.

In light of the above, it is important to obtain the electronic complex eigenvalues of a molecule as function of its nuclear coordinates. This function is defined as a complex potential energy surface (CPES), although only the real part of the complex eigenvalue is associated with the energy of the molecule. By using this CPES when solving the Schrödinger equation, coupling between electronic and nuclear coordinates upon autoionization is possible. This coupling may provide predictions of new observable phenomena and allow analysis of certain experimental data, meaning that there is a crucial need for CPESs both in theoretical fields and experimental fields. For example, see the most recent publications of Narevicius and his coworkers on cold molecular collisions[1–3] and the experimental and theoretical papers published on ICD.[4,5] The need for obtaining an accurate CPES is also explained in Ref. 6 where the calculated cross sections exhibit large sensitivity due to small changes in the CPES.

In order to obtain a PES, one needs to solve the electronic time-independent Schrödinger equation (TISE) within the framework of the BO approximation. However, a standard BO molecular Hamiltonian is Hermitian and its eigenvalues are real; hence only real PES can be calculated under the Hermitian regime. Thus, calculation of CPES in the Hermitian BO Hamiltonian is not trivial.

In order to overcome this problem, it is possible to impose outgoing boundary conditions (OBCs) on the asymptotes of the eigenfunctions. This implies that the obtained solutions of the TISE are the poles of the scattering matrix. These poles represent either bound or resonance states, where the asymptotes of these states can be described as:

$$\Psi_{bound/resonance}(\mathbf{r}_1,\ldots,\mathbf{r}_N)\xrightarrow{\vec{r}\to\infty}\sum_j d_j\mathcal{A}\Phi_j(\mathbf{r}_1,\ldots,\mathbf{r}_{N-1})\cdot\exp\left(+i\mathbf{p}_j\cdot\mathbf{r}_N/\hbar\right) \tag{1}$$

where N is the total number of electrons in the system under study, r_i is the coordinate of the ith electron, $\vec{r}$ is a vector of the coordinates of all the electrons, $\mathbf{p}_j$ is the component j of the momentum, $\mathcal{A}$ is the antisymmetrizer operator, and d_j is a weight coefficient.

For bound states, all components of the momentum, $\mathbf{p}_j$, get purely imaginary values, ie, $\mathbf{p}_j/\hbar = +i\mathbf{k}_j^{bound}$, where the wavevectors $\mathbf{k}_j^{bound}$ get real and positive values. As a result, at the asymptotes, the bound solutions under OBCs are of the following form:

$$\Psi_{bound}(\mathbf{r}_1,\ldots,\mathbf{r}_N)\xrightarrow{\vec{r}\to\infty}\sum_j d_j\mathcal{A}\Phi_j(\mathbf{r}_1,\ldots,\mathbf{r}_{N-1})\cdot\exp\left(-\mathbf{k}_j^{bound}\cdot\mathbf{r}_N\right)\overset{\vec{r}\to\infty}{\to}0. \tag{2}$$

It is thus clear that bound states are square-integrable functions.

The autoionization resonances are associated with eigenfunctions that diverge exponentially and therefore are not square integrable. In this case, all components of the momentum, $\mathbf{p}_j$, get complex values, where $Im[\mathbf{p}_j^{res}]<0$, ie, $\mathbf{p}_j^{res}/\hbar=\mathbf{k}_j^{res}$ where $\mathbf{k}_j^{res}=\mathbf{k}_{re,j}^{res}-i\mathbf{k}_{im,j}^{res}$. Therefore, at the asymptotes, the resonance solutions under OBCs become

$$\begin{aligned}\Psi_{res}(\mathbf{r}_1,\ldots,\mathbf{r}_N)\xrightarrow{\vec{r}\to\infty}\sum_j d_j\mathcal{A}\Phi_j(\mathbf{r}_1,\ldots,\mathbf{r}_{N-1})\cdot\exp\left(+i\mathbf{k}_{re,j}^{res}\cdot\mathbf{r}_N\right)\cdot\\ \exp\left(+\mathbf{k}_{im,j}^{res}\cdot\mathbf{r}_N\right)\overset{\vec{r}\to\infty}{\to}\infty.\end{aligned} \tag{3}$$

Knowing the behavior of the resonance wavefunction at the asymptotes, it is possible to calculate the autoionization resonances, within the BO approximation and without imposing OBCs. All one need to do is to operate a similarity transformation on the Hamiltonian which will bring the resonance exponential diverged functions back to the Hilbert space (for different types of similarity transformations see chapter 5 in Ref. 7).

It should be stressed here that in contrary to atomic calculations, when calculating molecules, the similarity transformations should take into account the singularity in the molecular BO Hamiltonian. In a molecular system, the electron–nucleus attractive potential terms are inversely proportional to the distance of any one of the electrons from the nuclei. Since the nuclei are held fixed in space within the BO approximation, it is clear that there are singularities in the electron–nucleus attractive potential terms whenever $|\mathbf{r}_i-\mathbf{R}_j|=0$ ($\mathbf{r}_i$ is the coordinate of the ith electron and $\mathbf{R}_j$ is the coordinate of the jth nucleus). As a result of these singularities, the electron–nucleus attractive potential terms are analytical functions of the electronic coordinates only inside a sphere of radius $\mathbf{R}_j$. Therefore, analytical dilation of the coordinates by *uniform* complex scaling (CS), $\mathbf{r}_i\to\mathbf{r}_i\exp(i\theta)$, is not applicable for the calculations of molecules, although it suppresses the exponential divergence of the resonance functions.[7]

In Section 2, we will briefly discuss different similarity transformations, which can be regarded as the basis for calculating molecular autoionization resonances. These transformations require severe modification of the standard codes, yet in some cases they have already been implemented in standard quantum chemistry packages (SQCPs). In Section 3 we will focus on new methods, developed by us, that enable calculating CPESs by using existing SQCPs without any modifications (CPES-from-SQCPs). We will focus on using available codes and commercial electronic structure packages rather than modifying them, since they are accessible by many researchers and are highly optimized and efficient. Such method developments are important since ab initio CPESs for polyatomic systems are lacking. Our preliminary results indicate that these polyatomic CPESs can generate cross-sections of ultracold collisions in excellent agreement with experiment. However, this is subject to a future publication. Finally, in Section 4 concluding remarks will be presented.

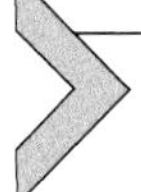

2. AB INITIO CPES BY USING EXISTING AND MODIFIED SQCPs—A BRIEF OVERVIEW

2.1 Complex Absorbing Potentials

Complex absorbing potentials (CAPs) are artificial potentials that are introduced into the molecular Hamiltonian in order to absorb scattered electrons. Doing so, CAPs bring the resonance states into the Hilbert space and avoid the singularity in the BO Hamiltonian that arises using other methods. However, since it is an artificial potential, the CAP provides accurate resonance positions and widths only in the limit in which the CAP effect goes to zero.

Since CAPs can be easily added into the molecular Hamiltonian, they enable the use of existing computational algorithms, which were originally developed for bound states. Therefore, electronic structure methods augmented with CAPs serve as a convenient approach for calculating molecular electronic resonances.[8–20]

In fact, there are a large variety of CAPs that can be used to calculate resonances. Few CAPs have been implemented within wavefunction-based electronic structure codes (see, for example, Refs. 10–13,16–20). In general the implementation of CAPs should be divided into two steps.[10–12,14,15] In the first step of the calculations, an *artificial* CAP term is added to the Hamiltonian in order to guarantee that the asymptotes of the resonance eigenfunctions decay to zero. In the second step, one attempts to remove this artificial effect of the CAP on the solutions of the TISE. However, it is important to note that many times only the first step is implemented.[13,16–20]

Eq. (4) describes the first step of the CAP implementation by introducing an artificial potential, V_{CAP}, into a physical (Hermitian and real) Hamiltonian, H_0,

$$H(\lambda) = H_0 - i\lambda V_{CAP}, \tag{4}$$

where λ is a strength parameter. CAPs are usually introduced only in the exterior region of the molecular system, eg, Eq. (5) represents a certain exterior CAP potential,

$$V_{CAP} = V_x + V_y + V_z; \quad V_{j(=x,y,z)} = \begin{cases} 0, & \text{if } |j| < r_j^{CAP} \\ (j - r_j^{CAP})^2, & \text{if } |j| > r_j^{CAP}, \end{cases} \tag{5}$$

where r_j^{CAP} is the onset of this potential. An exterior CAP holds two distinct advantages: unlike a uniform CAP, an exterior CAP does not absorb electrons in the interaction region, meaning it only absorbs emitted electrons and hence does not create a severe perturbation like a uniform CAP. In addition, an exterior CAP is closer to the exterior (or smooth exterior) scaling transformation, which is designed to avoid the singularities in the BO molecular Hamiltonian.[7,21–24]

It is relevant to note that introducing an exterior CAP makes the calculated resonance energies quite sensitive to the CAP onset.[8,10] The reason is that simple CAPs do not perfectly absorb but also generate some reflection. Therefore, the resonance wavefunction in the interior region is unphysically disturbed.[9] Obviously, the artificial reflection depends on the CAP parameters: its onset and strength. Hence, the CAP onsets need to be optimized somehow; an example for that is given in Section 3.1. As for the CAP strength effect; clearly the perturbation caused by CAPs becomes more pronounced as λ increases.[8,9] Ideally, we would like to take $\lambda \rightarrow 0$ since V_{CAP} itself is an artificial operator, and it may lead to strong perturbations that can result in improper resonance energies and discontinuous CPESs.[8,11] However, when a too small CAP strength is used, other problems emerge. Consequently, we are left with the question of how to identify the resonance energy when using a simple CAP and large λ's.

In order to resolve this question, the artificial effect of CAP on the solutions of TISE has to be removed, meaning a second step in the calculation is needed. In the exact limit, when using a complete and infinite basis set, by taking the CAP strength to zero one obtains $E(\lambda \to 0) \to E_{res}$, where E_{res} is the *exact* complex energy of a resonance state.[9,25] However, in the framework of finite basis set calculations it is very challenging to remove this effect.[10,26–28] In fact, using a finite basis set and a very small value of λ, the absorption of V_{CAP} becomes too weak. Thus, resonances cannot be represented with a finite basis set if λ becomes too small.[26] In practice, there is a critical value, λ_C, for which the exact and finite-basis-set results start to diverge as λ decreases. The value of λ_C depends on the set of basis functions used; the larger the basis set, the smaller λ_C is.[26] Thus, in the complete basis set limit $\lambda_C \to 0$. Overall, on the one hand, for finite bases a relatively large λ is needed in order to spatially restrict the wavefunction, on the other hand, large value of λ causes strong artificial reflections. This point is also discussed in length in Section 3.1 where we present a scheme which allows to take the desirable limit of $\lambda \to 0$.

For an additional reading regarding the use of CAPs in the study of molecular autoionization resonances see Refs. 8 and 9.

2.2 Reflection-Free CAPs

The reflection-free CAP (RF-CAP), unlike simple CAPs that are discussed above, provides accurate resonance positions and widths also when $\lambda \neq 0$ and even when the RF-CAP does not vanish in the interaction region. The RF-CAPs are universal and are calculated from a complex contour of integration, $F(\mathbf{r})$, when the Hamiltonian matrix elements are evaluated.[29] The imaginary part of $F(\mathbf{r})$ in the interaction region is as close to zero as one wishes, although it does not necessarily cover the entire interaction region. Outside the integration region, $F(\mathbf{r})$ behaves as $\mathbf{r}\exp(i\theta)$. When $F(\mathbf{r})$ is a smooth (analytical) function of the electronic coordinates, it is associated with the smooth-exterior-scaling (SES) transformation.[29]

In this case, the SES transformation is equivalent to adding to the Hamiltonian the RF-CAP of the following universal form:

$$\hat{V}_{RF-CAP} = \lambda[V_0(\mathbf{r}) + V_1(\mathbf{r})\partial_{\mathbf{r}} + V_2(\mathbf{r})\partial_{\mathbf{r}}^2], \quad (6)$$

where the RF-CAP-strength parameter, λ, is equal to unity and not $\lambda \to 0$ as in the standard CAP described in the previous section. V_0, V_1, and V_2 are complex potentials which vanish at the interaction region where the imaginary part of $F(\mathbf{r})$ also vanishes (not in the mathematical sense but in all significant digits which depend on the accuracy of the numerical calculations). In Refs. 30 and 31, an RF-CAP was obtained from an SES contour to calculate atomic and molecular autoionization Feshbach-type resonances. Moreover, in Ref. 31 the RF-CAP was implemented within configuration interaction (CI).

It should be stressed here that two conditions have to be satisfied in order to obtain the resonances by the RF-CAPs even when the CAP strength parameter $\lambda = 1$. The first condition is that the CAP is introduced in the noninteracting region where the electron–nuclei potential terms almost vanish. The second condition is that the electronic repulsion between an electron located outside the interaction region with all other electrons located in the interaction region is negligible. These conditions are required to avoid the singularity of BO Hamiltonian, as discussed earlier, and keep the Hamiltonian unchanged. For more explanation of these two conditions see Ref. 29.

Recently, we have shown that excellent results for the resonance positions and widths can be obtained even when these two conditions are *not* satisfied.[28] This can be achieved when the dependence of the resonance complex eigenvalues on λ is evaluated and then removed by the Padé approximant in which $\lambda \to 0$. In Section 3.1 the applications of this approach to the calculations of CPESs will be discussed in some more details.

2.3 Uniform Complex Basis Functions

The CAPs and RF-CAPs presented above enable the calculations of molecular autoionization resonances by avoiding the singularity in the molecular BO Hamiltonian. They manage to do so by imposing absorbing boundary conditions where the terms of the electron–nucleus potential vanish. Alternatively, in order to calculate molecular autoionization resonances, the complex basis functions (CBFs) approach was introduced by McCurdy and Rescigno.[21] The use of CBFs is highly connected to the analytical continuation of the Hamiltonian's matrix elements as presented by Moiseyev and Corcoran in Ref. 22. In Moiseyev's and Corcoran's work it was shown that although the molecular Hamiltonian within the BO approximation cannot be analytically dilated into the complex plane, the Hamiltonian matrix elements that are obtained by using Gaussian basis sets can be analytically dilated, meaning that instead of carrying out the uniform complex scaling transformation of the operator, ie, $\mathbf{r}_i \to \mathbf{r}_i\eta$ where $\eta = \alpha \exp(i\theta)$, one can carry out analytical continuation of the Hamiltonian matrix elements. Since the matrix elements using Gaussian basis functions are evaluated analytically, analytical dilation of the Hamiltonian matrix elements into the complex plane can be done by the use of CBFs. In other words, scaling the Gaussian exponential parameter by a phase factor of $1/\eta^2 = \alpha^{-2}\exp(-i2\theta)$ results in scaling the Hamiltonian matrix elements by a factor of η. In fact, the calculations presented in Ref. 28 confirm that the use of CBFs is equivalent to the use of an integration contour in the complex plane, where all the grid points lie on a straight line that is rotated into the complex coordinate plane by the angle θ. That is, the use of CBFs is equivalent to uniform complex scaling.

2.4 Partial CBFs—Scaling Only the Diffuse Basis Functions

As explained earlier, it is possible to avoid the singularities in the molecular BO Hamiltonian by carrying out analytical continuation of the Hamiltonian matrix elements rather than the operator itself.[22] Originally, this transformation was carried out by uniformly scaling all the Gaussian basis functions that were used in the calculations. Yet, under this analytical continuation, simultaneous description of both bound and resonance states becomes a computationally demanding procedure. The reason is that in order to calculate bound states in an accuracy which is comparable to an untransformed calculation, relatively large set of functions is required.[32–35] However, as explained in Ref. 36 there are an infinite number of analytical continuations that can be carried out for calculating resonances. All of them scale the asymptotes of the resonance states by $\eta = \alpha\exp(i\theta)$, and we can adopt a scaling procedure which reduces the basis set requirement.

One such analytical continuation is the mixed scaled basis set method, where only the most diffused functions are complex scaled.[37,38] This method was recently implemented within the Hartree–Fock approximation. However, this approach suffers from serious numerical and convergence problems.[39] This approach reproduces the correct asymptotic form of the Hamiltonian matrix elements governed by the complex scaled diffused basis functions. Moreover, similar to SES it does not tamper with the internal region governed by the tight unscaled basis functions.[23,24,29,40,41] That is, on the one hand a mixed basis set appropriately describes resonances, since the asymptotic behavior of resonances is determined by the most diffuse functions.[42] On the other hand, a mixed basis set allows for an accurate description of the bound state. Therefore, this approach reduces the basis set requirement for high accuracy description of both bound and resonance states.[35] An additional advantage is its capability to maintain a stable bound state energy over a wide range of complex scaling parameters, unlike the uniformed scaled basis set.[35] This is similar to the CAP approach, in which one places the absorbing potentials well outside the interaction region.

2.5 Summary

In this section we reviewed the current situation in the theoretical description of electronic resonance states. We discussed two main methods, CAPs and CBFs, which are the most promising approaches for explicit calculations of complex resonances and CPESs. In Ref. 28 these different methods were compared by using Gaussian basis functions, which are the most widely used basis functions in electronic structure calculations. The subject is of current interest as ab initio methods of computing resonances have accumulated and it is relevant to shed light on their differences and similarities. From this comparison it is clear that the different methods that were developed for the calculations of molecular resonances (and thereby CPES) are closely related to one another. In fact, Ref. 28 clearly shows how the CBFs method is related to the calculations of resonances by CAPs, and under which conditions the use of CBFs is equivalent to the use of CAPs for calculating the molecular resonances.

3. CPES FROM SQCPs—RECENT DEVELOPMENTS

In the following section we present two recent advances that were developed in our group in order to calculate metastable electronic states through existing SQCPs. Here we focus on using available codes and commercial electronic structure packages rather than modifying them, since they are accessible by many researchers and are highly optimized and efficient. In addition, they are also optimized for all kinds of electronically excited states, ie, for single-reference, multireference, and multiconfiguration states, and for different types of open-shell configurations. Finally, Hermitian (real) codes are computationally more cost-effective than explicit complex codes.

The first approach we present aims to improve the existing CAP-augmented equation-of-motion coupled-cluster (EOM-CC)[10–12] by removing the artificial effect of the CAP potential. The second approach presents a tool for evaluating complex resonance energies from standard, Hermitian, and real electronic structure codes. This approach can be based on any standard code out there, since it does not rely on any changes or modifications to a code. Hence, we find this approach very promising.

3.1 CPES by Removing (CAPs) via Padé[43]

As discussed in Section 2.1, there are various challenges in implementing CAPs in order to calculate resonances, most importantly, the removal of the artificial CAP. One way to overcome these challenges is to use an RF-CAP (Section 2.2), ie, a perfectly absorbing potential. The RF-CAP yields valuable resonance energy; however, it is a rather complicated scheme since it is a nonlocal operator.[29,31] Alternatively, Krylov and coworkers suggested a deperturbative correction, where the perturbation due to the finite-strength CAP is corrected using a first-order energy term.[10–12] The deperturbed-CAP approach, unlike the raw (uncorrected) CAP, provides smooth potential energy curves and lifetimes for diatomic systems.[11] However, this is merely a perturbative correction. Altogether, constructing a simple correction scheme for removing the total effect of the artificial CAP is desirable.

Such a scheme can exploit the fact that by using a complete basis set and taking the CAP strength to zero, one obtains E_{res}, which is the *exact* complex energy of a resonance state.[9,25] That is, if a function is fitted to the calculated $E(\lambda)$ values, it is possible to extrapolate it to $E(\lambda \rightarrow 0)$ and get E_{res}. For large λ's ($\lambda > \lambda_C$, where λ_C is a critical value), the results obtained using a complete basis set coalesce with the results obtained using a finite basis set.[26] Therefore, as long as the data, $E(\lambda)$, for the fitting are taken with $\lambda > \lambda_C$, the functions fitted at the finite basis set or exact limits are indistinguishable. Thus, by extrapolating to $\lambda \rightarrow 0$, a fitted function relying on a finite basis set yields the resonance complex eigenvalues that would be obtained if we were able to solve the TISE with OBCs (and without adding a nonphysical CAP).

In this section we suggest to use the Padé approximant in order to extrapolate to $\lambda \rightarrow 0$. We follow Refs. 26 and 28 that used this approach for a one-dimensional model system and apply it within a bona-fide electronic structure scheme. The input data for the extrapolation are calculated using the CAP-augmented EOM-CC with singles and doubles for electron attachment states (CAP-EOM-EA-CCSD)[10–12,14,15] implemented in Q-Chem.[44] After the energies for different and relatively large λ values are calculated, the analytical continuation is performed by fitting a ratio of polynomial (Padé) function, $E(\lambda) = P/Q$, using the Schlessinger point method[45] (for more details see Section 3.2.2). Then, we dilate the polynomial function to $E_C = E(\lambda \rightarrow 0)$ in order to get an estimation to the resonance complex energy. This removal scheme of the CAP effect via Padé is referred to as: rm-CAP-EOM-EA-CCSD (rm-CAP in short).

There is a delicate point that needs to be remembered when analytical continuation is applied. Since $[H_0, V_{CAP}] \neq 0$ the function $E(\lambda)$ has a singularity, meaning it is not analytical everywhere.[46] This singularity must occur at $\lambda = 0$, which represents a point of phase shift between positive and negative λs which correspond to absorbing and emitting complex potentials, respectively. Nevertheless, even with finite basis sets, the $\lambda = 0$ point represents the edge of the analytical regime of the function $E(\lambda)$, and we can get as close to it as we wish via Padé in order to have a good estimation of the complex resonance energy. Therefore, in practice, this singularity does not pose any problem.

The above approach was applied to the calculations of the $^2\Pi$ shape resonance of CO^- and excellent complex energies were obtained. Moreover, excellent results were also obtained using a uniform CAP, ie, a CAP that absorbs electrons in the interaction area. This demonstrated the strength of the scheme and renders the tedious effort of optimizing the onset redundant.

Table 1 compares the rm-CAP results, with optimal onset (r_j^{CAP}, see Eq. (5)) and without any onset ($r_j^{CAP} = 0$), to the experimental and the deperturbed (dp-CAP-EOM-EA-CCSD, dp-CAP in short) results. We have used the same basis set, aug-cc-pVTZ-3s3p3d(C), and onset, if used, as within dp-CAP (for details see Refs. 12 and 43). The rm-CAP results with an exterior ($r_j^{CAP} \neq 0$) and uniform ($r_j^{CAP} = 0$) CAPs are very close to each other and differ by a few hundredth of eV, which demonstrate the robustness of the rm-CAP method. The rm-CAP energy position is overestimated by about 0.5 eV; however, this is similar to the dp-CAP estimation. This relatively large error can be attributed to the size of the basis set, as demonstrated in Refs. 12 and 43. The calculated width is underestimated by about 0.2 eV, similar to the value obtained by dp-CAP. Overall, the rm-CAP results are in good agreement with experiment and with the dp-CAP scheme.

Table 1 Resonance Position (ReE) and Width ($\Gamma = -2\text{Im}E$) for the $^2\Pi$ Shape Resonance of CO^- in eV

	Re*E*	**Γ**
Experiment[a]	**1.5**	**0.8**
dp-CAP ($r_j^{CAP} \neq 0$)[b]	1.98	0.58
rm-CAP ($r_j^{CAP} \neq 0$)[c]	2.00	0.59
rm-CAP ($r_j^{CAP} = 0$)[c]	2.09	0.62

[a]Refs. 47–51.
[b]Ref. 12.
[c]Ref. 43.
The present rm-CAP calculations are performed with an exterior (onset≠0) and uniform (onset=0) CAPs. In addition, the dp-CAP calculations are presented along with the corresponding experimental values. All calculations use the same basis, aug-cc-pVTZ-3s3p3d(C), and onset if used (for details see Refs. 12 and 43).

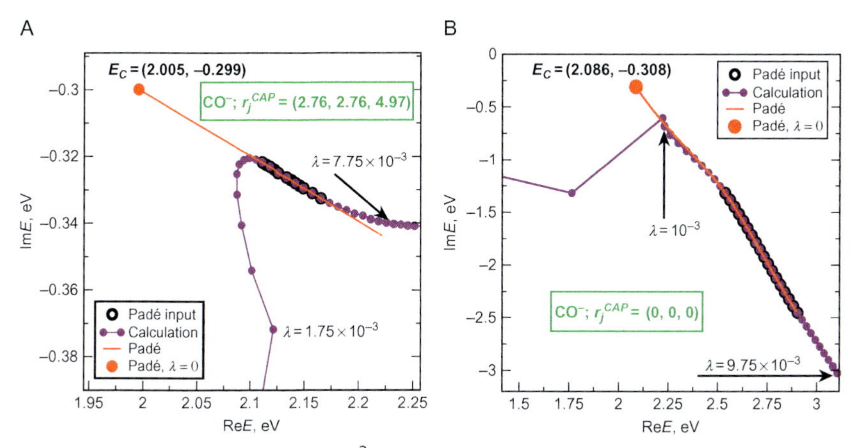

Fig. 1 The λ-trajectories of the CO^{-} $^2\Pi$ shape resonance in *purple*. (A) An exterior CAP with optimal $r^{CAP}_{j=x,y,x}$. (B) A uniform CAP, ie, $r^{CAP}_j = 0$. The *black circles* correspond to the input data used for generating the Padé trajectory (*red*). The resonance energy, E_C, estimated by analytic dilation to $\lambda \to 0$, is marked by a *full red circle*. For convenience, the E_C values are presented, in addition to two λ values that show the directionality of the trajectories.

Fig. 1 illustrates the way to compute and calculate the rm-CAP complex resonance energies. It presents λ-trajectory $[E(\lambda)]$ for CO^{-} with the optimized onset and without one. From this plot, particularly when no onset is used, it is clear that at some point the calculated energy trajectories (in purple) start to show abrupt changes from its "natural" path. The beginning of these abrupt changes is characterization of the critical λ_C, as discussed in Section 2.1. This implies that the calculated values with $\lambda < \lambda_C$ are meaningless, whereas the calculated $E(\lambda > \lambda_C)$ can be used for fitting an energy function, $E(\lambda)$. We use the data marked with black circles to generate such a function using the Padé method. These Padé trajectories, shown by a red line, reproduce the original data and are further dilated to $E(\lambda \to 0)$. These values, marked by red circles, are the estimated complex resonance energies, E_C.

Interestingly, we observe that when the optimal onset is applied, the calculated input values for the Padé are in the neighborhood of E_C (see Fig. 1A), whereas when a uniform CAP is applied the calculated input values are dramatically far from it (see Fig. 1B). This nicely demonstrates the advantage of the exterior-CAP potential over the uniform-CAP; the artificial perturbation generated by the uniform CAP potential is drastically strong. Moreover, it also shows the strength of the rm-CAP approach, which is not affected by the poor values calculated with $r_j^{CAP} = 0$, and yields excellent E_C values that are in agreement with the values obtained with the optimal onset.

An additional advantage of rm-CAP is its significantly lower computational cost as compared to that of dp-CAP. In the rm-CAP method only the right eigenvalue equation needs to be solved, whereas within the dp-CAP the left eigenvalue equation is also needed in order to construct the one-particle density matrix required for calculating the deperturbation. Overall, these features and the presented results make the rm-CAP a very promising approach for calculating resonance positions and widths.

3.2 CPES From a Single Stabilization Graph

An additional approach to calculate resonances is to carry out analytical continuation of eigenvalues from the stabilization graph into the complex plane. In the stabilization calculation, the eigenvalues are computed as a function of some parameters, $[E(\eta)]$, and later are dilated into the complex plane. For example, the eigenvalues can be calculated depending on the number of basis functions[52] or when a finite number of basis functions are scaled by a real factor.[53–55]

Every eigenvalue in the stabilization graph shows a different behavior, depending on its nature: an eigenstate with low or no curvature with respect to the scaling parameter implies that its wavefunction is mainly localized in space. Therefore, such a behavior represents a bound state. In contrast, an eigenstate with a high curvature with respect to the scaling parameter implies that its wavefunction is not localized in space, and therefore represents a continuum state. A resonance state is identified as a combination of both continuum and bound state, ie, its wavefunction is partially localized in space. Therefore, its eigenvalue exhibits low curvature with respect to the scaling parameter at some parts, and high curvature at other parts (see Fig. 2 for an example).

After identifying the resonance state as a function of the scaling parameter, calculations of all order derivatives with respect to the parameter will enable one to calculate its Taylor expansion series -

$$E(\eta) = E(\eta_0) + \sum_n a_n(\eta - \eta_0)^n. \tag{7}$$

If $E(\eta)$ is an analytical function of η it implies that the Taylor series expansion is converged, and η can get complex values in order to calculate the

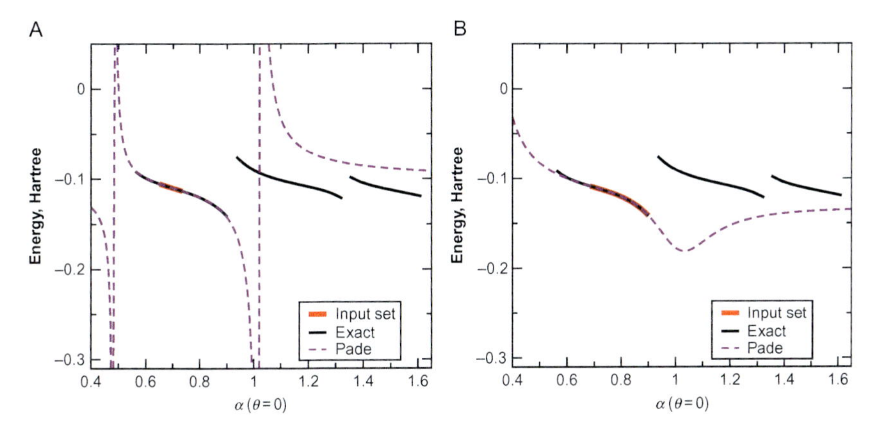

Fig. 2 Exact (*black*) and analytically dilated (*violet*) energy stabilization plots for the hydrogen molecular $1\sigma_u^2$ resonance state ($R = 1.4$ a.u.). The dilated graphs are generated via the Padé approximant using the input data marked in *red*. In (A) the input data are taken from the stable part of the stabilization, while in (B) the input data are taken from the high curvature regions. It is clear that while (A) reproduces the correct structure of the stabilization graph including the avoided crossings, (B) does not.

resonances energies. That is, using data from the stabilization graph, one can calculate the resonances energies, provided that $E(\eta)$ is an analytical function. However, accurately calculating high-order derivatives is not an easy task, and moreover, $E(\eta)$ is not necessarily an analytical function. Therefore, different approaches were explored during the years: in early 1981 Simons used the stabilization calculations to carry out a unitary transformation from the adiabatic energy levels to the diabatic presentation.[56] Doing so, crossings were exposed in the complex plane. These crossings represent branch points (BPs), and their exposure facilitated locating the nearby stationary points (the resonance energy). Later, a simpler approach was introduced by Thompson and Truhlar.[57] Under their approach, a single eigenvalue level from the stabilization graph, including both the stable part and the avoided crossing region, was analytically dilated into the complex plane. However, analytic continuation of a single root failed due to the existence of nonanalytic regions in the stabilization graph,[58] meaning that the avoided crossing region in the stabilization curve forbids such dilation since it is associated with BPs in the complex plane, ie, with nonanalyticity.

In order to overcome this obstacle, Jordan had suggested a method based on the BP structure.[59] Later this method was rigorously derived and modified by McCurdy and McNutt and was referred to as the multieigenvalue method.[58] Doing so, Jordan, McCurdy, and McNutt opened a new research direction for calculating resonances from stabilization graphs, as this multi-eigenvalue method is pursued until today.[60–62] This method takes into consideration the BP structure by using the following truncated characteristic polynomial and data from the avoided crossing region in the stabilization graph -

$$E^2(\eta) + p(\eta)E(\eta) + q(\eta) = 0. \tag{8}$$

Therefore, we refer to this method as the truncated characteristic polynomial (TCP) method (so called GPA in Refs. 60,61). Recently another approach to analytically dilate eigenvalues from the stabilization graph was suggested by Landau et al.[55] This approach avoids the need to calculate high-order derivatives as done in the Taylor expansion series. At the same time, it also avoids the nonanalytical regions in the stabilization graph. In this approach, the Padé approximant is used, and the eigenvalue obtained from the stable part of the stabilization graph is fitted as a function of a real scaling parameter to a ratio between two polynomials, like so -

$$E(\eta) = \frac{P(\eta)}{Q(\eta)}. \tag{9}$$

It is interesting to see that when the focus is not on the avoided crossing regions, but rather on the stable part of the stabilization graph, excellent results are obtained by analytically dilating the energy eigenvalue into the complex plane.[55] Moreover, these excellent results are obtained without using any assumption on the structure of the energy function, unlike the case in Eq. (8). We refer to this method as the resonances via Padé (RVP) method.

3.2.1 The TCP Method

According to McCurdy and McNutt, dilation of a single root is not possible since $E(\eta)$ is not an analytical function; therefore, within their method only the coefficients $p(\eta)$ and $q(\eta)$ in Eq. (8) are dilated into the complex plane. Moreover, Jordan and later McCurdy and McNutt use the fact that the avoided crossing areas are associated with BPs in the complex plane: since these areas result from the interaction of at least two roots of the characteristic polynomial, they set these areas as data points for their dilation into the complex plane.[58,59]

In practice, as Bentley and Chipman clearly explain,[63] a set of M data points from the avoided crossing in the stabilization graph is chosen.

Accordingly, a set of linearly depended equations is constructed. Each equation takes on the following form -

$$E_i^2 + p(\eta_i)E_i + q(\eta_i) = 0, \tag{10}$$

where η_i is one of the M data points for which E_i is an energy value taken from the avoided crossing region in the stabilization graph. $p(\eta_i)$ and $q(\eta_i)$ are taken to be polynomials of η_i in the order of N and $M - N$, respectively, where $N < M$. Therefore, Eq. (10) takes on the following form -

$$E_i^2 + E_i \cdot \sum_{j=1}^{N} p_j(\eta_i)^j + \sum_{k=1}^{M-N} q_k(\eta_i)^k = 0. \tag{11}$$

In order to find the coefficients of p and q, the array of these linear equations needs to be solved. Since these equations can be written in a matrix form, all one needs to assure, in order to find the relevant coefficients, is that this matrix is not singular. To do so, the curve from which the data points E_i is taken must be of high curvature, meaning that the data point, E_i, must be taken from the avoided crossing region and not from the stable part of the stabilization graph. In the next step, in order to find stationary points in the complex plane (resonances), one must find complex η for which $E'(\eta) = 0$, ie, using Eq. (10)[64]

$$E'(\eta) = q'^2(\eta) - p(\eta)p'(\eta)q'(\eta) + p'^2(\eta)q(\eta) = 0. \tag{12}$$

3.2.2 The RVP Method

Recently, Landau et al. showed that a single eigenvalue level that includes only the stable part of the stabilization graph is an analytical function.[55] Meaning that while a single eigenvalue level from the stabilization graph that includes the avoided crossing region is not an analytical function, a single eigenvalue level that includes only the stable part is an analytical function. Moreover, they demonstrated that using only the stable part, one can reproduce the entire stabilization graph (see Fig. 2), that is to say that these stable regions contain all the relevant information for analytical dilation into the complex plane.[55] This implies that one does not need to implement the BP structure in the analytical continuation as done by using the TCP method. On the contrary, the RVP method avoids the BP. In Ref. 55 the existence of an analytical path from the stabilization graph toward a complex stationary point was illustrated to bypasses the BP. Meaning that by using the RVP method one can always remain in an analytic area, and eventually converge to a stationary point in the complex plane. As explained in Ref. 55, the existence of such a path results from using a finite basis set, which is a must.

The RVP method resembles Moiseyev's and Corcoran's work, in which molecular resonances are evaluated by the molecular Hamiltonian matrix elements.[22] In their work, Moiseyev and Corcoran demonstrate that the matrix elements are analytical functions of the scaling factor, even though the operator is not, because the contour of their integration in the complex plane can be chosen to avoid the singular points.[22,24,32,37,65]

In practice, using the RVP method one generates an analytical approximation to $E(\eta)$ by the Schlessinger point method.[45] This method requires a set of M data points (η_i) and their corresponding values $[E(\eta_i)]$. The data points are taken from the stable region in the stabilization plot which excludes the avoided crossing. Once an energy function is fitted, an analytical continuation into the complex plane is performed by choosing η to be complex, ie, $\eta = \alpha e^{i\theta}$. Stationary points are identified by generating α- and θ-trajectories and looking for cusps in the complex plane.[66]

3.2.3 Proof of Concept—Autoionization Feshbach Resonance of the Hydrogen Molecule by the RVP Method

Calculating molecular resonance is a complicated task, which poses a challenge on the regular complex scaling methods.[22–24,37,65] Therefore, it is important to test the RVP method presented here on a molecular system such as the hydrogen $1\sigma_u^2$ molecular resonance. This Feshbach resonance was calculated at an internuclear distance of $R = 1.4$ a.u. for which there are several calculations available for comparison.

In Fig. 3 an optimal cusp for hydrogen $1\sigma_u^2$ molecular resonance is shown. This cusp was obtained through analytical continuation of the stable region in the stabilization plot of this resonance (marked in red in Fig. 2B).

In Table 2 the results are compared with other theoretical works. The results are in an excellent agreement with these works, particularly with the RF-CAP.[31] A good agreement with the complex CI and complex multiconfiguration self-consistent fields (CMCSCF) methods is also observed[67] (for details see Ref. 55).

3.2.4 CPES for Cold Collisions

The previous subsections of this review described a way to calculate molecular resonances at fixed geometries using SQCPs. A repetition of the above

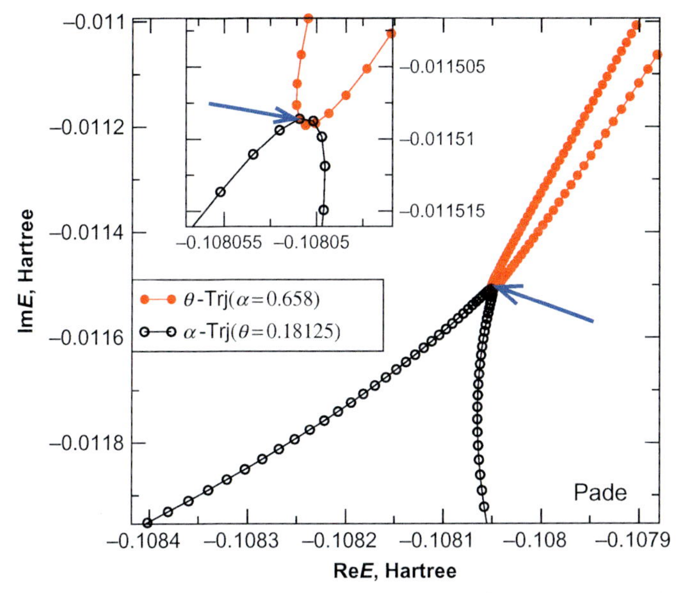

Fig. 3 α-Trajectory (*black*) and θ-trajectory (*red*) obtained from our analytical continuation scheme for the H_2 ($R = 1.4\dot{u}$.) $1\sigma_u^2$ autoionizing resonance. In this figure an obvious cusp is seen (*blue arrow*), indicating a stationary point at $\theta = 0.18125$ and $\alpha = 0.658$. The α-trajectory and θ-trajectory overlap at the cusp, as clearly shown in the *inset*.

Table 2 Position (Re*E*) and Width (Γ) of the $1\sigma_u^2$ Feshbach Resonance of H_2 ($R = 1.4$ a.u.) in Hartree, Re*E* is Presented With Respect to the H_2^+ Ground-State Energy (−0.56994 Hartree)

Method	Re*E*	Γ
Complex CI[a]	0.4630	0.0272
CMCSCF[a]	0.4638	0.0270
RF-CAP[b]	0.4615	0.0227
RVP[c]	0.4614	0.0232

[a]Ref. 67.
[b]Ref. 31.
[c]Ref. 55.

procedure and subsequent mapping of the corresponding resonance position and decay rate as a function of the geometry could generate a CPES. The CPES can be represented as,

$$V_{CPES}(R) = E(R) - \frac{i\Gamma(R)}{2}$$

where V_{CPES} is the complex potential, $E(R)$ and $\Gamma(R)$ are the resonance position and decay rate, respectively. CPESs are important to analyze and interpret certain experimental findings.[1–3] The strength of the above approach is seen in some preliminary results while evaluating the CPES for a polyatomic system. These computed CPESs are used to calculate cross-sections which are in excellent agreement with the measured cold molecular collision experiments.[68]

3.2.5 Summary

By using the TCP method, one can find stationary points in the complex plane (resonances), provided that the input is taken from the avoided crossing region. Meaning that the coefficients for the p and q polynomials must be computed when the input for the TCP method is from the avoided crossing area. In addition, using this approach, the stationary points are identified by solving an algebraic equation (Eq. (12)).

By using the RVP method, one can carry out analytical continuation of a single eigenvalue level into the complex plane, provided that the Padé approximant is used on the stable part of the stabilization graph, ie, the analytical part. Using this area, an analytical path to the stationary points that avoids the BP is assured, and there is no need for extra data from other eigenvalue levels. In this method, stationary points are identified by cusps in the complex plane using α- and θ-trajectories. That is to say, that by using this method one does not have to solve an algebraic equation in order to find the stationary points, but rather needs to explore the surrounding of each points. This approach enables one to gain more information on the stationary point and accurately choose the right one.

In light of the above, in Ref. 55 a new approach for calculating atomic and molecular autoionization resonances was presented. This approach facilitates calculating resonances in a very simple manner by using standard and available ab initio codes, which substantially lower the computational efforts in comparison with non-Hermitian electronic structure codes. However, this approach is limited to the calculations of narrow isolated resonances, since broad and overlapping resonances will not yield stabilization graphs. Therefore, these type of resonances should be calculated by other means, such as complex basis sets or by introducing CAPs into the molecular Hamiltonian.

4. CONCLUDING REMARKS

There is a growing interest in generating molecular electronic CPESs which are required in order to interpret experimental observations such as the recent cold molecular collisions.[1–3] These theoretical CPESs provide the energy positions and autoionization decay rates of a system as function of its geometry, $\mathbf{V}_{\mathbf{CPES}}(\{\mathbf{R_j}\}_{\mathbf{j=1,2,\ldots,N}})$. The real part of this potential stands for the energy position of the molecular system as function of the nuclear coordinates, ie, $\mathbf{E}(\{\mathbf{R_j}\}_{\mathbf{j=1,2,\ldots,N}}) = \mathrm{Re}[\mathbf{V}_{\mathbf{CPES}}(\{\mathbf{R_j}\}_{\mathbf{j=1,2,\ldots,N}})]$. The imaginary part of this potential is the autoionization decay rate of the molecular system as function of the nuclear coordinates, ie, $\boldsymbol{\Gamma}(\{\mathbf{R_j}\}_{\mathbf{j=1,2,\ldots,N}}) = -2\mathrm{Im}[\mathbf{V}_{\mathbf{CPES}}(\{\mathbf{R_j}\}_{\mathbf{j=1,2,\ldots,N}})]$. Since a transition from one geometrical configuration to another may result in dramatic changes in the autoionization decay rate, and since electron correlation plays an important role in evaluating CPESs,[6] it is crucial to be able to accurately calculate CPESs. Preferably, CPES should be calculated via ab initio methods, which provide accurate, robust, and reliable electronic eigenvalues that can be used to construct smooth CPESs.

In this review, based on our most recent research, two methods for describing resonance states, using state-of-the-art electronic structure, are discussed. Both of them are based on standard quantum chemistry calculations using commercial packages or in-house codes. In the first method, described here, we remove the artificial effect of V_{CAP}, which has been introduced into the original molecular Hamiltonian. In the second method, we use the stabilization graph in order to calculate CPESs. This method uses an even more standard quantum chemistry theory in order to calculate the CPESs, since it utilizes the Hermitian (real) Hamiltonian that was originally developed for the calculation of electronic bound (both ground and excited) states. Standard (Hermitian and real) codes are highly optimized; hence, this method is extremely efficient.

In our work, so far, we have used Q-Chem[44] that allows calculations of variety of open-shell multiconfiguration electronic states using the versatile implementation of the EOM-CC family of methods[69] at the singles and doubles level including triples corrections.[70,71] Furthermore, Q-Chem was recently implemented with CAP-augmented EOM-CC that was used for our first approach.[10–12]

The first method, describing the removal of the CAP perturbation, employs the CAP-EOM-EA-CCSD approach, which is suitable for calculating autodetachment for transient negative molecular ions. In this method, the artificial CAP introduces similar conditions to imposing OBCs that absorbs free electrons. However, since the CAP potential is artificial it creates unphysical perturbation that introduces errors into the calculations of resonance energies. Under the assumption that the resonance energy is an analytical function of the CAP-strength parameter, we fit a function to the energies calculated with different CAP strengths using the Padé approximant. Once an analytical energy function is found, we set the CAP-strength parameter to zero. Our preliminary calculations for the CO^- (and N_2^-, not shown) using rm-CAP show good agreement with experimental results and other theoretical schemes. Further, we discuss the advantages of our approach in terms of computational effort and theoretical aspects, in relation to the other CAP-removing method.[43]

In the second method presented here, the starting point is a stabilization graph calculation at every molecular geometry.[55] This calculation uses standard (Hermitian and real) quantum chemistry codes, where electronic energy levels are calculated as a function of a scaling parameter. The fingerprints of a resonance state are easily recognized in the calculated stabilization graph. The procedure continues by taking finite number of data points on a single stabilization graph, which are away from the avoided crossings, and fitting an analytical function to them using the Padé approximant. The ability of the Padé expression to describe (or reproduce) the entire multistabilization graphs is a numerical proof that the Padé algebraic expression is an analytical function of the scaling parameter. By setting *complex* values as the scaling parameter in the Padé algebraic expression [$\eta = \alpha \exp(i\theta_{CS})$], complex energies are obtained. Resonance energies are identified when the two scaling parameters (α and θ_{CS}) are varied to obtain stationary solutions at every geometry of the molecular system. Preliminary results, not shown, demonstrate the ability of the RVP method to yield an accurate ab initio polyatomic CPES. Its accuracy is validated by computing reaction cross-sections, thus, allowing direct comparison with experiment. Our pilot results show an excellent agreement with cold-chemistry experiments.[1–3]

The two methods described here for calculating CPESs from SQCPs open a door for a production-level calculation of shape and Feshbach-autoionization metastable electronic states and for neutral and anionic molecular systems, in their ground and excited electronic states.

ACKNOWLEDGMENTS

This research was supported by the I-Core: the Israeli Excellence Center "Circle of Light," by the Israel Science Foundation grants Nos. 298/11 and 1530/15.

Per-Olov Löwdin is one of the founders of Quantum Chemistry. One of us (N.M.) had the privilege to participate in the 1978 Sanibel Symposium; at that time N.M. was a postdoctoral fellow in the United States. Per-Olov Löwdin encouraged young scientist to particulate and represent their results in the international conferences which were conducted by him for many years. Among the first invitations N.M. got as an young scientist was to spend 1 month in Uppsala in 1984. N.M. had the opportunity to observe Per-Olov Löwdin and hear him during his lectures to the students and postdocs who participated in the Uppsala Summer School in Quantum Chemistry at that year. Besides being a great scientist, he was a great lecturer and had the "know-how" to attract the attention of young participants and influence them to get into research fields he felt strongly about since he believed they might contribute new research directions. Our contribution to this volume is a review on new methods for calculations of molecular resonances by complex scaling, and this chapter is dedicated to memory of Per-Olov Löwdin who is a role model to young scientists for his love for science and people.

REFERENCES

1. Henson, A. B.; Gersten, S.; Shagam, Y.; Narevicius, J.; Narevicius, E. Observation of Resonances in Penning Ionization Reactions at Sub-Kelvin Temperatures in Merged Beams. *Science* **2012**, *338*(6104), 234–238.
2. Lavert-Ofir, E.; Shagam, Y.; Henson, A. B.; Gersten, S.; Kłos, J.; Żuchowski, P. S.; Narevicius, J.; Narevicius, E. Observation of the Isotope Effect in Sub-Kelvin Reactions. *Nat. Chem.* **2014**, *6*(4), 332–335.
3. Shagam, Y.; Klein, A.; Skomorowski, W.; Yun, R.; Averbukh, V.; Koch, C. P.; Narevicius, E. Molecular Hydrogen Interacts More Strongly When Rotationally Excited at Low Temperatures Leading to Faster Reactions. *Nat. Chem.* **2015**, 7, 921–926.
4. Trinter, F.; Schöffler, M.; Kim, H.-K.; Sturm, F.; Cole, K.; Neumann, N.; Vredenborg, A.; Williams, J.; Bocharova, I.; Guillemin, R.; Simon, M.; Belkacem, A.; Landers, A. L.; Weber, T.; Schmidt-Böcking, H.; Dörner, R.; Jahnke, T. Resonant Auger Decay Driving Intermolecular Coulombic Decay in Molecular Dimers. *Nature* **2014**, *505*(7485), 664–666.
5. Gokhberg, K.; Kolorenč, P.; Kuleff, A. I.; Cederbaum, L. S. Site-and Energy-Selective Slow-Electron Production Through Intermolecular Coulombic Decay. *Nature* **2014**, *505*(7485), 661–663.
6. Scheit, S.; Averbukh, V.; Meyer, H.-D.; Moiseyev, N.; Santra, R.; Sommerfeld, T.; Zobeley, J.; Cederbaum, L. S. On the Interatomic Coulombic Decay in the Ne Dimer. *J. Chem. Phys.* **2004**, *121*(17), 8393.
7. Moiseyev, N. *Non-Hermitian Quantum Mechanics*. Cambridge University Press: Cambridge, 2011.
8. Santra, R.; Cederbaum, L. S. Non-Hermitian Electronic Theory and Applications to Clusters. *Phys. Rep.* **2002**, *368*(1), 1–117.
9. Muga, J. G., Palao, J. P.; Navarro, B.; Egusquiza, I. L. Complex Absorbing Potentials. *Phys. Rep.* **2004**, *395*(6), 357–426.
10. Jagau, T.-C.; Zuev, D.; Bravaya, K. B.; Epifanovsky, E.; Krylov, A. I. A Fresh Look at Resonances and Complex Absorbing Potentials: Density Matrix-Based Approach. *J. Phys. Chem. Lett.* **2013**, *5*(2), 310–315.
11. Jagau, T.-C.; Krylov, A. I. Complex Absorbing Potential Equation-of-Motion Coupled-Cluster Method Yields Smooth and Internally Consistent Potential Energy Surfaces and Lifetimes for Molecular Resonances. *J. Phys. Chem. Lett.* **2014**, *5*(17), 3078–3085.
12. Zuev, D.; Jagau, T.-C.; Bravaya, K. B.; Epifanovsky, E.; Shao, Y.; Sundstrom, E.; Head-Gordon, M.; Krylov, A. I. Complex Absorbing Potentials Within EOM-CC Family of Methods: Theory, Implementation, and Benchmarks. *J. Chem. Phys.* **2014**, *141*(2), 024102.
13. Sajeev, Y.; Ghosh, A.; Pal, S.; Vaval, N. Coupled Cluster Methods for Autoionisation Resonances. *Int. Rev. Phys. Chem.* **2014**, *33*(5–6), 397–425.
14. Jagau, T.-C.; Dao, D. B.; Holtgrewe, N. S.; Krylov, A. I.; Mabbs, R. Same but Different: Dipole-Stabilized Shape Resonances in CuF^- and AgF^-. *J. Phys. Chem. Lett.* **2015**, *6*(14), 2786–2793.

15. Jagau, T.-C.; Krylov, A. I. Characterizing Metastable States Beyond Energies and Lifetimes: Dyson Orbitals and Transition Dipole Moments. *J. Chem. Phys.* **2016**, *144*(5), 054113.
16. Ehara, M.; Sommerfeld, T. CAP/SAC-CI Method for Calculating Resonance States of Metastable Anions. *Chem. Phys. Lett.* **2012**, *537*, 107–112.
17. Ghosh, A.; Vaval, N.; Pal, S. Equation-of-Motion Coupled-Cluster Method for the Study of Shape Resonance. *J. Chem. Phys.* **2012**, *136*(23), 234110.
18. Ghosh, A.; Karne, A.; Pal, S.; Vaval, N. CAP/EOM-CCSD Method for the Study of Potential Curves of Resonant States. *Phys. Chem. Chem. Phys.* **2013**, *15*(41), 17915–17921.
19. Ghosh, A.; Pal, S.; Vaval, N. Interatomic Coulombic Decay in ($n = 2$-3) Clusters Using CAP/EOM-CCSD Method. *Mol. Phys.* **2014**, *112*(5–6), 669–673.
20. Krause, P.; Sonk, J. A.; Schlegel, H. B. Strong Field Ionization Rates Simulated With Time-Dependent Configuration Interaction and an Absorbing Potential. *J. Chem. Phys.* **2014**, *140*(17), 174113.
21. McCurdy, C. W., Jr.; Rescigno, T. N. Extension of the Method of Complex Basis Functions to Molecular Resonances. *Phys. Rev. Lett.* **1978**, *41*(20), 1364.
22. Moiseyev, N.; Corcoran, C. Autoionizing States of H_2 and H_2^- Using the Complex-Scaling Method. *Phys. Rev. A* **1979**, *20*(3), 814–817.
23. Simon, B. The Definition of Molecular Resonance Curves by the Method of Exterior Complex Scaling. *Phys. Lett. A* **1979**, *71*(2), 211–214.
24. Morgan, J. D.; Simon, B. The Calculation of Molecular Resonances by Complex Scaling. *J. Phys. B At. Mol. Phys.* **1981**, *14*(5), L167.
25. Riss, U. V.; Meyer, H.-D. Calculation of Resonance Energies and Widths Using the Complex Absorbing Potential Method. *J. Phys. B At. Mol. Phys.* **1993**, *26*(23), 4503.
26. Lefebvre, R.; Sindelka, M.; Moiseyev, N. Resonance Positions and Lifetimes for Flexible Complex Absorbing Potentials. *Phys. Rev. A* **2005**, *72*(5), 052704.
27. Sajeev, Y.; Vysotskiy, V.; Cederbaum, L. S.; Moiseyev, N. Continuum Remover-Complex Absorbing Potential: Efficient Removal of the Nonphysical Stabilization Points. *J. Chem. Phys.* **2009**, *131*, 211102.
28. Ben-Asher, A.; Moiseyev, N. On the Equivalence of Different Methods for Calculating Resonances: From Complex Gaussian Basis Set to Reflection-Free Complex Absorbing Potentials via the Smooth Exterior Scaling Transformation. *J. Chem. Theory Comput.* **2016**, *12*(6), 2542–2552.
29. Moiseyev, N. Derivations of Universal Exact Complex Absorption Potentials by the Generalized Complex Coordinate Method. *J. Phys. B Atomic Mol. Phys.* **1998**, *31*(7), 1431.
30. Sajeev, Y.; Sindelka, M.; Moiseyev, N. Reflection-Free Complex Absorbing Potential for Electronic Structure Calculations: Feshbach Type Autoionization Resonance of Helium. *Chem. Phys.* **2006**, *329*(1), 307–312.
31. Sajeev, Y.; Moiseyev, N. Reflection-Free Complex Absorbing Potential for Electronic Structure Calculations: Feshbach-Type Autoionization Resonances of Molecules. *J. Chem. Phys.* **2007**, *127*(3), 034105.
32. Rescigno, T. N.; McCurdy, C. W., Jr.; Orel, A. E. Extensions of the Complex-Coordinate Method to the Study of Resonances in Many-Electron Systems. *Phys. Rev. A* **1978**, *17*(6), 1931.
33. Kaprálová-Žďánská, P. R. A Study of Complex Scaling Transformation Using the Wigner Representation of Wavefunctions. *J. Chem. Phys.* **2011**, *134*(20), 204101.
34. Kaprálová-Žďánská, P. R.; Šmydke, J. Gaussian Basis Sets for Highly Excited and Resonance States of Helium. *J. Chem. Phys.* **2013**, *138*(2), 024105.
35. Landau, A.; Haritan, I.; Kaprálová-Žďánská, P. R.; Moiseyev, N. Advantages of Complex Scaling Only the Most Diffuse Basis Functions in Simultaneous Description of Both Resonances and Bound States. *Mol. Phys.* **2015**, *113*(19–20), 3141–3146.

36. Moiseyev, N.; Hirschfelder, J. O. Representation of Several Complex Coordinate Methods by Similarity Transformation Operators. *J. Chem. Phys.* **1988**, *88*(2), 1063–1065.
37. McCurdy, C. W.; Rescigno, T. N. Complex-Basis-Function Calculations of Resolvent Matrix Elements: Molecular Photoionization. *Phys. Rev. A* **1980**, *21*(5), 1499.
38. McCurdy, C. W.; Mowrey, R. C. Complex Potential-Energy Function for the $^2\Sigma_u^+$ Shape Resonance State of H_2^- at the Self-Consistent-Field Level. *Phys. Rev. A* **1982**, *25*(5), 2529.
39. White, A. F.; McCurdy, C. W.; Head-Gordon, M. Restricted and Unrestricted Non-Hermitian Hartree-Fock: Theory, Practical Considerations, and Applications to Metastable Molecular Anions. *J. Chem. Phys.* **2015**, *143*(7), 074103.
40. Nicolaides, C. A.; Beck, D. R. The Variational Calculation of Energies and Widths of Resonances. *Phys. Lett. A* **1978**, *65*(1), 11–12.
41. Gyarmati, B.; Vertse, T. On the Normalization of Gamow Functions. *Nucl. Phys. A* **1971**, *160*(3), 523–528.
42. White, A. F.; Head-Gordon, M.; McCurdy, C. W. Complex Basis Functions Revisited: Implementation With Applications to Carbon Tetrafluoride and Aromatic N-Containing Heterocycles Within the Static-Exchange Approximation. *J. Chem. Phys.* **2015**, *142*(5), 054103.
43. Landau, A.; Moiseyev, N. Molecular Resonances by Removing Complex Absorbing Potentials via Padé; Application to CO− and N_2^-. *J. Chem. Phys.* **2016**, *145*, 164111.
44. Shao, Y.; Gan, Z.; Epifanovsky, E.; Gilbert, A. T.; Wormit, M.; Kussmann, J.; Lange, A. W.; Behn, A.; Deng, J.; Feng, X.; et al. Advances in Molecular Quantum Chemistry Contained in the Q-Chem 4 Program Package. *Mol. Phys.* **2015**, *113*(2), 184–215.
45. Schlessinger, L. Use of Analyticity in the Calculation of Nonrelativistic Scattering Amplitudes. *Phys. Rev.* **1968**, *167*(5), 1411–1423.
46. Moiseyev, N.; Friedland, S. Association of Resonance States With the Incomplete Spectrum of Finite Complex-Scaled Hamiltonian Matrices. *Phys. Rev. A* **1980**, *22*(2), 618.
47. Ehrhardt, H.; Langhans, L.; Linder, F.; Taylor, H. S. Resonance Scattering of Slow Electrons From H_2 and CO Angular Distributions. *Phys. Rev.* **1968**, *173*(1), 222.
48. Zubek, M.; Szmytkowski, C. Calculation of Resonant Vibrational Excitation of CO by Scattering of Electrons. *J. Phys. B At. Mol. Phys.* **1977**, *10*(1), L27.
49. Zubek, M.; Szmytkowski, C. Electron Impact Vibrational Excitation of CO in the Range 1-4 eV. *Phys. Lett. A* **1979**, *74*(1), 60–62.
50. Jagau, T.-C.; Zuev, D.; Bravaya, K. B.; Epifanovsky, E.; Krylov, A. I. Correction to "A Fresh Look at Resonances and Complex Absorbing Potentials: Density Matrix-Based Approach" *J Phys, Chem. Lett.* **2015**, *6*(19), 3866.
51. Zuev, D.; Jagau, T.-C.; Bravaya, K. B.; Epifanovsky, E.; Shao, Y.; Sundstrom, E.; Head-Gordon, M.; Krylov, A. I. Erratum: "Complex Absorbing Potentials Within EOM-CC Family of Methods. Theory, Implementation, and Benchmarks". *J. Chem. Phys.* **2015**, *143*(14), 9901. (*J. Chem. Phys.* **2014**, *141*, 024102).
52. Holøien, E.; Midtdal, J. New Investigation of the $^1S^e$ Autoionizing States of He and H^-. *J. Chem. Phys.* **1966**, *45*(6), 2209–2216.
53. Taylor, H. S. Models, Interpretations, and Calculations Concerning Resonant Electron Scattering Processes in Atoms and Molecules. *Adv. Chem. Phys.* **1971**, *18*, 91–147.
54. Taylor, H. S.; Hazi, A. U. Comment on the Stabilization Method: Variational Calculation of the Resonance Width. *Phys. Rev. A* **1976**, *14*(6), 2071.
55. Landau, A.; Haritan, I.; Kaprálová-Zdánská, P. R.; Moiseyev, N. Atomic and Molecular Complex Resonances From Real Eigenvalues Using Standard (Hermitian) Electronic Structure Calculations. *J. Phys. Chem. A* **2016**, *120*(19), 3098–3108.
56. Simons, J. Resonance State Lifetimes From Stabilization Graphs. *J. Chem. Phys.* **1981**, *75*(5), 2465–2467.

57. Thompson, T. C.; Truhlar, D. G. New Method for Estimating Widths of Scattering Resonances From Real Stabilization Graphs. *Chem. Phys. Lett.* **1982**, *92*(1), 71–75.
58. McCurdy, C. W.; McNutt, J. F. On the Possibility of Analytically Continuing Stabilization Graphs to Determine Resonance Positions and Widths Accurately. *Chem. Phys. Lett.* **1983**, *94*(3), 306–310.
59. Jordan, K. D. Construction of Potential Energy Curves in Avoided Crossing Situations. *Chem. Phys.* **1975**, *9*(1), 199–204.
60. Chao, J.-Y.; Falcetta, M. F.; Jordan, K. D. Application of the Stabilization Method to the $N_2^-(1^2\Pi_g)$ and $Mg^-(1^2P)$ Temporary Anion States. *J. Chem. Phys.* **1990**, *93*(2), 1125–1135.
61. Falcetta, M. F.; DiFalco, L. A.; Ackerman, D. S.; Barlow, J. C.; Jordan, K. D. Assessment of Various Electronic Structure Methods for Characterizing Temporary Anion States: Application to the Ground State Anions of N_2, C_2H_2, C_2H_4, and C_6H_6. *J. Phys. Chem. A* **2014**, *118*(35), 7489–7497.
62. Jordan, K. D.; Voora, V. K.; Simons, J. Negative Electron Affinities From Conventional Electronic Structure Methods. *Theor. Chem. Acc.* **2014**, *133*(3), 1–15.
63. Bentley, J.; Chipman, D. M. Accurate Width and Position of Lowest ^1S Resonance in H^- Calculated From Real-Valued Stabilization Graphs. *J. Chem. Phys.* **1987**, *86*(7), 3819–3828.
64. Isaacson, A. D.; Truhlar, D. G. Single-Root, Real-Basis-Function Method With Correct Branch-Point Structure for Complex Resonances Energies. *Chem. Phys. Lett.* **1984**, *110*(2), 130–134.
65. McCurdy, C. W. Complex-Coordinate Calculation of Matrix Elements of the Resolvent of the Born-Oppenheimer Hamiltonian. *Phys. Rev. A* **1980**, *21*(2), 464.
66. Moiseyev, N.; Friedland, S.; Certain, P. R. Cusps, θ Trajectories, and the Complex Virial Theorem. *J. Chem. Phys.* **1981**, *74*(8), 4739–4740.
67. Yabushita, S.; McCurdy, C. W. Feshbach Resonances in Electron-Molecule Scattering by the Complex Multiconfiguration SCF and Configuration Interaction Procedures: The $^1\Sigma_g^+$ Autoionizing States of H_2. *J. Chem. Phys.* **1985**, *83*(7), 3547–3559.
68. Bhattacharya, D.; Landau, A.; Haritan, I.; Ben-Asher, A.; Pawlak, M.; Moiseyev, N. *Ab-initio* complex potential energy surface provides accurate cross-sections for ultracold atom-molecule collisions (to be submitted).
69. Krylov, A. I. Equation-of-Motion Coupled-Cluster Methods for Open-Shell and Electronically Excited Species: The Hitchhiker's Guide to Fock Space. *Phys. Chem.* **2008**, *59*(1), 433–462.
70. Manohar, P. U.; Krylov, A. I. A Noniterative Perturbative Triples Correction for the Spin-Flipping and Spin-Conserving Equation-of-Motion Coupled-Cluster Methods With Single and Double Substitutions. *J. Chem. Phys.* **2008**, *129*(19), 194105.
71. Manohar, P. U.; Stanton, J. F.; Krylov, A. I. Perturbative Triples Correction for the Equation-of-Motion Coupled-Cluster Wave Functions With Single and Double Substitutions for Ionized States: Theory, Implementation, and Examples. *J. Chem. Phys.* **2009**, *131*(11), 114112.

CHAPTER SIXTEEN

High-Resolution Quantum-Mechanical Signal Processing for in vivo NMR Spectroscopy

Dževad Belkić*[,1], Karen Belkić*[,†,‡]

*Karolinska Institute, Stockholm, Sweden

[†]School of Community and Global Health, Claremont Graduate University, Claremont, CA, United States

[‡]Institute for Health Promotion and Disease Prevention Research, University of Southern California School of Medicine, Los Angeles, CA, United States

[1]Corresponding author: e-mail address: dzevad.belkic@ki.se

Contents

Abstract

High-resolution quantitative signal analysis using the fast Padé transform (FPT) is applied to a specific problem (cerebral asphyxia) within magnetic resonance spectroscopy (MRS) for in vivo pediatric neurodiagnostics. Potential broader implications for the presented methodology are indicated for interdisciplinary research, including quantum chemistry. An iterative averaging procedure is introduced and validated, which could be automatically built-in, to provide denoised spectra, so vitally needed in clinical MRS. The full equivalence of nonparametrically and parametrically generated total shape spectra in the FPT is demonstrated. With subsequent parametric analysis,

Advances in Quantum Chemistry, Volume 74
ISSN 0065-3276
http://dx.doi.org/10.1016/bs.aiq.2016.06.004

exceedingly dense component spectra are reliably reconstructed, both with the mixture of absorption and dispersion components ("usual" mode) and by setting the reconstructed phases to zero, in order to eliminate interference effects ("ersatz" mode). Via the ersatz components, the consequences and extent of the said interference effect are distinctly visualized for every overlap of closely located resonances or hidden resonances. Practical implementation of Padé-optimized MRS from in vivo encoded time signals in the clinical setting is hereby demonstrated.

ABBREVIATIONS

Ace acetate
Ala alanine
ARMA autoregressive moving average
Asp aspartate
Au arbitrary units
BW bandwidth
Cho choline
Cr creatine
DFT discrete Fourier transform
FFT fast Fourier transform
FID free induction decay
FPT fast Padé transform
FWHM full width at half maximum
GABA gamma amino butyric acid
GE general electric
Gln glutamine
Glu glutamate
Glx glutamine plus glutamate
IDFT inverse discrete Fourier transform
IFFT inverse fast Fourier transform
Lac lactate
Leu leucine
Lip lipids
m-Ins myoinositol
MR magnetic resonance
MRI magnetic resonance imaging
MRS magnetic resonance spectroscopy
ms milliseconds
NAA *N*-acetyl aspartate
NAAG *N*-acetyl aspartyl glutamic acid
NMR nuclear magnetic resonance
PA Padé approximant
PC phosphocholine
PCr phosphocreatine
PE phosphoethanolamine
ppm parts per million
PRESS point-resolved spectroscopy sequence
s-Ins scylloinositol

SNR signal–noise ratio
SNS signal–noise separation
SVD singular value decomposition
Tau taurine
TE echo time
TR repetition time
Val valine

1. INTRODUCTION

From the onset of its development, signal processing as a huge interdisciplinary field, which was originated by Prony in 1797, has the rational polynomials in the form of the Padé approximant (PA), as the centrally important response function to external perturbations of general systems.[1,2] Prony considered that time evolution of all phenomena can ultimately be described by a linear combination of real-valued exponentials. Likewise, the universally valid quantum mechanics also predicts this latter formula, via the autocorrelation functions, except that the exponentials are complex-valued. Transforming this time domain ansatz to the frequency domain directly yields the PA, which was known to Prony nearly a century before Padé, who under the supervision of Hermite, did his PhD Thesis in 1892 on polynomial quotients and the related tables. By design, the PA is always exact for any experimental data whose energy or frequency parametrized description is naturally rooted in the class of rational polynomials. Such experimental data are ubiquitous in interdisciplinary research and applied fields. This leaves little doubt as to which signal processing model is the method of choice; the PA is interchangeably called the fast Padé transform (FPT) in signal processing.[1,2]

It is interesting to note that Löwdin rediscovered both the Prony method and the PA. In 1956,[3] while studying multielectron atoms and ions, he proposed an improved method using an analytical approximation to parametrize the numerically available self-consistent field wavefunctions. In the exact derivation of his analytical functions, Prony's sum of real-valued exponentials was obtained. In 1965,[4] as rightly pointed out by Brändas and Goscinski,[5] the PA was implicitly present in the Löwdin partitioning and inner projection method, which combines the perturbative and variational procedures.

The field of nuclear magnetic resonance (NMR) is one of the most remarkable examples of multidisciplinarity. There is no essential difference in the methodology between NMR in analytical chemistry and magnetic resonance spectroscopy (MRS) in medical diagnostics. The word "nuclear" has been dropped within the medical arena for MRS. In chemical laboratories, NMR and in medicine MRS differ in the instrumentation. Spectrometers for NMR are relatively small in accordance with the size of the suspended samples, while in medicine, the same scanners serve both magnetic resonance imaging (MRI) and MRS and must be sufficiently large to accommodate patients. Crucially, however, the problems germane to NMR are shared by MRS, especially regarding the mathematical analysis of encoded data. Methods developed herein are of general applicability and can be used for signal processing in many other fields that need not have any connection whatsoever with the magnetic resonance (MR) phenomenon itself. Apart from the practical implications of arriving at highly resolved spectra and images within MRS and MRI, introducing quantum mechanics into signal processing provides the fundamental framework of a complete theory of physics. Thereby, it becomes possible to directly relate arbitrary signals to the dynamics of the examined system and its time evolution described by the first principles of physics. For highly complex systems, the dynamics might be either unknown or unwieldy. Nevertheless, whenever there are experimental data, eg, time signals, quantum mechanics offers the possibility to look into the dynamics of the system and extract the parametrized full spectral information. This is possible due to the equivalence between time signals and autocorrelation functions. With this equivalence, the typical problem of nonlinear fitting of experimental data becomes, instead, a quantum-mechanical search for eigen spectra of the studied system. The obtained results reconstruct the unknown dynamics and interactions in the system which has undergone transitions due to the effects of perturbations, before generating the recorded time signals. Clearly, we are now speaking of an inverse problem in quantum mechanics, whereby one has experimentally measured data, from which the interaction potentials, or, more broadly speaking, the underlying causes are to be retrieved from the observed effects.[1,2]

One of the main challenges in MRS is to reliably solve the quantification problem. This problem amounts to unequivocally reconstructing the unknown, hidden, quantitative information from data encoded from patients for the purpose of diagnostics. In this context, one measures a quantity (time signal) from which no discernable clinical information is directly available. The reason is that the time signal appears only as heavily packed, multiple, damped sinusoidal oscillations. This immediately demands certain mathematical transforms of such measured data in order to extract the sought information. These transforms map the encoded data from the original time domain to a complementary and potentially more discernable domain called the frequency domain or the domain of spectra. The former and the latter domains are the subject of measurement and theory, respectively. It is in these computed frequency spectra that the desired clinical information becomes qualitatively apparent and teased out from the measured time signals. This is manifested by the emergence of a number of peaks (or resonances) in a spectrum. These peaks contain the sought information which is, in principle, more amenable to analysis and interpretation in the frequency than in the time domain, due to improved transparency.

The high-resolution quantum-mechanical method, the FPT, is applied herein to a specific problem within MRS for pediatric neurodiagnostics. We will now very briefly consider clinical aspects of that problem area with the main focus on cerebral asphyxia. This is a brain impairment due to oxygen deprivation usually at birth. Cerebral neurons exhibit a pronounced vulnerability to hypoxia which can be detected by MRS through a significant reduction of concentration of NAA, as will presently be shown in the Results section. By detecting the metabolic features of normal and pathologic tissues, MRS can substantially improve the specificity of MRI. Thus far, pediatric neurodiagnostics via MRS have been mainly based upon assessments of a few metabolites (molecules) and their concentration ratios. Among the most informative MR-visible metabolites is choline (Cho), whose resonant frequency is ~3.2 ppm (parts per million), reflecting phospholipid metabolism of cell membranes and being a marker for membrane damage, cellular proliferation, and cell density. The abundance and viability of neurons are indicated by nitrogen acetyl aspartate (NAA), resonating at ~2.0 ppm. Creatine (Cr), which resonates at ~3.0 ppm, is a marker of cerebral energy metabolism, and its concentration is usually stable after the first year of life.[6] Due to the predominance of anaerobic glycolysis, a lactate (Lac) doublet, centered at ~1.3 ppm, is expected in cerebral ischaemia/hypoxia, but Lac can also be observed in healthy newborns and in brain tumors.[6–8] With reperfusion after hypoxia, lipids (Lip), resonating near 1.3 ppm, often appear, as well.[6] Brain metabolite concentrations and their ratios assessed via MRS differ substantially according to the child's age. In newborns, myoinositol (m-Ins) is the dominant brain metabolite, while Cho is the largest peak in older infants. Further, Cr and NAA concentrations increase as the brain matures. Accordingly, with the child's age, Cho to NAA and Cho to Cr concentration ratios normally diminish.[9] Improved pediatric neurodiagnostics through MRS, particularly better possibilities to identify hypoxia/ischaemia and to clearly distinguish this from other pathology, is one rationale for the clinical focus of this paper. Another important clinical issue arises, with direct relevance to oncology. Namely, hypoxic regions may occur within tumors, and these indicate resistance to radiotherapy, as well as to chemotherapy. Hypoxia also drives genomic instability and is linked to progression with invasive/metastatic disease.[10] Thus, identifying these hypoxic regions via MRS could contribute to better cancer treatment planning not only for brain tumors but also for other malignancies. Such an important potential of MRS motivates efforts to glean maximal information from this diagnostic method. Critical to these efforts are not just improved technological advances, important as these may be, but also to explore what mathematical advances in signal processing could offer to MRS.[2] First, let us briefly examine the current signal processing method used in MRS.

1.1 Why Are the Current Signal Processing Methods Within MRS of Limited Value?

An averaged MRS time signal, ie, a free induction decay (FID)-digitized curve, has been encoded herein in vivo on a 1.5 T MR scanner from the brain of an 18-month-old child. The full metabolic content of the scanned brain tissue is reflected in the encoded time domain data, but this information is difficult to interpret from the FID, since its components are tightly packed in the harmonically oscillating and exponentially attenuated waveforms. When the time signal is mapped through mathematical transforms into the frequency domain, the resulting spectrum interpretation becomes possible. One such mapping is done automatically in MR scanners via the fast Fourier transform (FFT).

Although computationally speedy, the FFT is a low resolution signal processor, importing noise directly from the FID to the frequency domain. The Fourier spectrum is linear, as it is given by a single polynomial:

$$\text{FFT}: F_m = \sum_{n=0}^{N-1} c_n \exp(-2\pi i m n/N), \quad \text{IFFT}: c_n = \frac{1}{N}\sum_{m=0}^{N-1} F_m \exp(2\pi i m n/N). \quad (1)$$

Here, $2\pi\ m/T$ is the fixed mth Fourier grid frequency, and IFFT denotes the inverse FFT. Further, $\{c_n\}$ is the set of complex-valued time signal points, T is the total signal duration or total acquisition time, $T=N\tau$, where N is the total signal length, and τ is the sampling time (dwell time, sampling rate), which is the inverse of the bandwidth (BW). The variables $\exp(\pm 2\pi\ imn/N)$ are the undamped sinusoids and cosinusoids ($nm\tau/T=nm/N$). An FFT stick spectrum is constructed only on the fixed grid points with no resolution improvement by interpolation, nor can there be any extrapolation, ie, no information can be predicted past the last encoded FID point, c_{N-1}. Zero-filling of an FID yields merely the sinc-type wiggles on the baseline of an FFT spectrum. As is well-known, the speed of the FFT algorithm is gained for signals lengths of the composite form, $N=2^m$ ($m=1, 2, 3, \ldots$). When N is noncomposite, ie, any positive integer, the FFT and IFFT from (1) become the discrete Fourier transform (DFT) and the inverse DFT, namely, IDFT.

Since the FFT is nonparametric, it can produce only a total shape spectrum (envelope), but cannot generate spectral parameters needed for quantification which is the main goal of MRS. Typically, the FFT spectrum is fitted to a preselected set of Lorentzians, Gaussians, or their linear combination mimicking the Voigt profile. Metabolite concentrations are subsequently estimated via least square free-parameter-adjusting techniques. Thereby any number of preassigned peaks can be fitted to the given envelope (Lanczos' paradox).[1] Consequently, physical metabolites may be missed (under-modeling), while unphysical ones may be spuriously retrieved (over-modeling).[1,2] The waveforms from some informative metabolites for pediatric neurodiagnostics, eg, m-Ins, glutamate (Glu), glutamine (Gln)[a], and Lip rapidly decay, and are detected only at short echo times (TE)[11] at which fitting becomes even more troublesome. Recall, Lac at ~1.3 ppm generally appears with ischaemia/hypoxia. At short TE, there may be overlap of Lac with Lip which also resonates in the chemical shift region of ~1.3 ppm. Changes in TE can affect peak height ratios since spin–spin T_2*-relaxation times of various metabolites differ. Reliance upon metabolite ratios becomes even more problematic; eg, besides NAA at ~2.05 ppm, metabolites Glu, Gln within 2.1–2.5 ppm at short TE may contribute to the spectral area around 2.0 ppm.[12] Thus, assessing NAA levels at short TE becomes more difficult due to heavily overlapping resonances that cannot be reliably identified by the FFT followed by fitting. These considerations regarding the relation of metabolite concentration ratios and TE are particularly salient in pediatric neurodiagnostics.[7]

1.2 Advanced Signal Processing Method in MRS: Why Is the FPT Equivalent to Quantum-Mechanical Spectral Analysis?

Via advanced signal processing methods, such as the FPT, many of these problems can be overcome. Resolution and signal–noise ratio (SNR) in MRS are improved through better-equipped mathematics. As a quotient

[a] A joint acronym for both Glu and Gln is Glx.

of two frequency-dependent polynomials, P_K/Q_K, extracted from the examined FID, an envelope is generated by the FPT without limitation to a preassigned grid of the sweep frequencies ω. This yields interpolation by the FPT.[1] The presence of numerator (P_K) and denominator (Q_K) polynomials adds another degree of freedom to cancel noise from the Padé spectrum P_K/Q_K. Polynomial P_K suppresses noise via the "moving average," which is familiar from the autoregressive moving average (ARMA) process. The FPT and ARMA, which is from statistical mathematics, are equivalent.[1] Via polynomial Q_K, the FPT achieves extrapolation. The expansion coefficients of polynomial Q_K are identical to the prediction coefficients from the ARMA process. From these coefficients, new signal points $\{c_n\}$ can be computed for $n > N-1$ ($t > T$) to predict the FID data at times $t = n\tau$ ($n > N-1$) beyond the total acquisition time T.

Through exact quantum-mechanical spectral analysis (the quantification problem), the FPT directly quantifies MRS time signals, thereby yielding accurate quantitative information for many brain metabolites, as demonstrated in our controlled studies.[2] The solution to quantification contains K (model order), as well as the 4 K real-valued spectral parameters (K complex frequencies ω_k and K associated complex amplitudes d_k, $1 \le k \le K$) for K resonances. This complete parametrization of the encoded FID set $\{c_n\}$, via the sum of K physical resonances, is represented by the geometric progression:

$$c_n = \sum_{k=1}^{K} d_k \exp(i\omega_k \tau n), \quad \mathrm{Im}(\omega_k) > 0 \quad (0 \le n \le N-1). \tag{2}$$

As stated, this is an inverse problem because the FID set $\{c_n\}$ $(0 \le n \le N-1)$ is known and $\{\omega_k, d_k\}$ $(1 \le k \le K)$ are to be found. This harmonic inversion is linear in $\{d_k\}$ and nonlinear in $\{\omega_k\}$ with the exponentials in (2) being complex damped harmonics. At any real linear frequency ν, related to the angular frequency ω by $\omega = 2\pi\nu$, for the given Maclaurin expansion, the Padé spectrum or the response function $R(u)$ in, eg, its diagonal form is completely and uniquely determined by the rational function as a ratio of two polynomials P_K and Q_K of the common degree K:

$$R(u) = \frac{P_K(u)}{Q_K(u)}, \quad u = \exp(i\omega\tau). \tag{3}$$

For the encoded FID set $\{c_n\}$, polynomials P_K and Q_K are determined by solving a single system of linear equations from the imposed condition:

$$\sum_{n=0}^{N-1} c_n z^{-n} = \frac{P_K(u)}{Q_K(u)}, \quad u = z \quad \text{or} \quad u = z^{-1}. \tag{4}$$

The fundamental frequencies $\{\omega_k\}$ are extracted by rooting the denominator polynomial Q_K:

$$Q_K(u) = 0 \Rightarrow \text{Spectral pole frequencies} \quad \omega_k (1 \le k \le K). \tag{5}$$

Via the Cauchy analytical formula for the residues of P_K/Q_K, taken at $\omega = \omega_k$, the amplitudes $\{d_k\}$ are deduced. For a nondegenerate spectrum P_K/Q_K comprised of simple poles alone, the result is given by:

$$d_k = \frac{P_K(u_k)}{Q_K'(u_k)}, \quad Q_K'(u) = \frac{d}{du} Q_K(u), \quad u_k = \exp(i\omega_k \tau), \ 1 \le k \le K. \tag{6}$$

The kth metabolite concentration is computed from the reconstructed amplitudes $\{d_k\}$ and is usually corrected for the T_1-(spin-lattice) and T_2-relaxation times whenever T is too short for FIDs to decay to zero. The independent variable u used in Eqs. (3)–(6) denotes either variable $z = \exp(i\omega\tau)$ or $z^{-1} = \exp(-i\omega\tau)$ from the two versions of the FPT, the $\text{FPT}^{(+)}$ and $\text{FPT}^{(-)}$, inside and outside the unit circle, respectively, for the same input z- transform given by the lhs of (4).

The reason for which the FPT is a quantum-mechanical spectral analyzer is fourfold as per Eqs. (2), (4), (5) and (6). *First*, the form (2) for each data point in the time signal $\{c_n\}$ is equivalent to the quantum-mechanical auto-correlation function, $(\Phi_0|\Phi_n)$, defined by projecting the Schrödinger non-stationary state Φ_n of the system at time $n\tau$ onto the initial state Φ_0, with an asymmetric scalar product $(a\,|\,b) = (b\,|\,a)$. *Second*, the $z-$ transform is the quantum-mechanical finite-rank Green function $G(z^{-1}) = \sum_{n=0}^{N-1} c_n z^{-n}$ which, by construction (4), coincides with the spectrum P_K/Q_K in the FPT. *Third*, the fundamental frequencies $\{\omega_k\}$ from (5) in the FPT, are the same as the quantum-mechanical eigen frequencies from the stationary Schrödinger eigen value problem, $\Omega\Psi_k = \omega_k\Psi_k$, where Ω is a non-Hermitean system operator (complex "Hamiltonian"). This is because the corresponding Schrödinger secular equation (equivalent to the Schrödinger eigen value problem) is identical to the characteristic polynomial equation $Q_K = 0$ from (5) in the FPT. *Fourth*, the fundamental amplitudes $\{d_k\}$ from (6) in the FPT are equivalent to their quantum-mechanical counterpart $(\Phi_0|\Psi_k)^2$. This is due to the proof of uniqueness of the amplitudes $\{d_k\}$

for the same frequencies $\{\omega_k\}$. No matter how different any two given algorithms for generating the amplitudes might be, they must yield the same results for $\{d_k\}$ as long as the two reconstructed sets for $\{\omega_k\}$ coincide[1]. Overall, the advantage of using the FPT relative to directly solving the Schrödinger eigen value problem on a given set of basis functions, is twofold regarding: (i) the amplitudes and (ii) signal-noise separation. With respect to (i), the Padé amplitudes $\{d_k\}$ from (5) obviate altogether the computation of wave functions $\{\Psi_k\}$ needed for $d_k = (\Phi_0|\Psi_k)^2$ in quantum mechanics. Eigen frequencies $\{\omega_k\}$ from the said non-stationary Schrödinger equation are variational, in the sense that their estimates have no first-order errors. By contrast, eigen states $\{\Psi_k\}$ are non-variational and, thus, contain first-order errors that, in turn, deteriorate the accuracy of the ensuing amplitudes $(\Phi_0|\Psi_k)^2$. This is usually mitigated[1] by using a linear combination of a set of reconstructions for the amplitudes of the original type $(\Phi_0|\Psi_k)^2$. Regarding (ii), the FPT has the unique feature of separating signal from noise (i.e. disentangling genuine from spurious resonances) by way of identifying Froissart doublets, as the coincident spectral poles and zeros that are the solutions of the characteristic equations, $Q_K = 0$ and $P_K = 0$, respectively.

A critical problem is the presence of noise within in vivo encoded FIDs. Particularly with dense spectra, the number of genuine metabolites is a small percentage of the total number of retrieved resonances. In MR spectra from the brain, besides the reconstructed genuine metabolites, many more non-physical (spurious) resonances are generated. The FPT algorithm handles this obstacle via "signal–noise separation" (SNS) implemented by pole–zero coincidence which identifies spurious, Froissart doublets.[13]

Extensive validation has been performed for the FPT on MRS time signals encoded in vivo from normal human brain and pediatric brain tumor. Compared to the FFT, markedly improved resolution of total shape spectra is obtained by the FPT using in vivo MRS time signals from healthy human brain encoded on both high magnetic field (4 and 7 T) and clinical scanners (1.5 T). Even more striking is its parametric capability especially at high spectral density, where the FPT resolved the numerous, very closely overlapping resonances, many of which are diagnostically important, including cancer biomarkers.[1,2,14] Controlled studies applying the FPT to synthesized MRS time signals that were very similar to those encoded in vivo from the brain of a healthy volunteer at 1.5 T, provided proof-of-concept validation in that the spectral parameters were exactly reconstructed from which the concentrations of many metabolites were precisely computed, including metabolites for which chemical shifts differed only by 0.001 ppm or less.[2] Further proof of principle of the FPT was provided through an MRS study on the standard general electric (GE) phantom head,[15] where extensive examination of the convergence process confirmed the stability of the reconstructed spectral parameters. Statistical analysis, via "parameter averaging" in the $\mathrm{FPT}^{(+)}$ demonstrated the accuracy and precision of the reconstructed complex-valued fundamental frequencies $\{\omega_k^+\}$ and amplitudes $\{d_k^+\}$, including the dense regions of the spectrum, where very small and/or very closely overlapping resonances were found.

In the present study, we apply the FPT to in vivo MRS time signals encoded using a clinical 1.5 T scanner from a pediatric patient with cerebral asphyxia. The FPT and DFT are compared with respect to resolution performance on the total shape spectra. A new stabilization feature of the FPT through iterative averaging of computed envelopes is introduced and tested, with the goal of solving the critical bottleneck of all parametric methods, namely, marked sensitivity of spectral parameters to any change of the model order K. This will provide the basis for reliable generation of component spectra, whereby the closely overlapping resonances for metabolites known to be abundant in the brain can be disentangled. Further, comparisons between the nonparametric and parametric Padé reconstructions will be made to see whether they could yield the same total shape spectra. Such a test is one of the important consistency checks of the computed component shape spectra in the Padé parametric estimation. Improved pediatric neurodiagnostics through MRS, with better possibilities to identify hypoxia/ischaemia and to clearly distinguish this from other pathology, is the overall clinical aim of this investigation. Possibilities for wider applications within various aspects of in vivo MRS diagnostics could also be forthcoming, in particular, as noted, for improving cancer diagnostics and treatment planning. These concepts are of generic importance and, as such, hold promise for wide applications ranging from basic sciences to technology.

2. METHODS

2.1 Acquisition of MRS Time Signals

Using a 1.5 T GE clinical scanner at the Astrid Lindgren Children's Hospital in Stockholm, the MRS time signals, or FIDs, were encoded from the parietal–temporal brain region of an 18-month-old patient who had suffered cerebral asphyxia. Each of the encoded FID contains 512 data points, with BW = 1000 Hz, and the Larmor frequency $\nu_L = 63.87$ MHz corresponding to the magnetic field strength $B_0 = 1.5$ tesla ($B_0 = 1.5$ T). The sampling time was $\tau = 1/\text{BW} = 1$ ms. Single-voxel proton MRS with the point-resolved spectroscopy sequence (PRESS) was applied. The echo times were TE = 24, 136, and 272 ms, and the repetition time (TR) was 2000 ms. Altogether, 128-encoded FIDs were averaged to improve SNR. The Fourier analysis is carried out by zero-filling time signals, so as to double the original FID length, as per the usual practice. Therefore, after zero-filling once, the total signal length is $N = 1024$, so that $T = N\tau = 1024$ ms and, for consistency, this will be used for both the DFT and FPT. Water was partially suppressed through encoding via the standard spin-echo procedure. The Regional Ethics Committee, Karolinska Institutet (Dnr # 2007/708-31/1) stated that they found no ethical issues to preclude implementation of this research.

2.2 Reconstructions

As usual, a phase correction was made by multiplication of the encoded set $\{c_n\}(0 \le n \le 511)$ by $\exp(i\phi_0)$ via $c_n^{(0)} = c_n \exp(i\phi_0)$. Here, ϕ_0 is the zero-order phase which is chosen to be equal to -1.1831, 1.6058, and 1.7499 rad for TE = 24, 136, and 272 ms, respectively. These values of ϕ_0 represent the results of the calculation of the minimae of the real parts of the DFT spectra, $\min\{\text{Re}(F_m)\}_{m=0}^{511}$, with the originally encoded, raw, phase uncorrected time signal $\{c_n\}$. Once phase corrected in this way, we relabel the FID data set $\{c_n^{(0)}\}$ as $\{c_n\}$ and will use only the phase-corrected FIDs reannotated thereafter by $\{c_n\}$.

2.2.1 Nonparametric Reconstructions Through the DFT and the FPT

Via (1), the spectrum was reconstructed in the DFT at the fixed Fourier grid frequencies. The main focus will be on the variant $\text{FPT}^{(+)}$, where the variable u is taken to be z. Note that the same reconstructions from the $\text{FPT}^{(+)}$ were also obtained by the $\text{FPT}^{(-)}$. Nonparametric analysis via the $\text{FPT}^{(+)}$ was done first after the expansion coefficients of the polynomials P_K^+ and Q_K^+ were extracted from the time signal $\{c_n\}$ using (4). The ratio P_K^+/Q_K^+ is the complex-valued total shape spectrum. Its real and imaginary parts, $\text{Re}(P_K^+/Q_K^+)$ and $\text{Im}(P_K^+/Q_K^+)$, would be purely absorptive and dispersive, respectively, insofar as all the phases ϕ_k^+ of the reconstructed FID amplitudes $d_k^+ = |d_k^+| \exp(i\phi_k^+)$ were equal to zero, $\phi_k^+ = 0$ $(1 \le k \le K)$. In reality, for encoded MRS time signals, the phases ϕ_k of the FID amplitudes d_k are generally nonzero $(\phi_k \neq 0)$, due to dephasing in the course of encoding. Thus, the reconstructed values ϕ_k^+ are also such that $\phi_k^+ \neq 0$. Consequently, there is a mixture of absorption and dispersion lineshapes in both $\mathrm{Re}(P_K^+/Q_K^+)$ and $\mathrm{Im}(P_K^+/Q_K^+)$.

2.2.2 Parametric Signal Processing via the FPT$^{(+)}$

Since the Padé-based quantification is performed by polynomial rooting, the roots of the characteristic equations of the numerator (P_K^+) and denominator (Q_K^+) polynomials give the zeros and poles of the Padé spectrum, P_K^+/Q_K^+, respectively. The Padé spectrum is a meromorphic function since P_K^+/Q_K^+ has only polar singularities[1]: the zeros of Q_K^+ are the poles of P_K^+/Q_K^+. The roots $z_k^+ = \exp(i\omega_k^+\tau)$ of equation $Q_K^+(z) = 0$ reconstruct the fundamental frequencies ω_k^+ via $\omega_k^+ = \{1/(i\tau)\} \ln(z_k^+)$. Through (6), the amplitudes d_k^+ are generated.[1] The component spectra will be presented in different modes. A mixture of absorption and dispersion components appear in the "usual" component spectra, where the amplitudes $\{d_k^+\}$ $(1 \leq k \leq K)$ are all complex-valued because their phases ϕ_k^+ are nonzero, as stated. The "absorption" components frequently appear as skewed Lorentzians, especially in spectrally dense regions. By setting the reconstructed phases ϕ_k^+ to zero "by hand," ie, $\phi_k^+ \equiv 0$ $(1 \leq k \leq K)$, this interference effect is eliminated, ie, externally suppressed to produce pure absorptive Lorentzians (for visualization purposes only). This is the so-called ersatz mode of the component spectrum for the kth resonance:

$$\left(\frac{P_K^+(z)}{Q_K^+(z)}\right)_k^{\mathrm{Ersatz}} \equiv \frac{|d_k^+| z}{z - z_k^+} \quad \text{(Ersatz component } k\text{)}, \tag{7}$$

where $\{z_k^+\}$ is the set of zeros of the characteristic equation $Q_K^+(z) = 0$ with $z_k^+ = \exp(i\omega_k^+\tau)$. The "usual" mode of the component spectra is given by:

$$\left(\frac{P_K^+(z)}{Q_K^+(z)}\right)_k^{\mathrm{Usual}} \equiv \frac{d_k^+ z}{z - z_k^+} \quad \text{(Usual component } k\text{)}. \tag{8}$$

When proceeding from (8) to (7), setting $\phi_k^+ = 0$ means replacing $d_k^+ \equiv |d_k^+| \exp(i\phi_k^+)$ by $|d_k^+|$. Consequently, the real part of the total and component shape spectra of the ersatz form $\mathrm{Re}(P_K^+/Q_K^+)_k^{\mathrm{Ersatz}}$ from (7) is entirely in the absorption mode. The usual form $\mathrm{Re}(P_K^+/Q_K^+)_k^{\mathrm{Usual}}$ from (8) contains both absorption and dispersion modes.[15] For the kth component, the T_2*-relaxation time in the FPT$^{(+)}$ is denoted by T_{2k}^{*+} and relates to the imaginary part of the reconstructed complex frequency ω_k^+ or ν_k^+ via $T_{2k}^{*+} = 1/\{\mathrm{Im}(\omega_k^+)\} = 1/\{2\pi \mathrm{Im}(\nu_k^+)\}$. This quantity is used in the expressions for peak heights $(H_k^+)^{\mathrm{Usual}}$ and $(H_k^+)^{\mathrm{Ersatz}}$, respectively, as:

$$(H_k^+)^{\mathrm{Usual}} \equiv \frac{d_k^+}{D_k^+}, \quad (H_k^+)^{\mathrm{Ersatz}} \equiv \frac{|d_k^+|}{D_k^+}; \; D_k^+ = 1 - \exp\left(-\tau/T_{2k}^{*+}\right) > 0, \quad (9a)$$

$$\mathrm{Re}(H_k^+)^{\mathrm{Usual}} = (|d_k^+|/D_k^+)\cos(\phi_k^+) = (H_k^+)^{\mathrm{Ersatz}}\cos(\phi_k^+). \quad (9b)$$

The peak heights H_k^{c+} of an absorptive conventional Lorentzian, expressed directly in terms of ω instead of z as $|d_k^+|\{(\mathrm{Im}\omega_k^+)/\tau\}/\{(\omega - \omega_k^+)^2 + (\mathrm{Im}\omega_k^+)^2\}$, are given by $H_k^{c+} \equiv |d_k^+|/(\tau \mathrm{Im}\omega_k^+)$. For narrow resonances with long relaxation times, this latter result can also be deduced from (9a). Consequently, for small τ/T_{2k}^{*+}, the series expansion for $\exp\left(-\tau/T_{2k}^{*+}\right)$ gives $D_k^+ \approx 1 - \left(1 - \tau/T_{2k}^{*+} + \cdots\right) \approx \tau/T_{2k}^{*+} = \tau \mathrm{Im}\omega_k^+$. From (9a), it then follows that $|d_k^+|/D_k^+ \approx |d_k^+|/(\tau \mathrm{Im}\omega_k^+) = H_k^{c+}$.

The numerator (P_K^+) and denominator (Q_K^+) polynomials in Eqs. (7) and (8) are explicitly expressed by:

$$P_K^+(z) = \sum_{r=1}^{K} p_r^+ z^r, \quad Q_K^+(z) = \sum_{s=0}^{K} q_s^+ z^s. \quad (10)$$

Here $\{p_r^+\}$ and $\{q_s^+\}$ are the expansion coefficients with $p_0^+ \equiv 0$. For N_P even, it follows that $K = N_P/2$, where N_P is the partial signal length given by a part of the total FID length N. As per condition (4) for $u = z$, the expansion coefficients $\{q_s^+\}$ for the polynomial $Q_K^+(z)$ from (4) are extracted by solving the system of linear equations $\sum_{s=0}^{K} q_s^+ c_{s'+s} = 0$. Subsequently, the solutions $\{q_s^+\}$ are refined via singular value decomposition (SVD). When the set $\{q_s^+\}$ becomes available, the expansion coefficients $\{p_r^+\}$ in P_K^+ are computed simply from the analytical expression $p_r^+ = \sum_{r'=0}^{K-r} c_{r'} q_{r'+r}^+$. The free term, q_0^+ can be set to, eg, 1 or -1 without affecting the spectra or the spectral parameters $\{\omega_k^+, d_k^+\}$ $(1 \le k \le K)$ reconstructed by the FPT$^{(+)}$. Via (7) and (8), the envelopes in the ersatz and usual modes are given by the Heaviside partial fractions:

$$\left(\frac{P_K^+(z)}{Q_K^+(z)}\right)^{\text{Ersatz}} \equiv \sum_{k=1}^{K} \left(\frac{P_K^+(z)}{Q_K^+(z)}\right)_k^{\text{Ersatz}} = \sum_{k=1}^{K} \frac{|d_k^+|z}{z - z_k^+} \quad \text{(Ersatz envelope)}, \quad (11)$$

$$\left(\frac{P_K^+(z)}{Q_K^+(z)}\right)^{\text{Usual}} \equiv \sum_{k=1}^{K} \left(\frac{P_K^+(z)}{Q_K^+(z)}\right)_k^{\text{Usual}} = \sum_{k=1}^{K} \frac{d_k^+ z}{z - z_k^+} \quad \text{(Usual envelope)}. \quad (12)$$

The notational difference between the lhs of Eqs. (8) and (12) for the kth usual component $\left(P_K^+/Q_K^+\right)_k^{\text{Usual}}$ and the usual envelope $\left(P_K^+/Q_K^+\right)^{\text{Usual}}$, respectively, is the missing subscript k in the latter and likewise for the ersatz modes in (7) and (11). As mentioned, ersatz component spectra are useful because therein the lack of interference between the absorption and dispersion modes clearly reveals the overlap of closely located or hidden resonances. However, these ersatz spectra should not be over-interpreted, since there may not be a one-to-one correspondence between ersatz peak heights and the actual abundance of the metabolites. The latter information requires the final output list of the reconstructed parameters including the phases $\phi_k^+ \neq 0$, whenever they are retrieved as nonzero quantities from the encoded MRS time signal. Although there is a correspondence between the kth component resonances, $\left(P_K^+/Q_K^+\right)_k^{\text{Usual}}$ and $\left(P_K^+/Q_K^+\right)_k^{\text{Ersatz}}$, their full widths at half maxima (FWHM) are unequal. Consequently, the peak area of the same kth component differs in the usual and ersatz modes. Since the metabolite concentration is proportional to the peak area, only the usual components should be used to fully estimate those concentrations.

Note, that the derivation of the envelopes in the representation of the Heaviside partial fractions (11) and (12) makes use of the expression $\sum_{n=0}^{\infty}\left(z_k^+/z\right)^n = z/\left(z - z_k^+\right)$ where $|z_k^+/z| < 1$, under the assumption that the total length of time signal $\{c_n\}$ is infinite ($N=\infty$). In reality, time signals are finite ($N<\infty$) indicating that the latter series should be truncated at $n=N-1$, which gives $\sum_{n=0}^{N-1}\left(z_k^+/z\right)^n = \left[1-\left(z_k^+/z\right)^N\right]/\left(1-z_k^+/z\right)$. Overall, this implies that the peak heights from (9a) should be corrected for the factor $1-\left(z_k^+/z\right)^N$ taken at sweep frequency, $\nu = \text{Re}\left(\nu_k^+\right)$. Therefore, the corrected peak heights should read as follows:

$$
\begin{aligned}
&\mathrm{Re}\left(H_k^+\right)^{\mathrm{Usual}} \equiv \frac{|d_{k,T}^+|\cos\left(\phi_k^+\right)}{D_k^+}, \quad \left(H_k^+\right)^{\mathrm{Ersatz}} \equiv \frac{|d_{k,T}^+|}{D_k^+},\\
&|d_{k,T}^+| = |d_k^+| \left\{1 - \exp\left(-T/T_{2k}^{*+}\right)\right\}, \quad T = N\tau.
\end{aligned}
\tag{13}
$$

The "stability test" in the FPT consists of computing successive values of spectra of partial signal lengths N_{P} ($N_{\mathrm{P}}=2\,K$ for N_{P} even) for components and envelopes

$$
\left(\frac{P_{K+m}^+(z)}{Q_{K+m}^+(z)}\right)_k^{\mathrm{Y}} = \left(\frac{P_K^+(z)}{Q_K^+(z)}\right)_k^{\mathrm{Y}} \quad (m=1,2,\ldots), \tag{14}
$$

$$
\left(\frac{P_{K+m}^+(z)}{Q_{K+m}^+(z)}\right)^{\mathrm{Y}} = \left(\frac{P_K^+(z)}{Q_K^+(z)}\right)^{\mathrm{Y}} \quad (m=1,2,\ldots), \tag{15}
$$

respectively, where Y = Usual or Y = Ersatz. Only when convergence is reached, as N_{P} is systematically increased, are the fundamental frequencies and amplitudes $\{\omega_k^+, d_k^+\}$ acceptable. Envelopes from (15) can be computed nonparametrically and parametrically. In the former case, the spectrum is obtained by evaluating $P_K^+(z)/Q_K^+(z)$ for $(z) = \exp(2\pi i\nu\tau)$ at the chosen set of sweep frequencies ν. In the latter case, (11) and (12) are used for the ersatz and usual modes, respectively.

Herein, we will introduce and implement a novel procedure, through iterative averaging of spectra. The goal is to practically handle the stumbling block of harmonic inversion, namely, oversensitivity to changes in model order K.

3. RESULTS

3.1 The Encoded MRS Time Signals

The MRS time signals (with the zero-order multiplicative phase corrections, $e^{i\phi_0}$) encoded at TEs of 24, 136, and 272 ms with 512 data points are displayed in Fig. 1. The real parts of the encoded FIDs for these three TEs are depicted in the left column (A–C), respectively, with the imaginary parts on the right column (D–F). A magenta line is drawn across the abscissae, from which it is seen that at the end of the encoding (512 ms), some of these FIDs have not fully returned to their zero values. Moreover, the waveforms are asymmetric around the abscissae because the residual water peak is about 200 times more abundant than all the other metabolites.

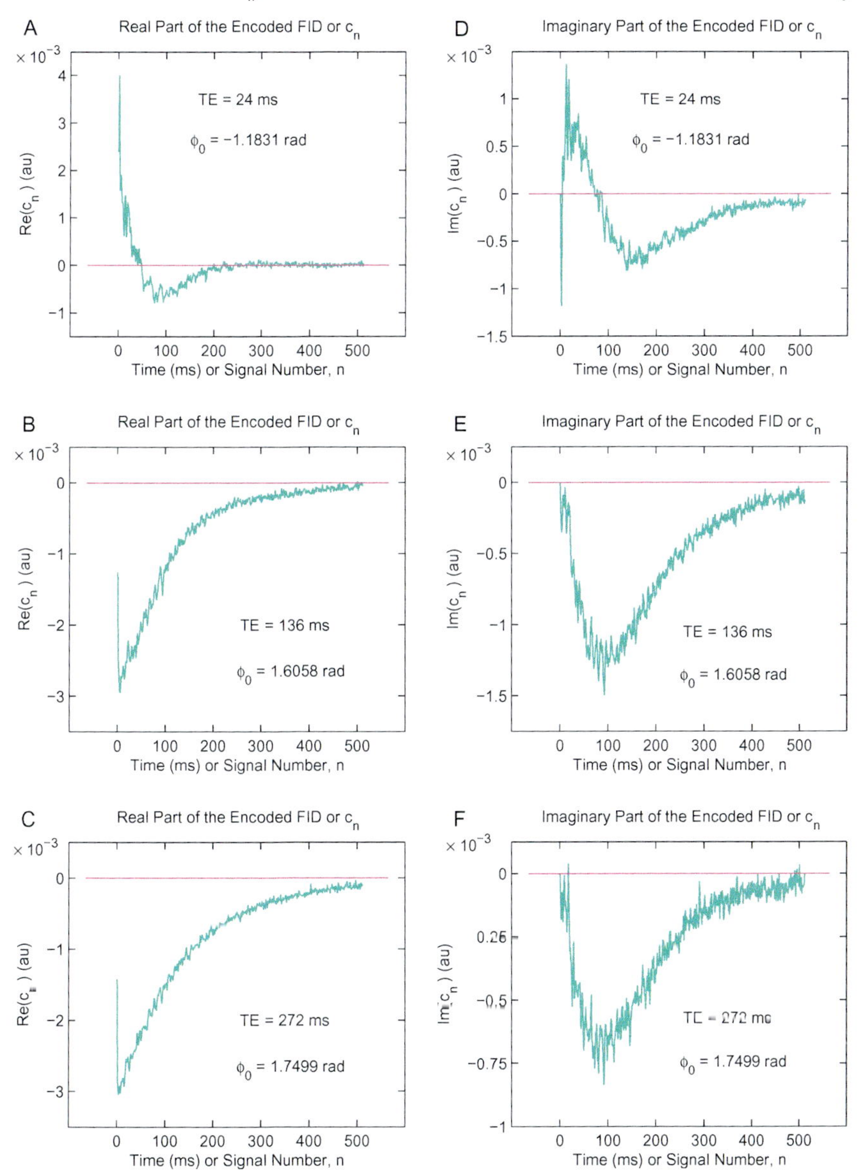

Fig. 1 The real part of the FID, Re(c_n), encoded (and corrected for the zero-order phase, ϕ_0) in vivo from the parietal–temporal brain region in an 18-month-old patient with cerebral asphyxia, using a 1.5 T MR scanner with 512 data points at echo times, TE = 24 ms (A), 136 ms (B), and 272 ms (C). The imaginary parts of the FID, Im (c_n), are shown for TE = 24 ms (D), 136 ms (E), and 272 ms (F). The *horizontal magenta lines* are drawn to guide the eye through the departures from the level of the zero-valued amplitudes in the oscillations of the FIDs.

3.2 Comparison of Total Shape Spectra Reconstructed by the DFT and Nonparametric FPT

Fig. 2 displays the total shape spectra reconstructed by the DFT (left column) and the nonparametric $\mathrm{FPT}^{(-)}$ (right column) for the three TEs: 24 ms (A and D), 136 ms (B and E), and 272 ms (C and F), between 0.75 and 4.25 ppm, at partial signal length $N_\mathrm{P} = 760$, where the encoded 512 data points were zero-filled to $N = 1024$. Several metabolites, especially those resonating above 3.5 ppm, eg, m-Ins, have decayed at longer TE, as seen on the decreasing maximum values of the ordinates. The baseline is also lowered at longer TE due to decay of the relatively immobile macromolecules. However, at TE = 272 ms a tall, narrow peak emerges, centered around 3.7 ppm. This corresponds to Gln and might reflect poor neurodevelopment.[7] At TE = 136 ms, the NAA peak at ~2.0 ppm is taller in the $\mathrm{FPT}^{(-)}$ than in the DFT. The inverted Lac doublet, as expected in hypoxic/ischemic conditions at TE = 136 ms, is seen with both reconstruction methods, but one of the two Lac peaks is more shortened in the DFT than in the $\mathrm{FPT}^{(-)}$. At TE = 272 ms, the Cho and NAA peaks are severely attenuated in the DFT, compared to the $\mathrm{FPT}^{(-)}$, and the serration at about 2.1 ppm can only be discerned in the $\mathrm{FPT}^{(-)}$. Overall, the resolution of the total shape spectra from the DFT is inferior to that of the $\mathrm{FPT}^{(-)}$, such that at all TEs, the peaks in the DFT are generally blunter and rougher, while many more structures are seen in the $\mathrm{FPT}^{(-)}$. It should be pointed out that, in general, and not only in Fig. 2, the $\mathrm{FPT}^{(-)}$ has a better resolution than in the Fourier envelope for the same number of the FID points. Conversely, by using a smaller number of the FID points, the $\mathrm{FPT}^{(-)}$ can achieve the same resolution as that in the Fourier envelope. Moreover, in contrast to the DFT, which can, on its own, generate nothing more than the total shape spectra, the FPT via parametric analysis yields further diagnostically important insights, as will be seen subsequently.

3.3 Iterative Averaging of Envelopes Through the FPT

We herein exploit a stabilization procedure within the FPT through iterative or adaptive averaging of nonparametrically computed envelopes. In the Padé rational functions:

$$\frac{P_{K+m}^{+}(z)}{Q_{K+m}^{+}(z)} = \frac{P_{K}^{+}(z)}{Q_{K}^{+}(z)} \quad (m = 1, 2, 3, \ldots), \tag{16}$$

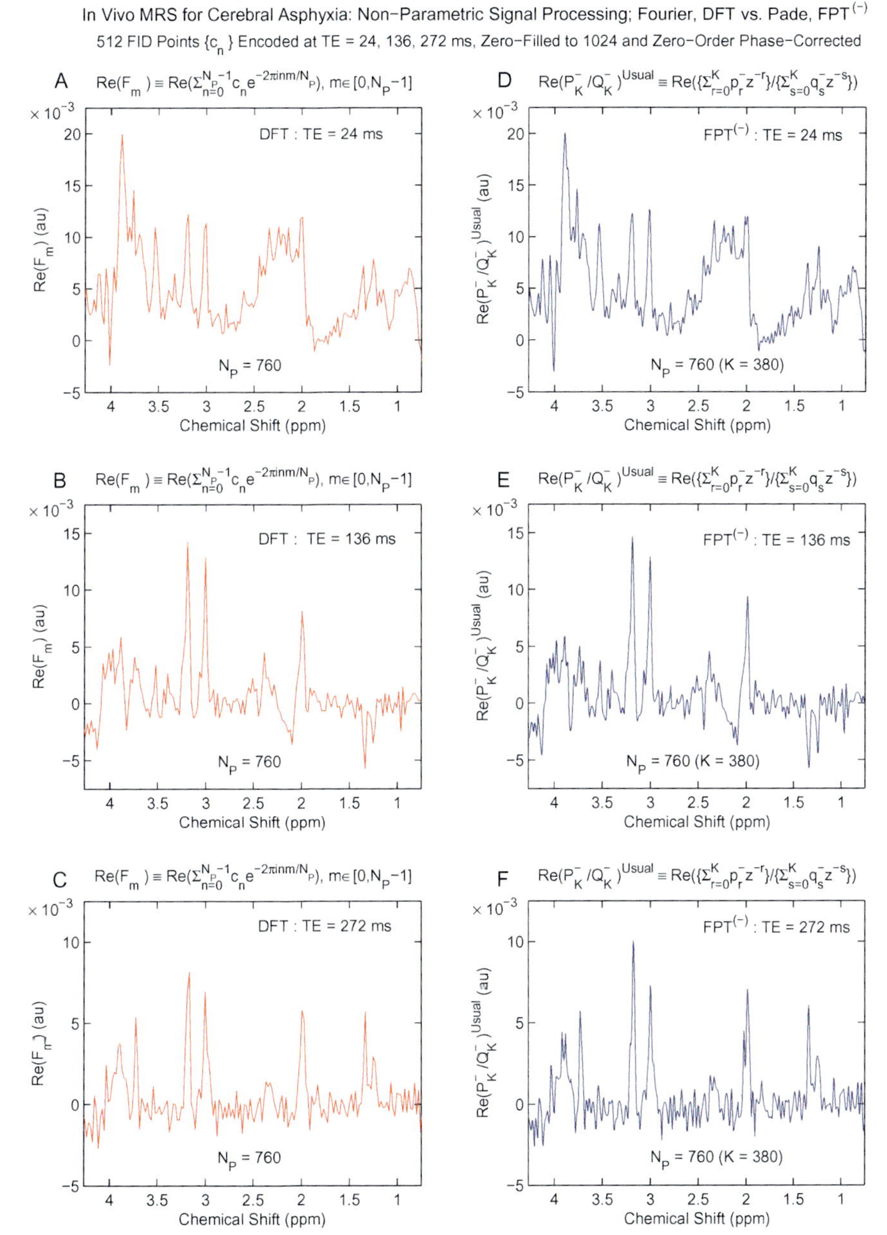

Fig. 2 The real part of the total shape spectra for TE = 24 ms (A and D), 136 ms (B and E), and 272 ms (C and F) for the chemical shift window between 0.75 and 4.25 ppm. *Left column* (A–C) corresponds to the DFT. *Right column* (D–F) corresponds to the nonparametric $FPT^{(-)}$. In both cases reconstruction was carried out on the original FID set $\{c_n\}$ encoded in vivo from the parietal–temporal brain region in an 18-month-old patient with cerebral asphyxia, using a 1.5 T MR scanner with 512 data points (ϕ_0−corrected), with zero-filling to $N = 1024$, and subsequently truncated to partial signal length $N_P = 760$.

all the spurious resonances will cancel out with stabilization for systematically and gradually increased polynomial degree $K+m$ ($m=1, 2, 3, \ldots$). The mechanism for this is rooted in pole-zero cancelations. This occurs because spurious resonances exhibit coincidence or near-coincidence of their poles and zeros. These confluences (known as Froissart doublets) make such spurious resonances markedly unstable, especially for changes in the model order K. Each envelope $P^+_{K+m}(z)/Q^+_{K+m}(z)$ ($m=1, 2, 3, \ldots$) will show different spuriousness. This is evidenced in Fig. 3A, where the real parts of 31 usual envelopes $\mathrm{Re}\left(P^+_K/Q^+_K\right)^{\mathrm{Usual}}_{\mathrm{It:1}}$ are displayed for $K=385, 386, \ldots, 415$ from the FID encoded at TE$=136$ ms. Therein, numerous noise-like spikes are observed. Subsequently, the arithmetic average is taken using the complex 31 envelopes. The result, denoted by $\left\{\mathrm{FPT}^{(+)}\right\}^{\mathrm{Usual}}_{\mathrm{Av:1}}$, is shown in (B) via $\mathrm{Re}\left\{\mathrm{FPT}^{(+)}\right\}^{\mathrm{Usual}}_{\mathrm{Av:1}}$, where a "clean" spectrum is seen. In this average spectrum, the stable structures remain, but the spikes are gone and/or greatly diminished. Hereafter, the subscripts It:m and Av:m ($m=1, 2, 3, \ldots$) denote the Iteration (It) and Average (Av) with m being their number. The metabolite assignments are given in Fig. 3B.

Taking the arithmetic average of a precomputed sequence of the retrieved envelopes (31 envelopes in Fig. 3) is seen to be a simple and powerful way to stabilize the process of shape estimations. Thus, for example, Fig. 3A displays some sharp and narrow spikes pointing to sensitivity of estimation to model order K within a number of frequency bands, collectively denoted as α: (0.75, 0.9) ppm, (1.4, 1.9) ppm, (2.5, 2.9) ppm, (3.25, 3.6), and around 4.25 ppm. On the other hand, the same Fig. 3A shows that a large number of spectral structures are manifestly insensitive to changes in the model order K, judging from their almost identical reproduction in all the 31 envelopes within several complementary frequency bands, jointly denoted by β: (0.9, 1.4) ppm, (1.9, 2.5) ppm, (2.9, 3.25) ppm, and (3.6, 4.2) ppm, containing among other molecules, diagnostically important metabolites (Lip, Lac), (NAA, Glx), (Cr, Cho), and (m-Ins, Cr; Cho), respectively. Fig. 3B proves that the average envelope simultaneously eliminates and/or significantly reduces the instability (spuriousness) in the α-band and confirms the stability (genuineness) of the β-band. Altogether, this stabilizes the shape estimation in the α–β bands, ie, at all the chemical shifts on the abscissa in Fig. 3B and the same holds true throughout the entire Nyquist range (-3.2, 12.2) ppm. The underlying mechanism of this outcome is the ability of the arithmetic average to damp the unstable (noise-like) peak heights by a significant factor, which is less than or equal to $\sqrt{31} \cong 5.6$ for Fig. 3. The present procedure of "spectra averaging" is

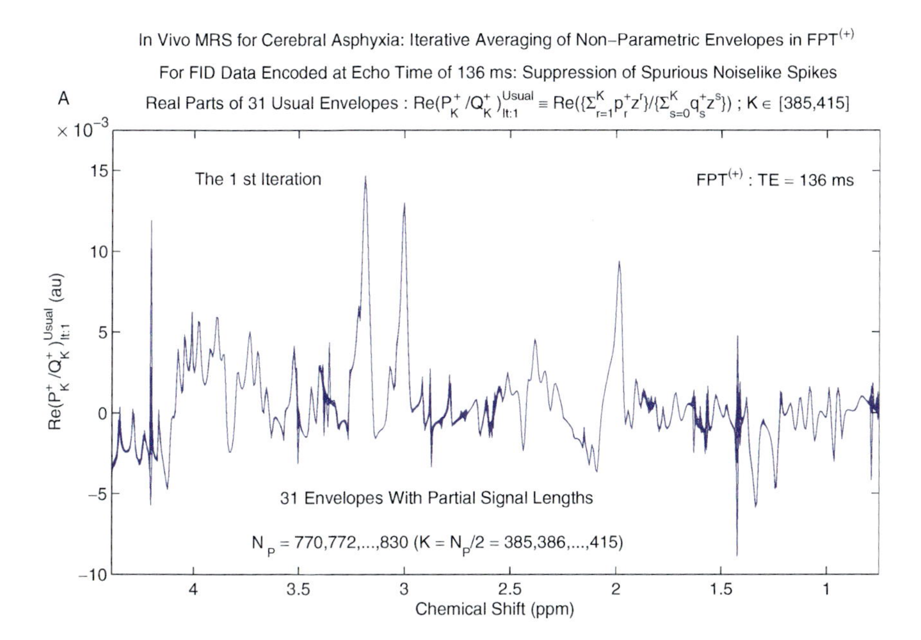

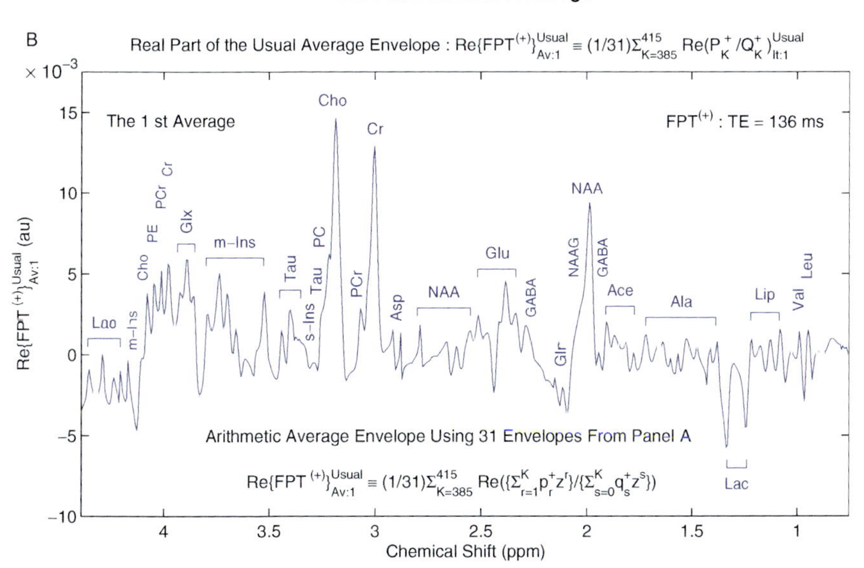

Fig. 3 The real parts of 31 usual envelopes $\mathrm{Re}\left(P_K^+/Q_K^+\right)^{\mathrm{Usual}}_{\mathrm{It:1}}$ are plotted for $K=385$, 386, …, 415 ($N_\mathrm{P}=2\,K=770, 772, \ldots, 830$) from the FID (with 512 original data points, zero-filled to $N=1024$) encoded in vivo from the parietal–temporal brain region in an 18-month-old patient with cerebral asphyxia, using a 1.5 T MR scanner at TE = 136 ms (A). Numerous noise-like spikes are observed. These 31 envelopes are averaged as denoted by $\mathrm{Re}\left\{\mathrm{FPT}^{(+)}\right\}^{\mathrm{Usual}}_{\mathrm{Av:1}}$ (B), where a "clean" spectrum is seen. In this averaged spectrum, the stable structures remain, but the spikes are gone and/or substantially reduced. The metabolite assignments are shown in (B). See the list of abbreviations and acronyms for their full names.

the frequency domain equivalent of the well-known SNR-improving "FID averaging" in the time domain.

Fig. 4 shows the nonparametrically (B) and parametrically (C) generated envelopes in the $\mathrm{FPT}^{(+)}$ at $K=512$. The FID used in (B) and (C) was generated by the IFFT-based inversion of $\{\mathrm{FPT}^{(+)}\}_{\mathrm{Av:1}}^{\mathrm{Usual}}$ whose $\mathrm{Re}\{\mathrm{FPT}^{(+)}\}_{\mathrm{Av:1}}^{\mathrm{Usual}}$ is from (A). It is seen from (B) and (C), that there is full equivalence of the Padé estimation using nonparametric and parametric analysis. Moreover, the spectra from (B) and (C) coincide with (A). This verification is crucial since, in turn, it validates the correctness of the reconstructed FID which will be subsequently subjected to Padé-based quantification. The same results from (B) and (C) for $K=512$ have also been obtained for the middle of the interval of the K values, ie, $K=400=(385+415)/2$. The case with $K=512$ confirms that the extrapolation feature of rational polynomials of the PA type remains valid beyond the originally selected interval for K. By contrast, when ordinary polynomials are used to approximate a given function in a selected interval, they perform poorly outside of that interval.

The next step is to check the robustness of the stability of the average envelope. To this end, the complex amplitude $\{\mathrm{FPT}^{(+)}\}_{\mathrm{Av:1}}^{\mathrm{Usual}}$ is subjected to the IFFT to generate a new FID. The $\mathrm{FPT}^{(+)}$ is then applied to this reconstructed FID, and the new 31 envelopes for $K\in[385, 415]$ are displayed in Fig. 5B alongside the average envelope from the first iteration repeated in Fig. 5A. The 31 new envelopes from (B) have markedly fewer spikes, and, moreover, these are of much smaller peak heights compared to the corresponding spurious structures in Fig. 3A. Once again, averaging is performed, and this time for the 31 envelopes in Fig. 5B. This yields the envelope for the second arithmetic average, shown in Fig. 5C. It is seen that the average envelopes in Fig. 5A and C are nearly indistinguishable.

Further, the envelope from the second complex arithmetic average whose $\mathrm{Re}\{\mathrm{FPT}^{(+)}\}_{\mathrm{Av:2}}^{\mathrm{Usual}}$ is in Fig. 5C is subjected to the IFFT to generate yet another FID, from which 31 new envelopes $\mathrm{Re}(P_K^+/Q_K^+)_{\mathrm{It:3}}^{\mathrm{Usual}}$ for $K=385, 386, \ldots, 415$ are obtained, with exceedingly sparse spikes of very small amplitudes as seen in Fig. 6C. To see the systematic elimination and/or suppression of spikes, Fig. 6A and B recapitulate, respectively, Figs. 3A and 5B. Comparing Fig. 6A–C, the reduction in the size and abundance of spurious structures through this iterative averaging procedure is clearly observed. Fig. 6D displays the three average envelopes using the 31 envelopes from (A–C), and this demonstrates their very close agreement. The latter three envelopes are color coded, in coherence with the like colors

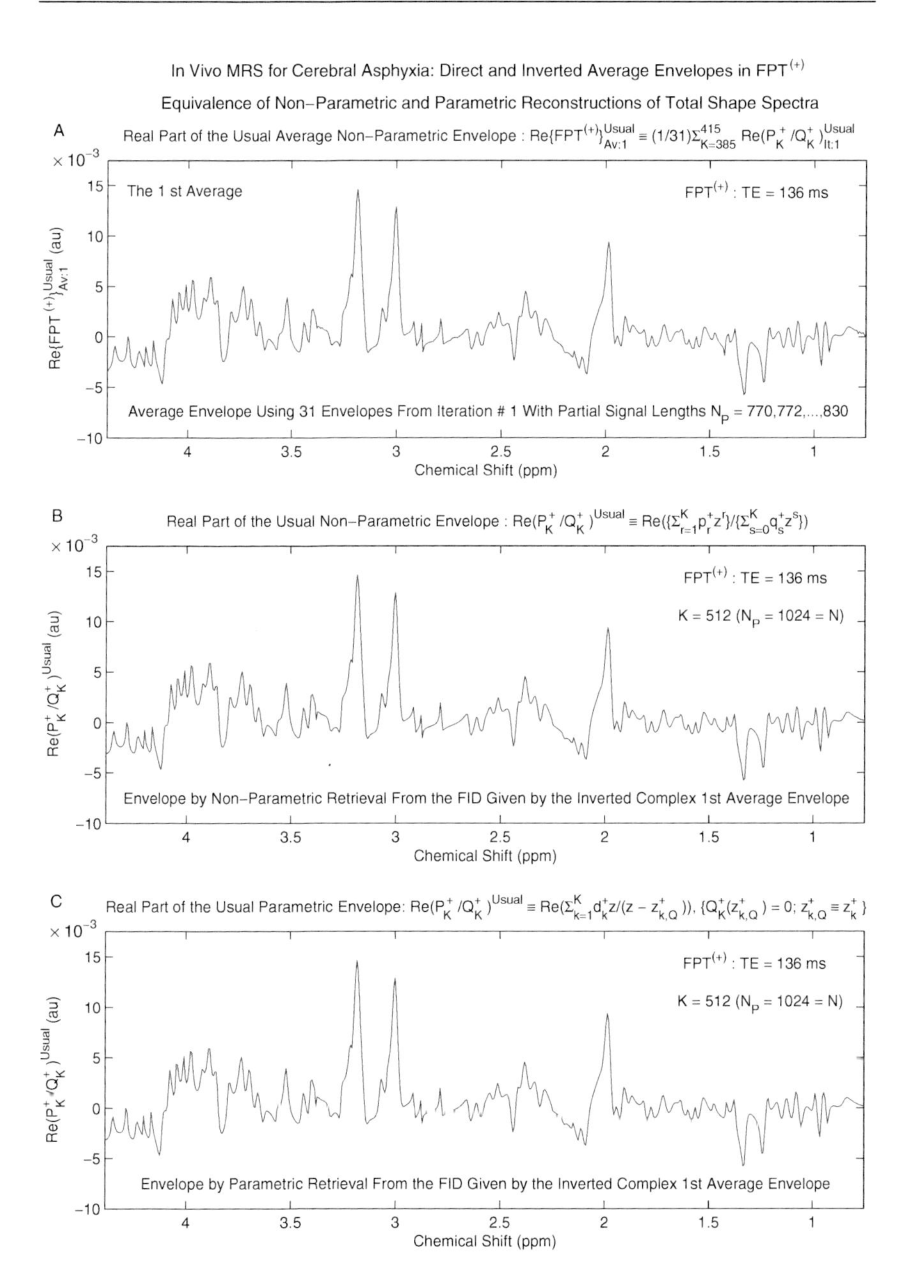

Fig. 4 The first average envelope (A) together with the nearly identical nonparametric (B) and parametric (C) envelopes in the $\text{FPT}^{(+)}$. In both cases reconstruction was carried out on the FID generated by inverting (via the IFFT) the first complex average envelope whose real part is in (A).

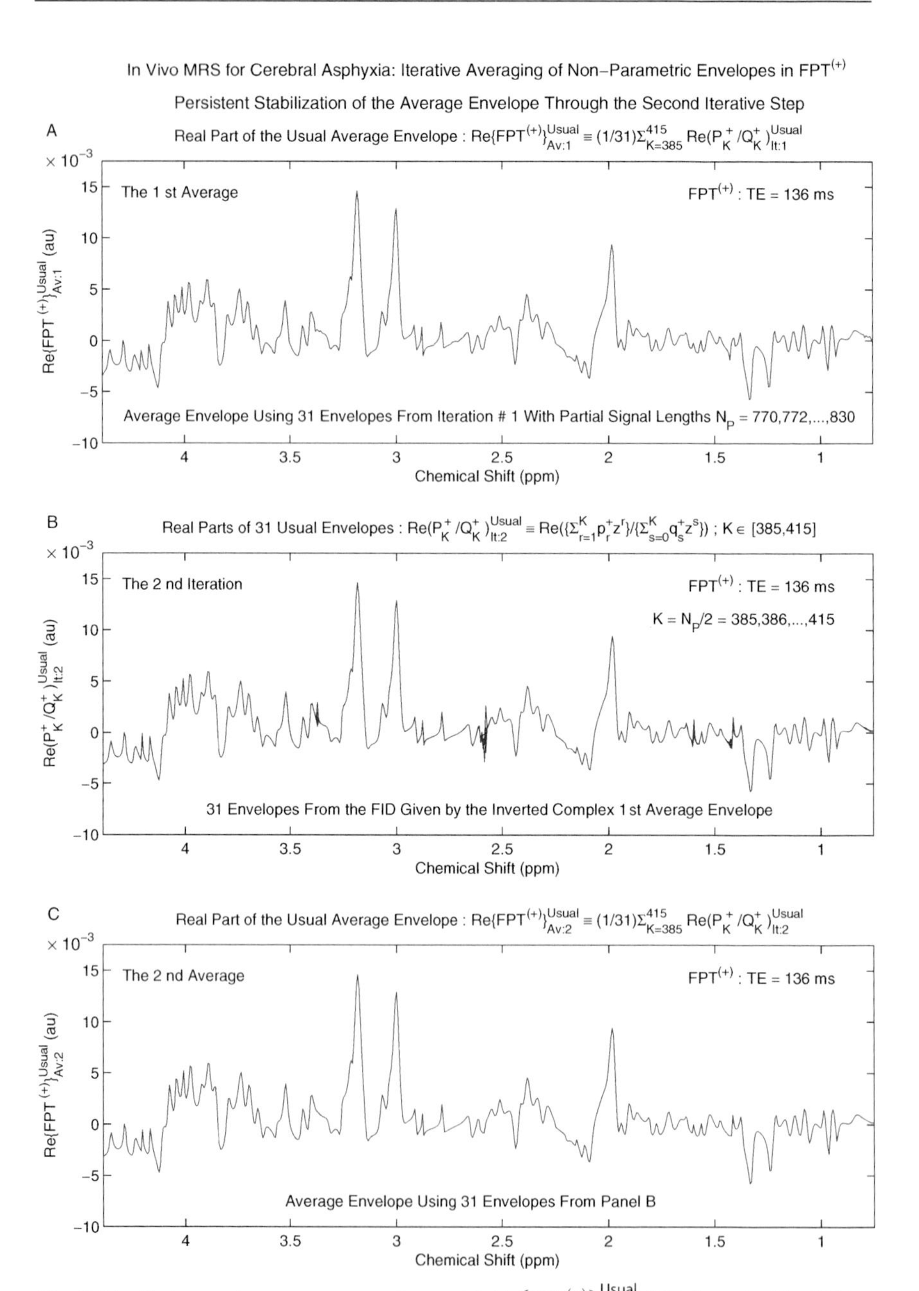

Fig. 5 The first complex average envelope with $\mathrm{Re}\left\{\mathrm{FPT}^{(+)}\right\}^{\mathrm{Usual}}_{\mathrm{Av:1}}$ (A) is inverted by IFFT to obtain a new FID to which the $\mathrm{FPT}^{(+)}$ was applied to generate 31 envelopes for $K \in [385, 415]$ (B) from which the second arithmetic average $\mathrm{Re}\left\{\mathrm{FPT}^{(+)}\right\}^{\mathrm{Usual}}_{\mathrm{Av:2}}$ (C) was produced.

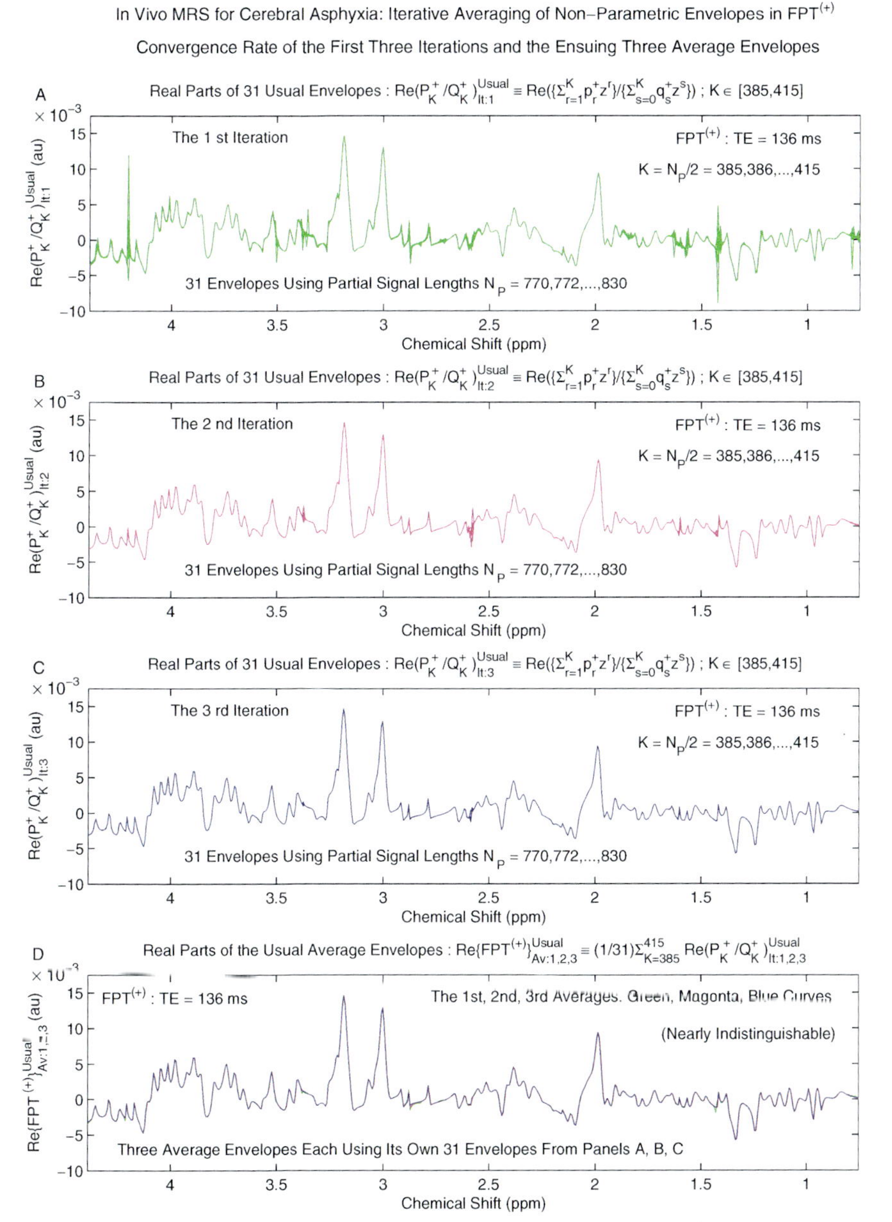

Fig. 6 The convergence rate of the first three iterations $\mathrm{Re}\left(P_K^+/Q_K^+\right)_{\mathrm{It:1,2,3}}^{\mathrm{Usual}}$ (A), (B) and (C), respectively, is shown herein. The third iteration $\mathrm{Re}\left(P_K^+/Q_K^+\right)_{\mathrm{It:3}}^{\mathrm{Usual}}$ generates almost identical envelopes, with a slight fluctuation only near 1.4 and 2.6 ppm. This shows convergence even without averaging the 31 spectra. The three envelopes $\mathrm{Re}\left\{\mathrm{FPT}^{(+)}\right\}_{\mathrm{Av:1,2,3}}^{\mathrm{Usual}}$ on (D) averaged over the spectra from (A–C) are in very close agreement, as seen by nearly complete coincidence of the *green*, *magenta* and *blue* curves.

of curves in (A–C), as green, magenta, and blue curves, respectively, and it is seen that their averages in (D) are nearly indistinguishable. This establishes the sought robustness of stability of the average envelopes generated by the expounded iterative averaging of the reconstructed spectra.

3.4 Parametric Reconstruction (Quantification) Through the FPT

Using the FID from the third complex average envelope, with its real part shown in Fig. 7A at TE = 136 ms, the usual component spectra are reconstructed by the parametric $\mathrm{FPT}^{(+)}$ and displayed in (B). Therein many absorption spectra are seen with admixtures of dispersion components since, as stated, the amplitudes $\{d_k^+\}$ are all complex-valued with their nonzero phases $\phi_k^+ (1 \le k \le K)$. The Lac doublet centered at ~1.3 ppm is inverted below the baseline (180 degrees out of phase) due to the effect of J-coupling[14] at TE of 136 ms. The bottom of Fig. 7C depicts the super-resolution of the ersatz component spectra, where the numerous closely overlapping resonances are distinctly identified. Here, the "positive" peaks (pointed upwards) of both Lac resonances are due to using $|d_k^+|$ in (7). Several underlying and closely overlapping resonances can be seen in association with the most prominent peaks such as NAA at ~2.0 ppm, Cr at ~3.0 ppm, and Cho at ~3.2 ppm.

Also shown in (B) and (C) are the peak heights for the usual and ersatz component spectra, computed from the analytical expressions for $\mathrm{Re}(H_k^+)^{\mathrm{Usual}}$ and $(H_k^+)^{\mathrm{Ersatz}}$ given in (13). It can be seen in (C) that peak heights $(H_k^+)^{\mathrm{Ersatz}}$ match the tops of the displayed resonances. As to peak heights $\mathrm{Re}(H_k^+)^{\mathrm{Usual}}$ in (B), there are two patterns. The first shows matching of the peak heights and the tops of the purely absorptive resonances, eg, Cr and Cho at ~3.0 and 3.2 ppm, respectively. In the second pattern for dispersive lineshapes, the peak heights are slightly displaced from the tops of the resonances; these departures depend upon the angle ϕ_k^+ of the amplitudes $\{d_k^+\}$. The location of the open circles denoting peak heights for dispersive resonances is closer to the dominant of the two lobes (recall that the dispersive lineshapes have two lobes).

The mentioned concept of SNS, with the underlying pole–zero coincidence known as Froissart doublets, is demonstrated in Fig. 8, which displays the results of quantification in the $\mathrm{FPT}^{(+)}$. Fig. 8A shows the Argand plot as the imaginary vs real frequencies. Therein, a complete separation is observed

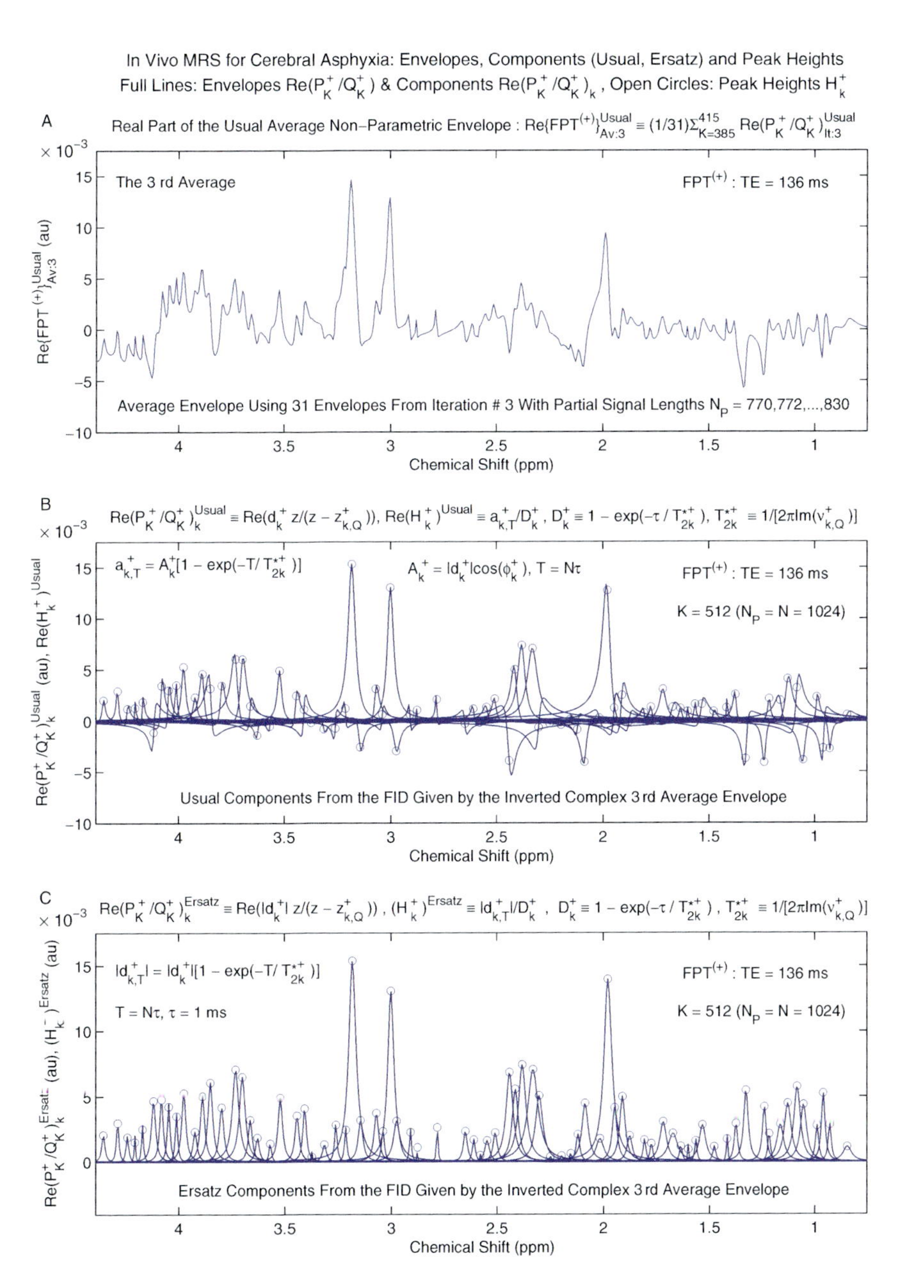

Fig. 7 The FID from the IFFT-based inversion of the third complex average envelope with its real part in (A) is subjected to the $\mathrm{FPT}^{(+)}$ to produce the usual and ersatz component spectra in (B) and (C), respectively. In (B), the usual components mix the absorption and dispersion lineshapes. Ersatz component spectra (C) show numerous closely overlapping positively oriented resonances that are distinctly identified, all in the absorption mode. Peak heights computed from the analytical expressions in (13) are depicted as *open circles*.

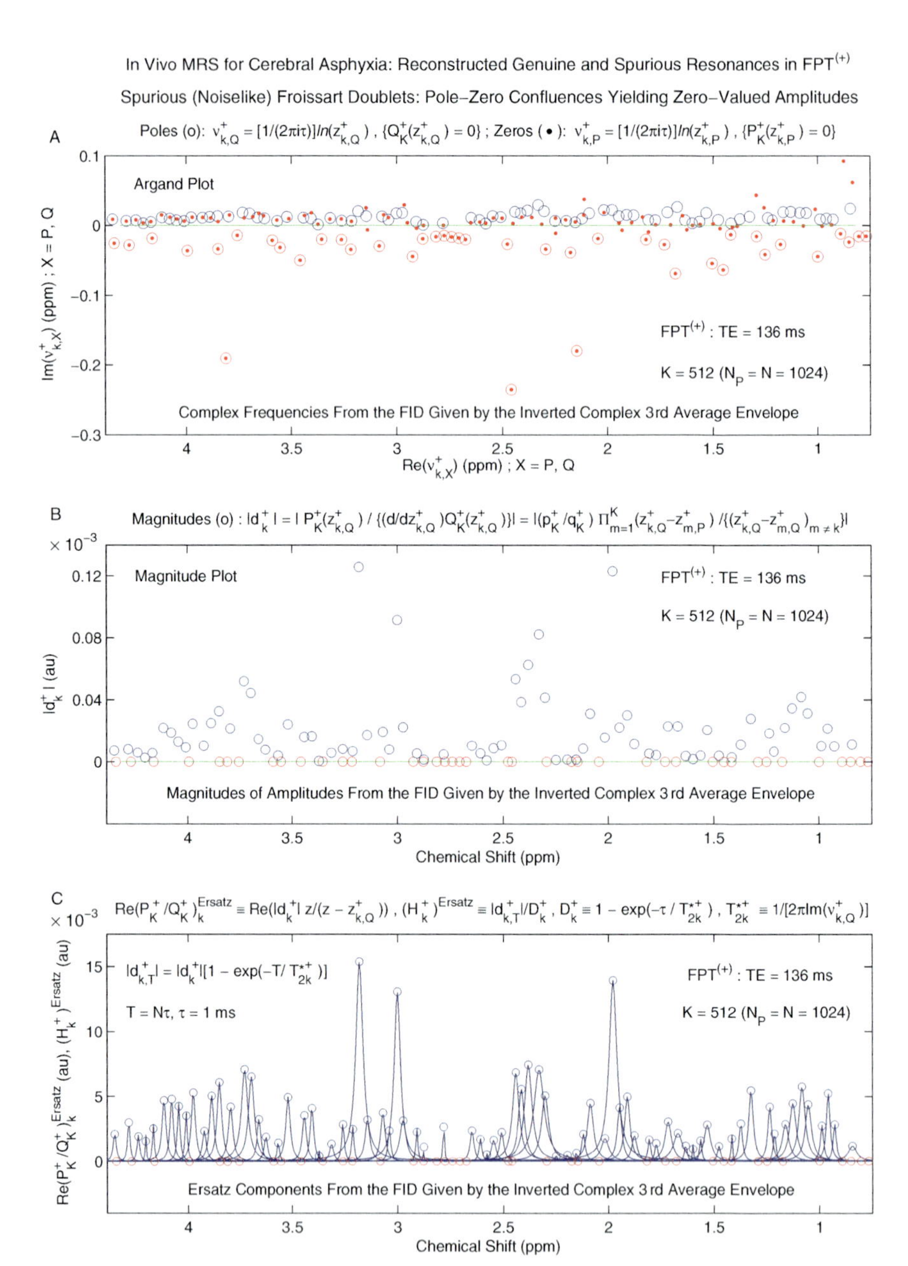

Fig. 8 Illustration of signal–noise separation, SNS. (A) The Argand plot in the complex frequency plane, with a complete separation of genuine from spurious frequencies. The twofold signature of a spurious resonance is pole–zero coincidence (confluence of *open circles with dots*) and zero-valued magnitudes in (A) and (B), respectively. Zero-valued magnitudes imply zero-valued peak heights, as seen in (C).

of genuine from spurious frequencies that lie in the positive and negative imaginary frequency regions, respectively. An auxiliary green line shown therein facilitates observation of this separation. Poles and zeros are symbolized by open circles and dots, respectively, that completely coincide for spurious resonances. These pole–zero coincidences result in full annihilation of the amplitudes for spurious resonances, as evidenced in (B) and (C), for magnitude plots and ersatz component spectra, respectively. Observe that poles and zeros can occasionally be very close for genuine resonances (eg, the Lac quartet around ~4.25 ppm). However, these near pole–zero coincidences are not Froissart doublets due to their stability against changes in model order K, as presently illustrated for $K\in[385, 415]$. Note that circles for genuine and spurious resonances are depicted in blue and red colors, respectively, for a more distinct separation in Fig. 8.

In the Results section, poles and zeros, denoted by $z_{k,Q}^{+}$ and $z_{k,P}^{+}$, are the solutions of the characteristic equations $Q_K^{+}(z)=0$ and $P_K^{+}(z)=0$, respectively. The second subscripts Q and P in $z_{k,Q}^{+}$ and $z_{k,P}^{+}$ indicate the association with the numerator and denominator polynomials, respectively. Note, in the title of (B) from Fig. 8, the expression $d_k^{+}=P_K^{+}(z_{k,Q}^{+})/[(d/dz_{k,Q}^{+})Q_K^{+}(z_{k,Q}^{+})]$ is a shortened notation for $d_k^{+}=P_K^{+}(z_{k,Q}^{+})/\left[(d/dz)Q_K^{+}(z)\right]_{z=z_{k,Q}^{+}}$. Here, for pole–zero coincidence $(z_{k,Q}^{+}=z_{k,P}^{+})$, it follows $P_K^{+}(z_{k,Q}^{+})=0$ and, thus, $d_k^{+}=0$. This is one of the two signatures of Froissart doublets (zero-valued amplitudes). Moreover, in (B), the same amplitude d_k^{+} is given by the equivalent canonical form $d_k^{+}=(p_K^{+}/q_K^{+})\prod_{m=1}^{K}(z_{k,Q}^{+}-z_{m,P}^{+})/(z_{k,Q}^{+}-z_{m,Q}^{+})_{m\neq k}$. Here, in the numerator $(z_{k,Q}^{+}-z_{m,P}^{+})$, we can have $m=k$, in which case $d_k^{+}=0$, and this also signifies Froissart doublets.

4. DISCUSSION AND CONCLUSIONS

The present investigation provides further evidence of the super-resolution capabilities of the FPT. At short, intermediate and long echo times, TE = 24, 136, and 272 ms, respectively, the FPT generated better resolved lineshapes compared to the DFT. The novelty brought here is in a procedure termed "spectra averaging" in the frequency domain, as a counterpart to the SNR—improving "signal averaging" in the time domain.

The advantages of Padé-processing with regard to its stabilization capabilities are herein clearly demonstrated. This is achieved by the newly introduced envelope-averaging procedure which in each iteration progressively suppresses all the highly sensitive spurious structures.

In our previous study,[16] we posed the question: Given that with realistic levels of noise, pole–zero coincidence is often either lacking or not complete, how can one then be certain which of the resonances are spurious and which are genuine? A key part of the answer to that query was to perform "the stability test," since spurious resonances are always unstable with even a minor change in either the noise level or model order. In the present study, we move that line of reasoning a critical step further, by taking advantage of the possibility to iteratively/adaptively transform back and forth from the frequency to the time domain, such that ultimately a spectrum is generated that is free from unstable, ie, spurious structures. It should, moreover, be reemphasized that particularly for dense spectra, such as those from the brain, convergence entails a highly overdetermined system of linear equations. The iterative averaging procedure, which is presently validated, could thus be automatically built-in, to remove the spurious noise-like spikes and thereby generate the "clean" spectra so vitally needed in clinical MRS. Inversion of the complex iteratively averaged total shape spectrum (envelope) gives the new time signal which is the starting point for Padé-based parametric analysis (quantification).

It was also demonstrated herein that nonparametric and parametric Padé reconstructions yield identical average envelopes. This, in turn, retroactively validates the process of extraction of quantitative information via the fundamental pairs $\{\omega_k^+, d_k^+\}$ reconstructed parametrically by the $\mathrm{FPT}^{(+)}$.

As yet another crossvalidation, the same envelopes are obtained by the two algorithmically different versions of the FPT, namely the $\mathrm{FPT}^{(+)}$ and $\mathrm{FPT}^{(-)}$ in the complementary domain of the complex harmonic variable, inside and outside the unit circle, respectively. The more stringent SNS capabilities of the $\mathrm{FPT}^{(+)}$ become particularly salient in this framework[17]. This multiple crossvalidation is the prerequisite for estimation reliability which is of utmost importance to clinical diagnostics by way of MRS. All the signal processors are plagued with marked sensitivity to changes in the otherwise unknown model order K, which is the number of the component metabolites in the present analysis. Spectra averaging can dramatically suppress this sensitivity and, thus, stabilize shape estimations, as well as the subsequent quantification.

Given the extremely high density of MR spectra from the brain, going beyond the total shape spectra to analyze the spectral components is absolutely essential. Clearly, no fitting procedure could ever provide reliable reconstruction of the actual number of peaks underlying, eg, the prominent

NAA at ~2.0 pm, Cr at ~3.0 ppm or Cho at ~3.2 ppm. The chemical shift region around 1.3 ppm where Lac and lipids overlap is particularly important for identifying hypoxia. Here, the "ersatz" component spectra reveal that besides the Lac doublet, several other peaks are closely overlapping even at a TE of 136 ms, after a number of the broad lipid resonances have already decayed. It can thus be seen that concentration ratios of these major metabolites based on conventional Fourier processing and postprocessing via fitting would be completely unreliable, as would any attempt to actually quantify these peaks without first correctly determining their true numbers.

From the present paper, together with our previous, related work, it can be concluded that Padé-optimized MRS applied to in vivo encoded time signals in the clinical setting is entirely feasible. Practical implementation is facilitated by the iterative averaging procedure which is herein introduced and validated. This latter procedure could be straightforwardly and automatically programmed to provide stable estimation of spectra and the fundamental parameters (complex frequencies and complex amplitudes), the long-sought goal so vitally needed in clinical MRS. Further applications of spectra averaging for in vivo MRS time signals encoded from the human brain and ovary have very recently been reported in Refs. 18 and 19.

Overall, the area addressed in the present study belongs to the category of inverse problems. Their intrinsic difficulty is in mathematical ill-conditioning or, equivalently, ill-posedness. This obstacle stems from the lack of continual dependence of observables on the independent variables (time $t=t_n=n\tau$ in the harmonic inversion problem). The most severe consequence of this hurdle is nonuniqueness of the solution of the inversion problem. Here, ill-conditioning is manifested in obtaining vastly different solutions for even the slightest changes of the input data (noise level, truncation of the total acquisition time, etc.). The FPT is herein demonstrated to robustly solve the harmonic inversion problem for the encoded time signals heavily corrupted with noise. On the level of envelopes, this ill-conditioning is observed by sensitivity of lineshapes to changes in model order K. This raises the question: when do we stop with systematic increase in K? We show that the presented iterative averaging of envelopes computed for a sequence of values of K is able to remarkably stabilize the total shape spectra. Subsequently, from a given stable arithmetically averaged complex spectrum, the time signal is reconstructed and subjected to quantification. The result is the set of regularized spectral fundamental parameters (complex frequencies and complex amplitudes) that represent the unique solution to the harmonic inversion problem.

By reference to "system theory," the importance of parametrization of a general complex system is in capturing the essence of the system's performance and dynamics through a relatively small set of the dominant features. In quantum mechanics this, and, in fact, the entire information is held by the twofold equivalent methodologies, called the Schrödinger eigenvalue problem and resolvent spectrum. Both concepts invariably produce the frequency or energy spectrum in the form of the Heaviside partial fraction decomposition which exactly sums up to the quotient of two polynomials, ie, to the FPT. The parameters in the partial fractions are the fundamental frequencies and amplitudes that hold the full information about the studied, generic system. This is how quantum mechanics parametrizes any system using its completeness relation. The fact that the quantum-mechanical spectrum of a general system is the unique ratio of two polynomials makes the FPT the method of choice for quantitative description (via parametrization) of the system's performance. Critical to robust performance of systems is their stability which is, in turn, governed by the stability of their dynamical parameters. The iterative averaging procedure of the present study is of key importance in achieving this stability. These features have no borders in terms of their applicability across disciplines.

ACKNOWLEDGMENTS

This work was supported by the King Gustav the 5th Jubilee Fund, the Marsha Rivkin Center for Ovarian Cancer Research, and FoUU through Stockholm County Council to which the authors are grateful.

REFERENCES

1. Belkić, Dž. *Quantum-Mechanical Signal Processing and Spectral Analysis*; Institute of Physics Publishing: Bristol, 2005.
2. Belkić, Dž.; Belkić, K. *Signal Processing in Magnetic Resonance Spectroscopy with Biomedical Applications*; Taylor & Francis: London, 2010.
3. Löwdin, P.-O.; Appel, K. Studies of Atomic Self-Consistent Fields. Wave Functions for the Argon-Like Ions and for the First Row of the Transition Metals. *Phys. Rev.* **1956**, *15*, 1746.
4. Löwdin, P.-O. Studies in Perturbation Theory. X. Lower Bounds to Energy Eigenvalues in Perturbation-Theory Ground State. *Phys. Rev.* **1965**, *139*, A357.
5. Brändas, E.; Goscinski, O. Variation-Perturbation Expansions and Padé Approximants to the Energy. *Phys. Rev. A* **1970**, *1*, 552.
6. Dezortova, M.; Hajek, M. 1H MR Spectroscopy in Pediatrics. *Eur. J. Radiol.* **2008**, *67*, 240.
7. Roelants-Van Rijn, A. M.; Van Der Grond, J.; De Vries, L. S.; Groenendaal, F. Value of 1H-MRS Using Different Echo Times in Neonates with Cerebral Hypoxia-Ischemia. *Pediatr. Res.* **2001**, *49*, 356.

8. Sijens, P. E.; Oudkerk, M. 1H Chemical Shift Imaging Characterization of Human Brain Tumor and Edema. *Eur. Radiol.* **2002**, *12*, 2056.
9. Kreis, R.; Ernst, T.; Ross, B. D. Development of the Human Brain: In Vivo Quantification of Metabolite and Water Content with Proton Magnetic Resonance Spectroscopy. *Magn. Reson. Med.* **1993**, *30*, 424.
10. Rundqvist, H.; Johnson, R. S. Hypoxia and Metastasis in Breast Cancer. *Curr. Top. Microbiol. Immunol.* **2010**, *345*, 121.
11. Peet, A. C.; Lateef, S.; MacPherson, L.; Natarajan, K.; Sgouros, S.; Grundy, R. G. Short Echo Time 1H Magnetic Resonance Spectroscopy of Childhood Brain Tumors. *Childs Nerv. Syst.* **2007**, *23*, 163.
12. Kaminogo, M.; Ishimaru, H.; Morikawa, M.; Ochi, M.; Ushijima, R.; Tani, M.; Matsuo, Y.; Kawakubo, J.; Shibata, S. Diagnostic Potential of Short Echo Time MR Spectroscopy of Gliomas with Single-Voxel and Point-Resolved Spatially Localized Proton Spectroscopy of Brain. *Neuroradiology* **2001**, *43*, 353.
13. Belkić, Dž. Exact Signal-Noise Separation by Froissart Doublets in the Fast Padé Transform for Magnetic Resonance Spectroscopy. *Adv. Quantum Chem.* **2009**, *56*, 95.
14. Belkić, Dž.; Belkić, K. Improving the Diagnostic Yield of Magnetic Resonance Spectroscopy for Pediatric Brain Tumors Through Mathematical Optimization. *J. Math. Chem.* **2016**, *54*, 1461.
15. Belkić, Dž.; Belkić, K. Quantification by the Fast Padé Transform of Magnetic Resonance Spectroscopic Data Encoded at 1.5T: Implications for Brain Tumor Diagnostics. *J. Math. Chem.* **2016**, *54*, 602.
16. Belkić, Dž.; Belkić, K. Strategic Steps for Advanced Molecular Imaging with Magnetic Resonance-Based Diagnostic Modalities. *Technol. Cancer Res. Treat.* **2015**, *14*, 119.
17. Belkić, Dž.; Belkić, K. How the Fast Padé Transform Handles Noise for MRS Data from the Ovary: Importance for Ovarian Cancer Diagnostics. *J. Math. Chem.* **2016**, *54*, 149.
18. Belkić, Dž.; Belkić, K. Iterative Averaging of Spectra as a Powerful Way of Suppressing Spurious Resonances in Signal Processing. *J. Math. Chem.* **2016**, http://dx.doi.org/10.1007/s10910-016-0693-9.
19. Belkić, Dž.; Belkić, K. In vivo Magnetic Resonance Spectroscopy for Ovarian Cancer Diagnostics: Quantification by the Fast Padé Transform. *J. Math. Chem.* **2016**, http://dx.doi.org/10.1007/s10910-016-0694-8.

INDEX

Note: Page numbers followed by "*f*" indicate figures, and "*t*" indicate tables.

M

Printed in the United States
By Bookmasters